CIM Technology
Fundamentals and Applications

by

Russell Biekert

Contributing Authors:

David Berling
Richard J. Evans
Donald G. Kelley
Dale E. Palmgren
Michael W. Pelphrey
Joe Richardson
David L. Thorson
W. Van Twelves, Jr.

Publisher

The Goodheart-Willcox Company, Inc.
Tinley Park, Illinois

Library of Congress Catalog Number
International Standard Book Number 1-56637-426-X

2 3 4 5 6 7 8 9 10   98   01 00 99

Library of Congress Cataloging in Publication Data

Biekert, Russell.
CIM technology: fundamentals and application / by Russell Biekert; contributing authors, David Berling. . . [et al.].

    p. cm.
    Includes index.
    ISBN 1-56637-426-X
    1. Computer integrated manufacturing systems.
I. Berling, David. II. Title.
TS155.6.B54 1998
670'.295--dc20   92-32085
                                       CIP

Cover illustration: Giddings & Lewis' latest cell manager, Navigator™, is a user-friendly interface, windows-based operation with real-time system monitoring.

Contents

Introduction

This textbook was written to help you understand the concepts involved in CIM (Computer-Integrated Manufacturing) and to provide a foundation for applying those concepts in actual industrial situations. The authors of the 13 chapters in the book bring to bear a great breadth of experience and depth of knowledge in the many specializations that contribute to successful implementation of CIM. Each is a specialist in the topics about which he writes. They bring together, in one volume, an exciting blend of industrial and academic backgrounds. Thus, **CIM Technology: Fundamentals and Applications** provides a means of conveying that knowledge to those who will be implementing CIM or moving into established CIM environments in a few short years.

CIM is important in a variety of areas from business planning, to purchasing procedures to customer support for a product. Although this book focuses on some of the most widely understood, high-payback applications of CIM, the concepts of Computer-Integrated Manufacturing can be applied to other areas in a manufacturing enterprise (or even nonmanufacturing businesses), where sharing of information between people can be improved with computer tools.

CIM is—in literal terms—using **C**omputer systems to **I**ntegrate a **M**anufacturing enterprise. Computer tools, combined with the appropriate CIM data architecture, permit communications between applications, processes, and users.

CIM is defined differently by different users, and can be implemented in varying and increasing degrees of complexity. For many companies, improving shop-floor communications is the primary goal. Others extend the degree of integration to encompass communication between engineering and manufacturing functions. The ultimate benefit of CIM is the improvement of communication and control of information flow to all aspects of an enterprise.

The need for Computer-Integrated Manufacturing develops within a given company as a result of growth in both the size and complexity of the company's operations. The need to share information between people and company functions is met by the use of computer tools. At the same time, the growth of technology has played a role in creating the need for CIM. The availability of computers, the widespread use in industry of CAD systems, databases, and numerically controlled production machines have made essential the efficient sharing of information.

Communication of data is made possible by networks that connect computers, data collection devices, and NC machine tools. The interfacing of the various devices is often made easier by working with a consultant specializing in systems integration. Beyond hardware problems, implementation of CIM may require overcoming difficulties posed by the very structure of the company, as well as developing justification to win the support of company decisionmakers. Often, cultural change is needed if an organization is to take advantage of CIM. Without management and user support, CIM applications have little chance for success. Departments and individuals in a company must learn to work together instead of optimizing their own processes at the expense of others.

Russell Biekert
Phoenix, AZ

Introduction to Computer-Integrated Manufacturing

by David L. Thorson

Key Concepts

- ☐ Computers are enabling technology for CIM. They provide a level of control over production processes not by manual means.
- ☐ Integration involves development of a single database that can be shared by all related applications, using a common interface.
- ☐ CIM is more than the simple interconnection of shop floor machinery and computers. In the fullest sense, it can involve every aspect of the manufacturing enterprise, from the front office to the shipping dock.
- ☐ CIM will have different definitions for different users, but can be applied to any business where the control, use, and distribution of information is important.

Overview

The purpose of this book is to help you understand CIM (Computer-Integrated Manufacturing) and provide a foundation for its application. This chapter will introduce you to the basic concept of CIM as a means of using *computer systems* to integrate a manufacturing enterprise. It sets the stage for the rest of the book, which covers in greater depth the applications and benefits of CIM, including details of the automation techniques enabled by CIM.

David L. Thorson is Project Leader, CIM Support, for Garrett Engine Division, Allied Signal Aerospace Company in Phoenix, Arizona.

CIM IN DESIGN, SCHEDULING AND CONTROL

Computer-integrated manufacturing (CIM) is important in a variety of areas, from business planning to purchasing to customer support. The areas most often recognized as needing help in the form of CIM are product and process design, production scheduling and production control. These are also perhaps the areas with the greatest payback from applying CIM. As a result, these are also the areas most supported by research, consulting services and commercial products.

APPLICATIONS OF CIM IN AUTOMATION

In the manufacturing area, the focus of CIM shifts from planning applications to production hardware applications. Even the best plans are of little use if they cannot be carried out, and CIM applications in a manufacturing shop help to magnify the benefits of planning through better control of how those plans are executed.

Computer interfaces and networks ensure information is sent to the right machines and operators. At the same time, computers collect feedback from shop operations, providing more accurate information for the next round of planning processes. Information on scrap rates, completed parts quantities, broken or misaligned machines, unavailable tooling, process capabilities (such as a machine's ability to hold tolerances), and operator performance all help to fine-tune shop schedules. Information on product quality and producibility directly affects the initial design or redesign of products and the processes to make them, Fig. 1-1.

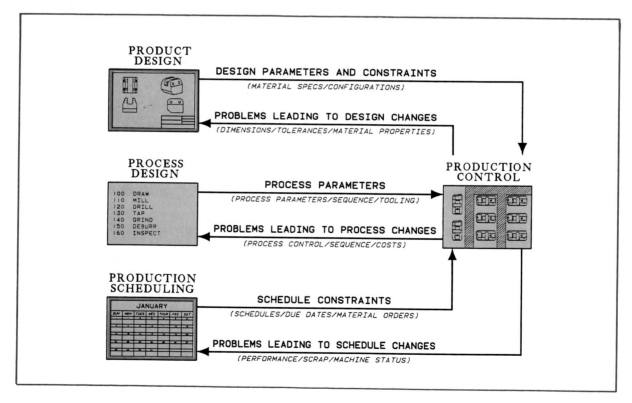

Fig. 1-1. *In a CIM environment, feedback from the shop operations helps to refine planning and scheduling processes.*

Without CIM to provide intelligent and accurate communication between these areas, designers and planners are working with an incomplete picture. Product designs will continue to be "thrown over the wall" to manufacturing, with no effective means for understanding how decisions made in each area impact each other. Opportunity for huge savings in production cost, time, scrap and field maintenance are lost.

Computers in production control

Computers, the enabling technology for CIM, provide a level of control over production processes that is not possible through manual means. A robot like the one shown in Fig. 1-2, for example, can be programmed to perform tasks with repeatable accuracy. A robot can spray an even coat of paint or molten metal across a complex contoured surface, simultaneously moving its spray head through several axes of motion. At the same time, the table holding the part being sprayed can move through several other axes of rotation and translation. Table movements are perfectly synchronized with the motions of the robot.

GARRETT ENGINE DIVISION/ALLIED SIGNAL AEROSPACE COMPANY

Fig. 1-2. A production robot like this one can simultaneously move through several axes of motion. The workpiece is held on a positioning table that moves in a coordinated manner with the robot.

Introduction to Computer-Integrated Manufacturing 7

In addition to controlling the robot's motion, a computer can monitor and adjust spray flow rates to account for differences in temperature or in the thickness of the sprayed material. This tightly controlled process can produce better quality parts with less waste and in less time than a human operator could achieve. It involves use of one or more computer languages, communication between computers and actuators on the robot, and feedback responses from visual or tactile sensors, Fig. 1-3. All the information must be processed at a high rate of speed to give the mechanical systems a reasonable chance to adapt to slight changes in the process.

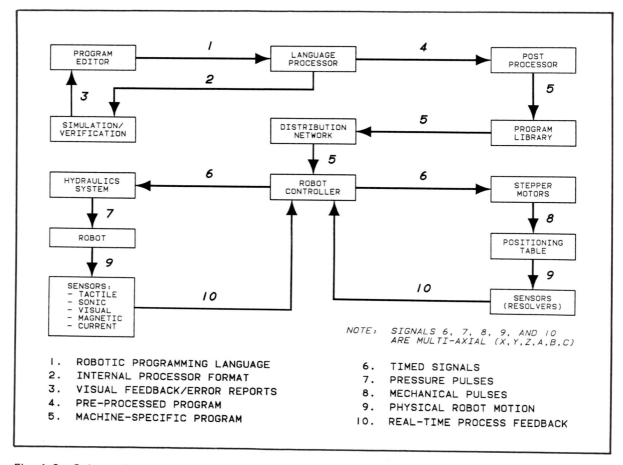

Fig. 1-3. Schematic representation of the processes involved in programming and coordinating movements of a multi-axis robot and a positioning table.

WHAT IS CIM?

Before you attempt to understand how to use CIM, it's helpful to know more about it. Some common questions about CIM are:

Is CIM an "off-the-shelf" software package bought from a computer company?

Not in the foreseeable future. Many software suppliers have (or are planning) systems that are designed to communicate freely with each other, often through a common database or information storage area. Hardware suppliers are way ahead in this area: agreed-upon standards in the computer industry generally allow one supplier's computer to exchange packets of information with a computer from another supplier. Unfortunately, software standards have not progressed as rapidly: it is often difficult to make software programs residing within the same computer share desired information.

Does CIM look the same for every company?

No, and it probably is not the same even for different users within the same company. Every company is concerned with unique information. Even such a seemingly common piece of information as a part number will vary widely between companies. Almost every company producing a product deals with such numbers, but uses them in many different ways. Some companies use part numbers that are serially assigned; others employ coded letter or number combinations to define product line, part type, supplier, or other relevant information. In some systems, a unique part number is assigned to every component; in others, suffixes are applied to a single part number to identify different variations. To be useful across all companies, a generic "CIM system" would have to handle all these different types of part numbers and more. A given company may need to track hundreds of *data elements* (distinct types of information), each with as many possibilities for interpretation. Even within a single company, users may interpret "part number" in different ways. For example, Engineering may establish an "official" part number that appears on drawings. Manufacturing, however, may make that single part from several components, each with a unique "phantom" part number that Engineering never knows about.

Is CIM really nothing more than wires connecting computers and machines together?

No, it is much more than physical connections. Besides wires and communication protocols (the low-level "languages" that tell one device how to exchange data with another), CIM brings together the right kinds of information under the right level of control. It provides an organizational structure for the enterprise that permits different departments and users to take advantage of the often-overlooked resource called "data."

"C" + "I" + "M"

Simply put, CIM is the use of **C**omputer systems to **I**ntegrate a **M**anufacturing enterprise. CIM provides the tools to enable the use of organizational programs such as *Total Quality Management, Continuous Improvement, Concurrent Engineering, Design for Manufacturability, Design for Assembly,* and the back-to-basics concept of "Do it right the first time." In very small organizations these efforts and tools may not be needed. In such smaller companies, people can adequately comprehend the overall business, design, and production systems and organizations, and the complex interactions between them. (See Fig. 1-4.) But most organizations quickly outgrow their ability to survive by such "seat of the pants flying."

Integrating information and organizations will decrease the *logistical* size of a company, making it appear to be small again—at least from the management, administration, and information-sharing viewpoints. The goal of CIM is to provide the computer applications and communications needed to bring about the integration (with matching organizational changes) that will allow a company to take advantage of these new capabilities. (See Fig. 1-5.)

"C" = Computer

CIM requires a tool to integrate functions and information within a manufacturing enterprise. The scope of information in a large organization, and the speed at which it is generated, makes ineffective such manual information systems as index cards, file cabinets, phone calls, company mail, or even "sneaker-net" (the physical conveying of information from place to place).

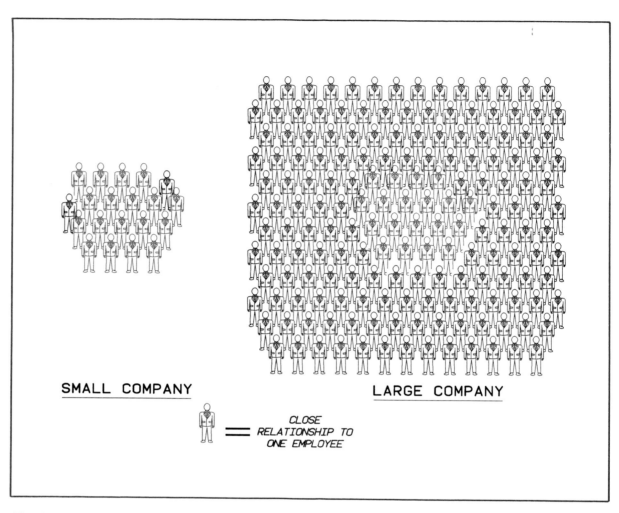

SMALL COMPANY

LARGE COMPANY

CLOSE RELATIONSHIP TO ONE EMPLOYEE

Fig. 1-4. *The larger the company, the smaller the percentage of employees that will be involved in a close relationship with any given individual. This affects the employee's comprehension of the overall operation, as well as affecting communication and coordination of efforts.*

The "C" in CIM specifies a *computer* as the most promising tool for handling an explosion of information. The "C" doesn't specify which computer platform to use, which software packages to run, or even which database technology to build on.

Enabling tool. The computer is an ***enabling tool***: it provides the potential for making information available, for protecting it, and for helping people create, summarize, and analyze it. For example, the search capabilities of a computer are far faster and more sophisticated than any manual method. The computer makes it possible to search through many thousands of pieces of stored information (data), while seeking complex relationships among them. The resulting information then can be sent throughout the company at high speed. More importantly, in terms of CIM, the computer serves as a consistent and convenient vehicle for sharing information and communicating ideas, Fig. 1-6.

Information flow. In most manufacturing organizations, information has traditionally flowed from product concept to marketing to design to analysis to manufacturing and quality assurance to scheduling to production to field service, as shown in Fig. 1-7. The small amounts of feedback information

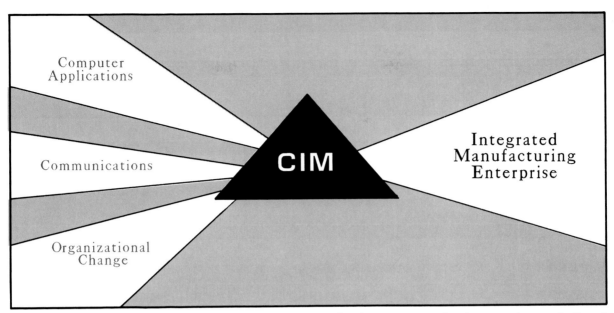

Fig. 1-5. The basic CIM process blends computer applications, communications, and organizational change into an integrated manufacturing enterprise.

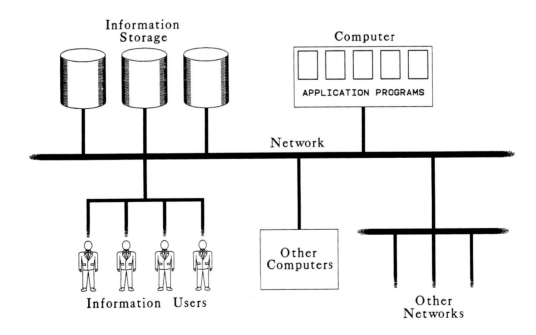

Fig. 1-6. The computer is the vehicle for storing information and sharing it with other computers and users through communication networks.

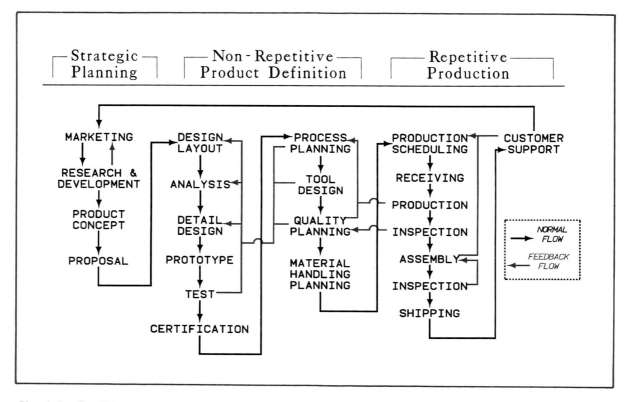

Fig. 1-7. *Traditional information flow in a company is primarily one-way, from product development through design and production operations to customer support functions. Feedback is present, but is not strongly emphasized.*

flowing upstream were often ignored, since they were not part of the formal information network.

To be most effective, each department in a manufacturing business must know how its decisions affect areas both downstream and upstream. For example, if designers understand how tolerances on certain features make a product difficult to produce, they can seek alternate design solutions. Without accurate feedback from the shop, many design decisions are little more than guesses.

Computers provide a means to store descriptive information, such as case studies, solutions to problems, or business process goals, in ways that allow the information to be easily retrieved. A computer can make it feasible, when faced with a production problem, to find out how similar problems were solved. Other information might include the impact the solution made on the organization in terms of cost, schedule, and quality, and what similar parts might benefit from a solution of the same type.

Information Management. Access to information cannot provide far-reaching benefits if the data is not stored properly. Without alphabetically sorted entries, a dictionary would be practically useless. A library of books and periodicals becomes most useful when an index system is used to catalog the contents by title, subject, author, date, and description.

A computer *could* store vast amounts of information in the order it was received, with no index to the data other than date and time of receipt. Clearly, most manufacturing applications would need to access information by other criteria, such as due date, part number, reason code, material type, vendor code, or where-used (the assembly or product on which a component or tool is used). Fig. 1-8 shows the data elements to which typical manufacturing operations must have access.

Application Functions	Shared Information Database	Part Number	Material Type	Part Family	Due Date	Reason for Change	Where-Used	Prod. Release Date	Cost per Piece
Design Change	X	X	X		X	X		X	
Process Change	X	X	X		X			X	
Inventory Control	X		X	X			X	X	
Process Design	X	X	X		X	X	X	X	
Materials Ordering	X	X	X				X	X	
Accounting	X							X	
Requirements Planning	X		X	X		X	X		
Proposals		X	X	X	X	X		X	
Production Scheduling	X	X	X	X			X		

Fig. 1-8. Data elements to which various planning and manufacturing functions must have access.

Information must be managed at a strategic level; that is, data definitions and structures (a *data architecture*) must be carefully planned with both short- and long-term needs in mind. Just as it becomes more difficult to reorganize a library's book index as the collection of books grows, a system's data architecture becomes increasingly expensive to change as the volume of stored data grows. Proper security procedures or access authorization must also be established. These procedures are needed to prevent changes in controlled data by unauthorized users or the wrong applications, and to limit user access to proprietary or classified information.

Strategic information management includes providing for data integrity. A sound data architecture describes the format and possible content of each type of information to be stored. It should not be possible to enter text (letters and numbers) in a location planned for numbers only. Where it is important to screen data for correct values, applications should enforce correctness by comparing all inputs against a master list, then rejecting unacceptable values.

A purchasing system that permits entry of a fictitious part number or supplier number will lead to data integrity problems, because sooner or later, a person will hit the wrong key. Typing part number GF133*62*, instead of GF133*26*, on a purchase order will have these results:

- Part GF13362 stays in inventory, because no assembly ever needs it.
- Final assembly is delayed because Part GF13326 was required, but never ordered.
- part GF13326 is delayed until more material can be ordered.

A major part of strategic information management is understanding user and business information requirements, Fig. 1-9. A data architecture supported by the biggest, fastest computer does no good for a company unless the correct information is directly stored or can be constructed from data relationships.

Information stored in a computer must also be managed at a tactical level, Fig. 1-10. Day-to-day operations must ensure that adequate space for data storage is maintained and that information is copied to backup storage media in case the primary storage should fail. Computer hardware must be supported and upgraded to meet expanding needs. Computer system performance must be measured to watch for lengthening of *response time* (the time needed for a computer to respond once a user enters some information) and increases in network traffic (information flow) that could cause unacceptable delays.

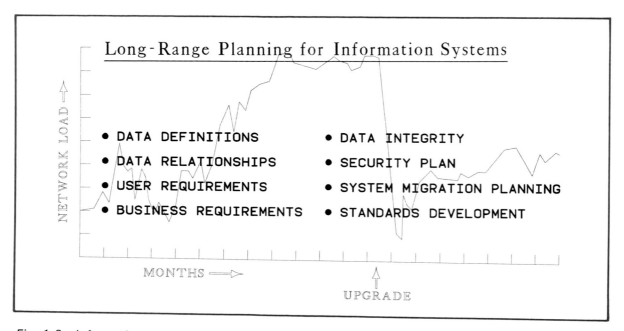

Fig. 1-9. *Information management at the strategic level involves planning for both short-term and long-term needs of the system.*

"I" = Integrated

The "I" in CIM represents the *integration* of manufacturing functions and information, using the enabling power of computers.

Integration vs. interfacing. Most "integration" being done today in manufacturing operations is really "interfacing." *Integration*, in terms of computer systems, means two or more software applications share the same information from the same source. This is not the same as *interfacing*, where data needed by two systems is duplicated, but can be sent between them through some transfer or translation strategy.

As an example, consider the personal computer program *Symphony*®, developed by Lotus Development Corporation. *Symphony* was one of the first truly integrated software packages for the personal computer. Data in memory could be viewed as a spreadsheet by part of the application, as a database by another, or as a word processing document from a third perspective. Each of these applications used the *same* data in a different way; users were not required to translate the data between applications. Users could be more productive when they had need to use their data in different ways.

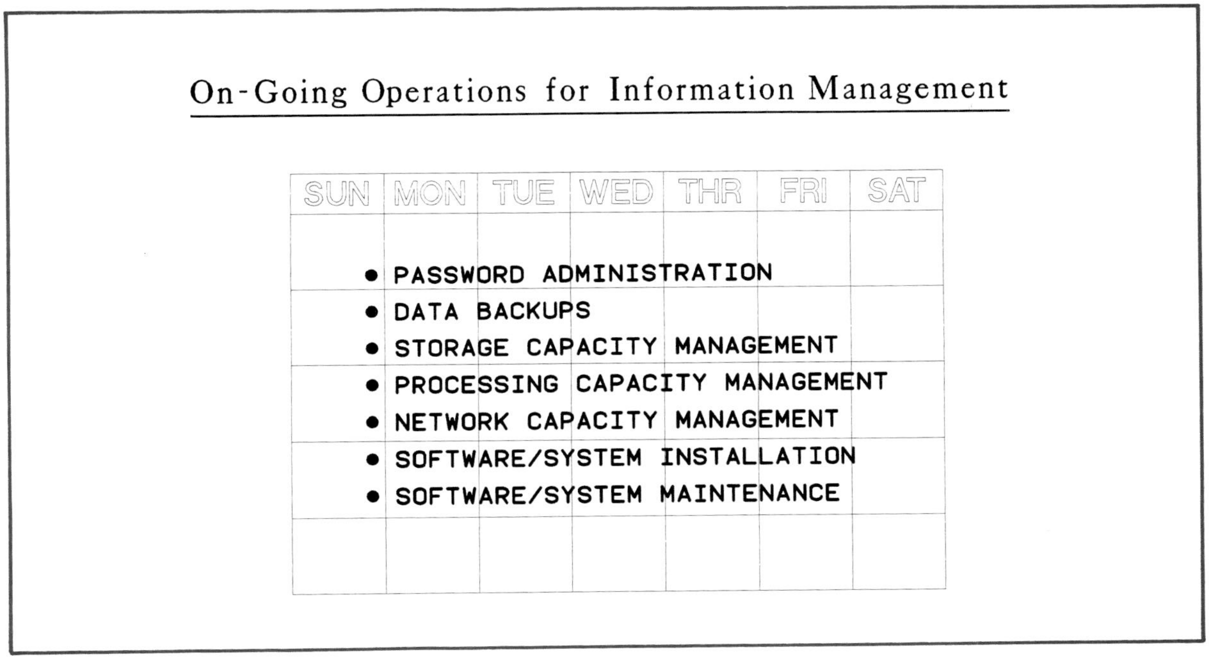

On-Going Operations for Information Management

SUN	MON	TUE	WED	THR	FRI	SAT
• PASSWORD ADMINISTRATION						
• DATA BACKUPS						
• STORAGE CAPACITY MANAGEMENT						
• PROCESSING CAPACITY MANAGEMENT						
• NETWORK CAPACITY MANAGEMENT						
• SOFTWARE/SYSTEM INSTALLATION						
• SOFTWARE/SYSTEM MAINTENANCE						

Fig. 1-10. Managing information at the tactical level involves meeting day-to-day operating requirements and closely monitoring system performance.

Of course, users who only needed a word processor could find more powerful or easier-to-use software designed specifically for that purpose. By giving up some word-processing power in moving to an integrated software package, they could trade for improved productivity in other applications. Taken separately, the main applications of Symphony compared poorly to more powerful and narrowly focused software packages. Taken as whole, however, the integrated package provided a solution much stronger than the sum of its parts.

This same philosophy applies, on a more grand scale, to a manufacturing enterprise. In moving toward a CIM environment, each department may have to give up some of its capabilities. Integrating applications within the enterprise means creating a single database to be shared by all related applications. This database can be in one physical location, or can be spread among various computer systems with a single "logical view" of the data (data elements are defined once and stored once, then software logic can access the data as needed, regardless of its physical location). To keep the database manageable, most departments will have to give up some of the information they use that is not important to the overall scheme of things in the company. Worse yet, some data thought to be essential by a department may exist only to enable a *workaround* (a means of overcoming a limitation or achieving a result that a system was not intended to provide) in a system not designed to handle processes that outgrew it.

By keeping data in one place (physically or logically), storage needs are reduced, data integrity (accuracy) is improved, and time or effort are not wasted in translating or transporting data from one application to another. Another important benefit is that users of different applications have a time-consistent (simultaneous) view of company information. In contrast, if some overnight update process is used to add one application's data to another's separate database, users of each application will see different information on screen until the overnight transactions are completed. Even if the cross-application update process occurs every ten minutes, there is still a good chance decisions will be based on old or incomplete information, Fig. 1-11.

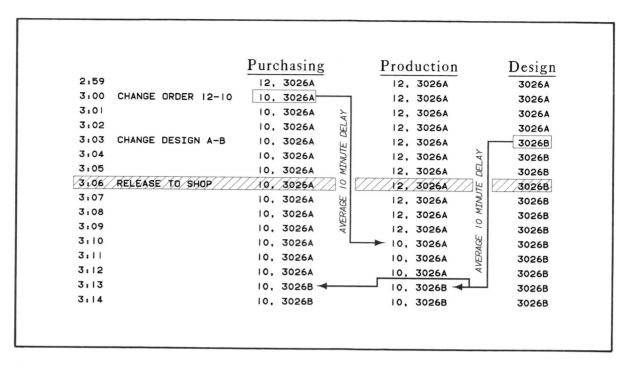

Fig. 1-11. Time delays can cause problems in updating database files. In the situation shown here, a 10-minute delay in updating transactions across different databases results in the wrong quantity of the wrong part being released to the shop. In large companies, the updating delay can be considerably longer, even overnight.

Software development in such an environment is quicker and more certain because the data is understood. This means that developers will not waste time trying to figure out which of two systems has the correct definition for lot number, scrap cost, or order status. As they move from one project to another, they will be working with the same set of data definitions. Users will no longer worry why reports from two different systems give different results when they requested the same information. This also leads to reduced maintenance for data translators (which may need to be revised each time either of the applications they translate between is changed). While there *is* a cost involved in standardizing data storage definitions and locations, and although the system may no longer provide *all* the data users had before, there is tremendous potential in redesigning systems and data definitions for integration.

Shared information. A *data dictionary* defines how and where information is stored. The data dictionary allows software applications to consistently access information they need, even if the physical format or location of the data changes. Although a data dictionary is commonly thought of as buried somewhere deep in the computer system (useful only to programmers and smart applications), the concept can readily be extended to benefit a wider audience, Fig. 1-12. If users can refer to a data dictionary, they can find out what information is available within the company, and how to use it.

To be most useful, this extended data dictionary should contain descriptive text explaining the purpose of each data element. It also should include information on typical values, where the data comes from, who controls it, and what applications use it or update it. This helps an information user make certain the correct data elements are used in applications and reports. For the information user (designer, process planner, scheduler, or even programmer), the data dictionary provides visibility into the processes related to those data elements.

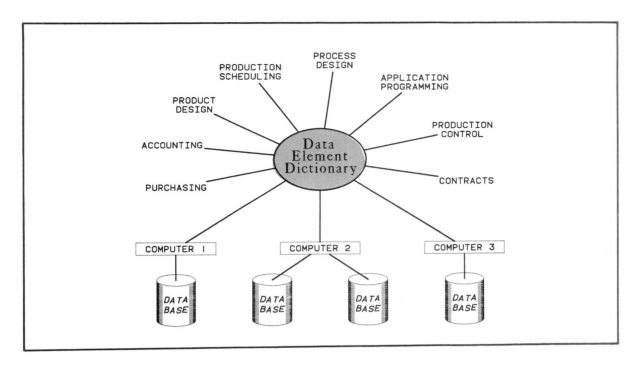

Fig. 1-12. A data element dictionary defines how information will be stored and where it will be stored. It provides a single, consistent view of data and data definitions for all users.

Information shared between applications must be stored in a physically or logically centralized database supporting those applications. In a CIM environment, it may seem that all company information must reside in one database. This is impractical, however, for such reasons as storage space, access time, disparate data types, security, and connectivity among various computer systems.

Fortunately, there are several groups of data that can be separately stored as a compromise. Personnel data, involving salary and addresses, has little use in a product definition process. CAD data models can be stored independently of labor reporting information. Of course, personnel information such as phone numbers or department names is of value in a design review process, where a reviewer may want to contact a designer to understand design intent, or where a designer may have questions on manufacturability and must determine the right person to ask. A CIM database designer must choose carefully which information is best to keep in one place, and which information can be duplicated without a serious loss of data integrity. Chapter 2 introduces techniques for linking data relationships across separate databases.

Shared functionality. One final way to improve data sharing is to use the same systems for different functions where appropriate. An example is an electronic signoff and review process for routings, software development, plant layouts, documentation, and part drawings. These functions may seem dissimilar on the surface, but all require read/write security, secured signoffs, review capability (perhaps with markup or comment capability), a distribution mechanism, and a way to track changes (including control of versions, revision history, and release status).

At the lowest level, a common system must be able to display CAD drawings and layouts, program listings, or perhaps shaded images used in customer documentation. The details of the process data are different for each user group, but the process management and control data can be the same. The same software system can support all these processes, if it allows high-level menus and screens to be tailored to each process. Similar data elements for each process can be related across processes.

Users of processes related this way will learn a common system; system developers can support different parts of the company with the same technology, reducing development and maintenance time. (See Fig. 1-13.)

User interfaces seen in Apple's Macintosh® System, and operating systems such as Microsoft's *Windows* and the multi-vendor *X-Windows,* are examples of this: each application is based on the same low-level functions (base services), so that similar processes are done the same way in each application. Some examples of such base services are selecting documents or files, choosing menu options, scrolling, or copying data between applications. Developers create this set of base services one time, and users learn how to use it one time, no matter what the high-level application will accomplish.

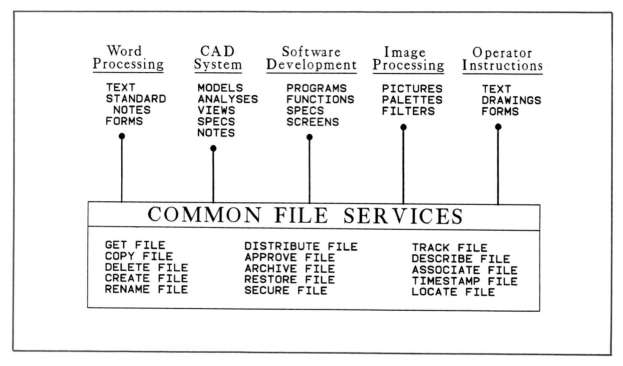

Fig. 1-13. *Various applications can share common base services. These applications all use file-oriented data that can be managed by the same tools across applications.*

"M" = Manufacturing

The "M" in CIM is often misleading. The word "Manufacturing" makes it easy to assume that CIM is only useful in a manufacturing shop area, that it is simply the interconnection of shop floor machinery and computers. This view was taken by many pioneers of CIM before they realized that its real impact was in areas *beyond* the shop floor. If manufacturing problems are *designed into* a product, then improvements in production processes can recover from those problems to only a limited extent. Information—including designs, plans and schedules—is most useful when it is accurate from the start and not when it must be patched by other users in downstream processes. *Manufacturing*, in the CIM sense, is best understood as the collection of departments, processes, and functions that enable an enterprise to create a product, Fig. 1-14.

Production control. Production Control is responsible for performing detailed shop scheduling and for handling schedule changes as shop events alter schedule assumptions. In short, the job of

Production Control is keeping the shop running as close to the overall production schedule as possible. This requires awareness of errors and delays, and knowledge of capacity, operator and machine capability, operation sequences, and priorities. In a smaller shop involving use of only a few machines or processes, the operators or their supervisor might handle this manually.

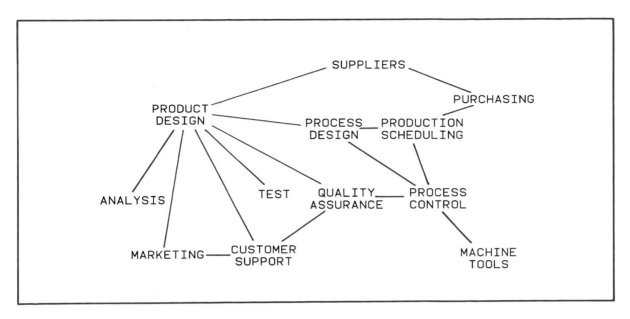

Fig. 1-14. Computer-integrated manufacturing can be defined as a collection of processes, departments, and functions that enable a company to produce a product. Each line in this graphic represents a potential CIM application focus.

In larger shops, however, this information is difficult to obtain, analyze, and understand quickly enough to solve problems and keep production on schedule. No one person can hope to comprehend all the activities in the shop at once. Information must be gathered efficiently. This usually calls for electronic data communications between machines and computers. Information from several sources must be compared and analyzed, calling for consistent storage formats and locations for data. Finally, shop information must be shared between higher-level scheduling systems and the data collection/distribution/execution systems at the shop level, Fig. 1-15.

Production scheduling. Establishing dates for shipping finished products, assembling components, fabricating components, ordering raw materials and tooling, and managing inventories, is the responsibility of Production Scheduling. Another key duty of Production Scheduling is planning which components will be assembled into a final product, based on inventory, schedule, and customer requirements.

There are several approaches in use for Production Scheduling:
- **MRP II** (Manufacturing Resource Planning) back-schedules activities from the final ship date of a product, based on lead time (manufacturing or ordering time) of assemblies, components, and raw materials.
- **MRP I** (Manufacturing Requirements Planning) is similar to MRP II, but considers a smaller organizational scope and does not consider production capacity when scheduling.
- **JIT** (Just-In-Time) eliminates work-in-process (WIP) inventory by scheduling arrival of parts and assemblies for an operation at the time they are needed and not before. By careful planning

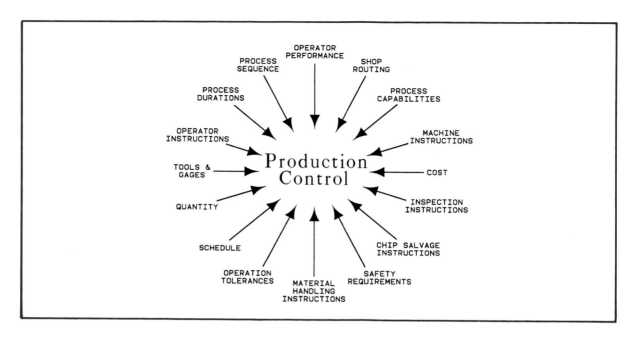

Fig. 1-15. Information needed to help Production Control keep the shop running as close to the overall production schedule as possible.

with suppliers, supplies of materials and purchased parts are received only as required to keep production on schedule. The supplies go directly to the first operation, eliminating receiving inventory.

Clearly, these and other scheduling approaches must deal with a high volume of information from a large number of sources. To be accurate, schedulers need reliable, up-to-date shop information, including feedback on scrap rates and performance to planned production. Production scheduling systems need planning information on process sequence, duration, and location. The parts list or **bill of material** that defines which parts (and which versions of those parts) to use has its basis in the Product Design area. Only a carefully applied CIM strategy will guarantee the correct information is shared between systems at the right times and with the right interpretation. (See Fig. 1-16.)

Process design. A *process design* describes the sequence of operations to make each component and/or assemble components into a final product. Details of each operation include instructions for operators, instructions for machine tools (Numerical Control or "NC" programs), instructions for any necessary inspection, and time standards. Time standards are estimates of process times, move times, inspect times, set-up times, and so on. There are callouts for tools and gages, machine tool requirements, and perhaps references to specifications on processes, material handling, chip salvage, cutting fluids, safety, and other topics.

The goal of Process Design is to define a method, or *process plan*, for making component parts and assemblies that meet defined design criteria. Process Design also must provide enough information to Production Scheduling and Production Control to allow a realistic schedule for following the process plan. Again, information created in one area is used by a wide variety of external systems. Feedback from Production Scheduling and Production Control will help to improve process planning accuracy, Fig. 1-17.

Product design. Defining a product—what it should look like, what it should do, how long it should last, how it should work, and how well it should meet a defined need—is the task of *Product Design*. This function has, in the past, been considered separately from the other manufacturing concerns

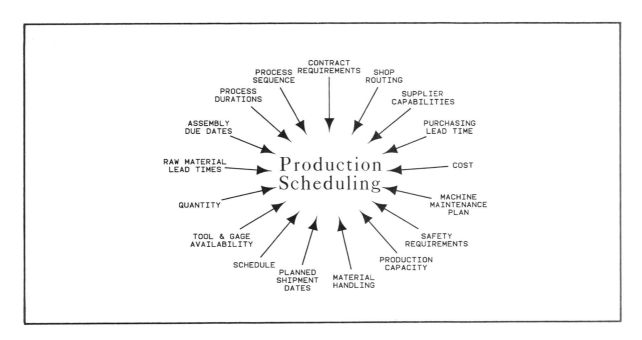

Fig. 1-16. To establish key dates for each step of the manufacturing process, Production Scheduling needs information on a variety of topics.

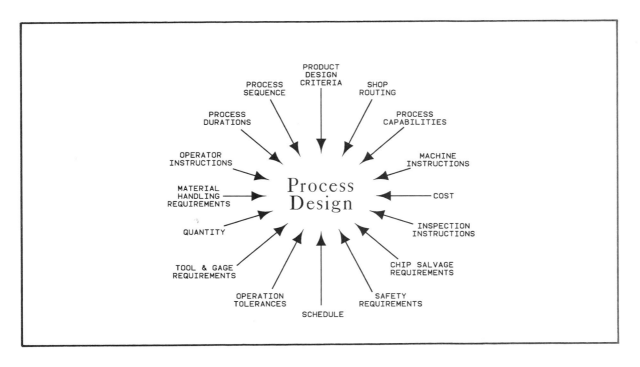

Fig. 1-17. Process Design defines the methods to be used to manufacture a product. Information needed by Process Design is shown.

described above. However, to be most effective, it must interact fluently with the other areas of the manufacturing enterprise. *Producibility* (how easy or difficult a product or component will be to manufacture) and production costs must be considered. So must other design tradeoffs, such as performance and quality.

The design process (including analysis and testing) creates information useful to other manufacturing areas, including parts lists (a starting point for Production Scheduling) and geometric models (used for NC programming and component visualization). A CIM strategy would promote access to this information as required. Even unreleased data can be of value for planning and "early warning" purposes, if managed properly. The CIM goals of minimal data translation and single storage of data, independent of its use, will ensure data accuracy throughout the organization. (See Fig. 1-18.)

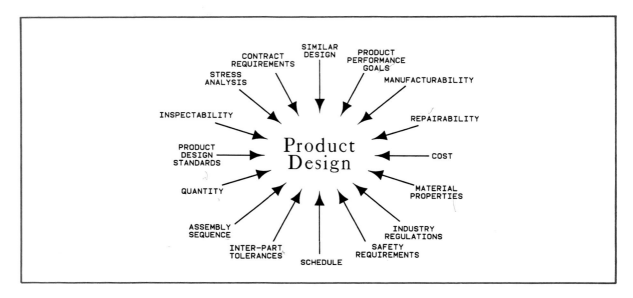

Fig. 1-18. To be effective, Product Design must interact with other areas on the manufacturing enterprise. Information inputs to Product Design are shown.

Manufacturing enterprise. The areas just described (Production Control, Production Scheduling, Process Design, and Product Design) are all described in greater depth in later chapters. Other areas can and should be considered in an overall CIM plan. The "M" in CIM must go beyond the shop floor, but it is difficult to decide where to stop. Accounting, Marketing, Purchasing, Sales, Research & Development, Facilities Planning, and other internal departments all can benefit from access to information created, stored, and used in a CIM environment. Each of these areas has information that can be fed back to the CIM environment; information that would enhance the effectiveness of other departments. CIM can even be extended to suppliers and contractors. For a typical company, though, the first step must be to get its own house in order.

DIFFERENT DEFINITIONS FOR DIFFERENT USERS

As you read this book, keep in mind that CIM concepts can be applied to almost any business where the control, use, and distribution of information is important. Such information can concern product definition, schedules, procedures, standards, costs, training materials, product documentation, proposals, or other topic areas. CIM will have a different definition for each type of business, and even for different companies of the *same* business type.

CIM becomes more beneficial as greater amounts of information are shared between more organizations in a company. Each group in a company may correctly view CIM as effectively communicating and controlling information between itself and other areas immediately upstream or downstream in the process flow. This may provide a reasonable starting point for improvements, but benefits will increase by expanding the scope of CIM efforts. Be forewarned, however: larger projects require greater organizational commitment, more resources, and more time. They also open the organization to greater risk.

CIM definitions will evolve over time. After a company tackles smaller, simpler integration issues, there will remain those issues spanning a broader spectrum of departments, policies, or hardware. The knowledge and confidence gained in early CIM efforts will provide increased leverage in solving such larger, more complex problems.

Shop communications

The primary CIM goal, for many companies, is improving shop floor communications. Computer networks and intelligent machine controllers pass data, such as *NC* (numerical control) programs directly to the correct machine. In *manual* machining environments, operator skill was the limiting factor in controlling the shape of parts turned on an engine lathe. Tracer lathes improved the repeatability of complex patterns by letting a cutter follow a pattern traced by a sensor.

Numerical control lathes eliminated the need for a pattern. Contours could be defined as a set of mathematical curves; a perfectly machined physical pattern was no longer needed before beginning a production run. (See Fig. 1-19.)

Fig. 1-19. The CNC controller on the left is PC technology-based. The full-feature control pendant on the right allows the operator to roam free of the main control station while still maintaining access to all major control functions and readouts.

Each advance in automation introduces new problems. An NC lathe needs an NC program, which is normally written **off-line** (away from the machine) by an experienced programmer. Paper or Mylar® tape is punched with program data, and filed in a tool crib. An operator checks out the tape when needed, carries it to the machine, loads the tape, and tells the machine to read it. This process takes time away from production and is highly susceptible to tape-reading errors and tape damage (from being stepped on and torn). Also, it is possible to mix up tapes as they are returned to or checked out of the tool crib.

A computer network and the appropriate application software can solve this problem, providing error checking, high transfer rates, and NC program tracking. Called **DNC** (Distributive Numerical Control), this type of software is a good beginning CIM application. As with other technological advances, DNC introduces new problems. Solving these problems often paves the way for additional integration efforts.

Problems arising from the introduction of DNC include providing computer storage space for NC programs, backing up program data, ensuring sufficient network flow rates and availability, and providing additional operator training. Since NC programmers work in a different, higher-level language than the resulting NC programs used by NC machine controllers, and since they may work on a different computer system altogether, Fig. 1-20, there is the additional problem of tracking

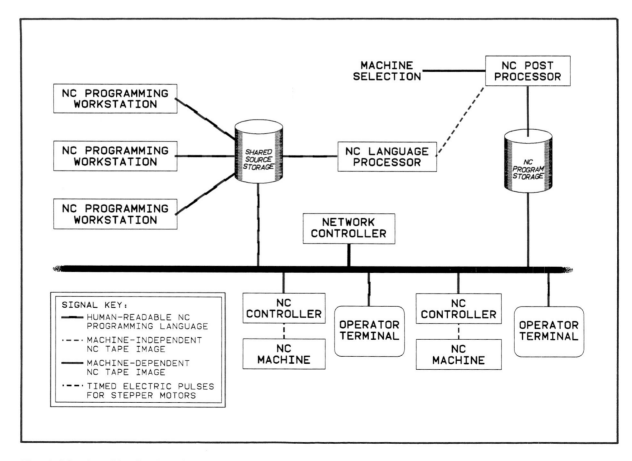

Fig. 1-20. In a Distributive Numerical Control (DNC) network, programmers work in higher level languages than the binary patterns read by the NC controllers.

24   Computer-Integrated Manufacturing

machine-readable NC programs to the source programs or geometric models on which they are based. These controllers convert binary instructions into timed electronic pulses to move stepper motors on a machine the desired direction and amount.

Recurring processes

In manufacturing, *recurring processes* are those surrounding the ordering, scheduling, production, inspection, and shipping of products; for a given item, the same series of events will take place each time it is produced. CIM in recurring processes is usually directed at improving accuracy of information and providing tools to analyze the data for process improvement. CIM applications take advantage of common information across processes, such as similar manufacturing problems that may be caused by equivalent, difficult-to-produce geometric features. Products or components with similar processing requirements may be made from similar process plans, reducing the planning effort and tooling costs. Also, *statistical process control* or SPC (the science of measuring a process to predict, as soon as possible, trends away from the expected process) can base its measurements across processing of *similar* features on *dissimilar* parts to build a stronger statistical base for process trend analysis. (See Fig. 1-21.) Collection, analysis and retrieval of this information would be difficult without a CIM architecture.

Non-recurring processes

In manufacturing, events that take place a limited number of times for a product or component, no matter how many times that part is produced, are known as *non-recurring processes*. These include process planning, tool design, NC programming, product design, inspection plans, and make-versus-buy decisions. CIM can aid non-recurring processes by reducing process lead times and re-using information from similar efforts.

Non-recurring lead times are the delays needed to design, analyze, prototype, and test a design, as well as the delays needed to plan processes, program NC tapes, and develop tooling and inspection plans. Lead times may also include time to order tooling or new machines, time to set up material handling equipment, or any other delays from the conception of a product up to the point it is ready for production. Obviously some lead time is necessary and unavoidable, but every week lost is a week a competitor can use to win market share. CIM can reduce non-recurring lead times by speeding communications, automating sign off and review processes, and automating the functions of drawing control (storage, revision control, access security, searching) and configuration management (tracking which components are used in a final product, and comparing this to the planned assembly).

The biggest impact of CIM for non-recurring processes comes from re-using information learned from similar and related efforts. Product designers can reject decisions that lead to geometry or other specifications that have historically caused problems in production, assembly, inspection, and repair.

Arbitrary decisions are routinely made in most organizations. This occurs because the persons making the decisions have no access to relevant information others possess, information that would shape decisions leading to products and processes more beneficial to the organization and its customers. *Information sharing* is the basis on which CIM is built. Computers can store and retrieve information in numerous ways, but the appropriate techniques must be used. One such technique, called *group technology* (GT), consists of taking advantage of the *similarities* in what you are trying to improve. GT can be applied to parts, tools, production processes, management processes, and machines or people skills, Fig. 1-22.

By classifying production problems into groups of similar problems, each larger group can be studied for a *common* solution (rather than a separate solution for each problem). If a group of problems is caused by tolerance restrictions on certain kinds of geometry, then designers can retrieve this information by classifying the geometry of the component they are designing. If the new component has features and tolerances matching those of a group already in the database, designers can get to this information quickly. The knowledge captured in Production is effectively (and consistently) shared with Design. A problem solved one time will not need to be solved again; it will not add lead time to every new project that comes along.

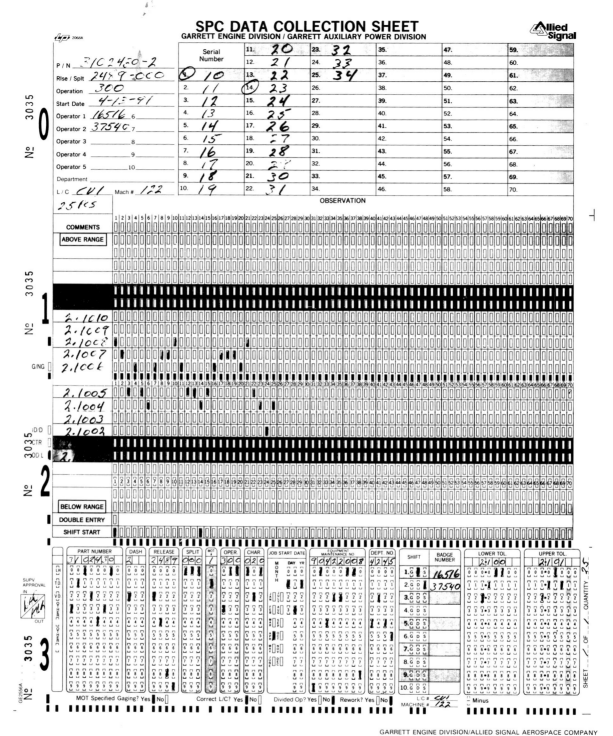

Fig. 1-21. An example of a data collection sheet used for Statistical Process Control. The machine-readable form doubles as an X-bar control chart for predicting process trends.

GARRETT ENGINE DIVISION/ALLIED SIGNAL AEROSPACE COMPANY

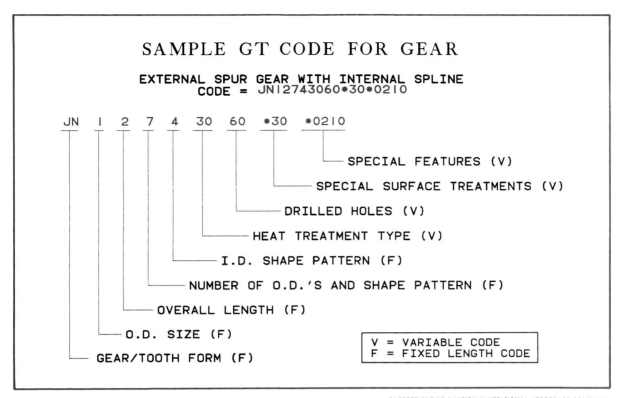

Fig. 1-22. A Group Technology (GT) classification scheme for gear parts.

Engineering/manufacturing communication

The simple concept of sharing a geometric model between product and process design groups can have a tremendous impact on both lead times and product quality. A blueprint, the traditional output of Design and input to Manufacturing Planning, is subject to interpretation (the more complex the part, the more open the interpretation). Inaccuracies can occur in transcribing a blueprint into the necessary NC and CMM (*coordinate measuring machine*) programs, *comparator* charts, tooling, and processes, Figs. 1-23 and 1-24. Such inaccuracies lead to scrap, rework, engineering changes, and more lead time. If a single geometric model can be used by computer-aided tools such as CAD systems for producing drawings, coordinates, and programs, then accuracy can be maintained throughout the product definition processes.

Other users

A CIM environment allows product and process design information to be shared with other departments. For example, news of a pending design change becomes visible to Purchasing as soon as evaluating and incorporating the change into a design begins. This could prevent ordering a large quantity of raw materials that would be made obsolete by the change. In a more traditional environment, Purchasing might not find out about the change until after it is evaluated, approved, incorporated, and released to Manufacturing. By that time, the obsolete raw material could already be in inventory.

As another example, a documentation group might be able to share assembly drawings made for Manufacturing, instead of recreating similar views from blueprints to send to customers.

Fig. 1-23. A Coordinate Measuring Machine (CMM) permits precise measurements for comparing a machined surface with a geometric model. The digital readouts provide the exact location, in Cartesian coordinates (X,Y,Z axes), of point where the probe tip touches the object's surface.

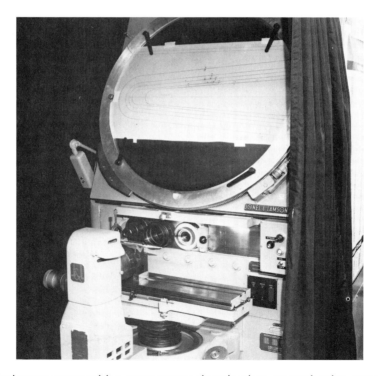

Fig. 1-24. An optical comparator with a comparator chart in place on projection screen. The comparator is used by placing the item to be compared (in this case, a machined cutting tool) between a light source and a lens. The shadow of the object's profile is optically magnified and shown on the projection screen. The overlay taped to the projection screen permits an in-scale visual comparison of the shadow profile to the drawing.

Improving communication through CIM

CIM applications can begin in an isolated corner of the organization, but quickly grow as more and more people understand the concepts and potential benefits. No group within an organization can survive by itself, or contribute to the organization without using information from other groups and creating information for other groups to use. The most effective way to improve a process in a given department (such as Design Review) is to improve the interaction with other departments. CIM provides the tools to improve communication and control of information everywhere in an enterprise.

SUMMARY

CIM is a means of using *computer systems to integrate a manufacturing enterprise.* The scope of CIM ranges from product design, process design, product scheduling and control, to advanced integrated functions within a production facility. It is important that all functions of a company be part of a CIM plan. Functions from business planning and strategic planning to processing to customer support should be included.

There is no single definition of CIM because CIM is designed to fit the needs and applications of a specific situation. Thus, each company will implement CIM in a slightly different fashion.

CIM requires a tool to integrate functions and information within a manufacturing enterprise. That tool is the computer or the *C* in CIM. A computer provides the potential to make information available, to secure it, and to allow people to analyze it. Information can then be sent throughout the company at high speeds with great data integrity.

CIM provides the tools to enable organizations to implement programs such as *Total Quality Management, Continuous Improvement, Concurrent Engineering, Design for Manufacturability, Design for Assembly,* and a back-to-basics approach concept to *"do it right the first time."* This is considered the integration tool or the *I* in CIM.

The *M* in CIM is rather misleading because the word *manufacturing* makes it easy to assume that it is intended for use only in manufacturing or for integrating the operation of shop floor machinery. The *M* in CIM is actually the connection of all manufacturing-related functions linked in a computer network in an integrated manner. Production planning and scheduling, production control, process design, and purchasing, as well as all manufacturing processes and equipment, must be integrated in a total CIM system.

IMPORTANT TERMS

bill of material
comparator
Computer-Integrated Manufacturing
coordinate measuring machine
data architecture
data dictionary
data elements
DNC
enabling tool
Group Technology
integration
interfacing
JIT (Just-In-Time)
machine controllers
manufacturing

MRP I
MRP II
NC (Numerical Control)
non-recurring lead times
non-recurring processes
off-line
process design
process plan
producibility
product design
recurring processes
response time
Statistical Process Control
Strategic Information Management
workaround

QUESTIONS FOR REVIEW AND DISCUSSION

1. Explain how three of the following departments in an organization might share information with the more traditional CIM areas of Production Control, Production Scheduling, Process Design, and Product Design:
   a. Accounting
   b. Marketing
   c. Purchasing
   d. Sales
   e. Research & Development
   f. Facilities Planning

2. Explain how you might apply aspects of CIM in one of:
   a. A process industry, such as producing rolls of newsprint.
   b. A service industry, such as a fast-food chain.
   c. A large amusement park.
   d. A fire department with several stations in a large city.

3. Describe the differences in applying CIM to a high volume production facility, such as an automotive plant, to a low-volume operation, such as building a space shuttle.

4. How would CIM help a small "Mom & Pop" job shop with twenty employees? Can the application of CIM be effectively scaled down to meet the needs of such a shop? Consider computers, automation, networks, and databases. Will CIM concepts help this shop to deal with the larger companies it supplies?

5. Which of the following are non-recurring processes? If the answer depends on the type of product, state your assumptions:
   a. Ordering raw materials.
   b. NC programming.
   c. Stress analysis.
   d. Final inspection.
   e. Receiving inspection.
   f. Change request processing.
   g. Production scheduling.
   h. Time standard application.
   i. Production learning curves.

6. For each of the following types of businesses, name the processes that correspond to the traditional manufacturing functions of product design, process design, production scheduling, and production control.
   a. Personal computer software developer (for example, Microsoft).
   b. New car dealership.
   c. College or university.
   d. Amusement park.

7. When obtaining user requirements for a CIM application design, how can you avoid capturing those requirements that are needed only to enable process workarounds already in use?

2

Computer Systems and CIM

by David L. Thorson

Key Concepts

- ❏ Increased complexity of manufacturing operations leads to the need for integration through sharing of data.
- ❏ Development of powerful computers has made possible such vital-to-CIM applications as CAD, CNC machine control, and management information systems.
- ❏ Selection of a network depends on the computers used and the specific data transmission needs of an operation.
- ❏ Systems integration is the process of interfacing different types of computer hardware and making sure different software applications can *talk to each other.*
- ❏ To justify CIM to management, a workable cost/savings measurement plan is a necessity. CIM implementation also must overcome reluctance of departments to share data and resources.

Overview

This chapter first traces the development of the need for Computer-Integrated Manufacturing as a company's operations grow and become more complex. It then explores the central role of computer systems technology in both creating the need for CIM and making it possible. Numerical control programming, Computer-Aided Design, the importance of sharing data between departments and protecting the integrity of that data, networking, and systems integration are all covered as a basis for understanding the total CIM environment. Finally, this chapter introduces additional topics not covered elsewhere, but important to an overall understanding of CIM and its implementation.

David L. Thorson is Project Leader, CIM Support, for Garrett Engine Division, Allied Signal Aerospace Company in Phoenix, Arizona.

CHANGING NEEDS CALL FOR NEW METHODS

To better understand why CIM is needed in a manufacturing enterprise, it is helpful to review a typical sequence of events as a company grows and expands. In small shops, it is common for the same person to design and plan the manufacture of a part or an entire product. Someone else may order materials and standard parts, such as washers and fasteners. A tooling person may set up machines and order or make necessary tooling. An operator may have the responsibility of making a part, given no more than a blueprint or even a simple sketch, Fig. 2-1. There is good communication between the people in various parts of the organization, because they work under one roof and see each other every day. They work together on problems and can often fill in for each other as the job demands.

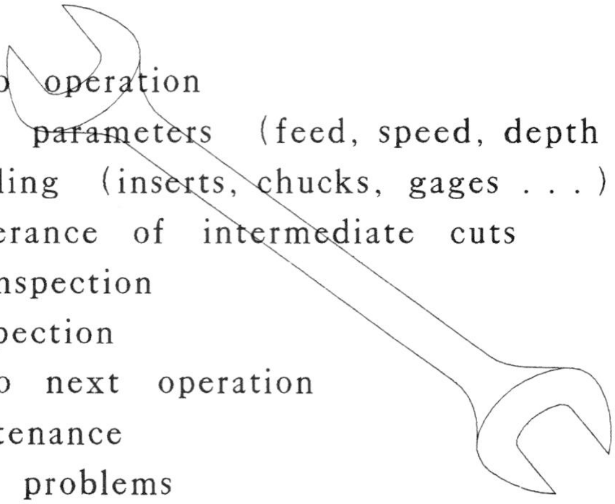

- Machine setup
- Move stock to operation
- Determine cut parameters (feed, speed, depth of cut . . .)
- Determine tooling (inserts, chucks, gages . . .)
- Determine tolerance of intermediate cuts
- First-article inspection
- In-process inspection
- Move parts to next operation
- Machine maintenance
- Report design problems

Fig. 2-1. In a job shop or small manufacturing operation, an individual may have a variety of responsibilities. Typical activities carried out by a machinist in a small shop are listed.

As product complexity and diversity and production volumes grow, so does the organization. There are now too many parts in the product line for an individual operator to be able to make them all effectively. The task is divided into more manageable chunks, and a person hired to take care of each subset of the overall requirements. As product complexity grows, operators specialize in different machines and part types, because no single person can hope to know how to do everything required, Fig. 2-2.

COMPLEXITY FORCES DIVISION OF LABOR

Eventually, designers and manufacturing planners find it impossible to keep close ties with the increased number of people. While a handful of close friendships had worked well in the past, the people performing these product and process definition functions must grow as well to handle the work load. Complexity encourages diversity and again, forces division of labor. Now, different peo-

```
        Tool  Engineer : Machine  setup
    Material  Handling : Move  stock  to  operation
Process Planner/NC Programmer : Determine  cut  parameters
 Tool  Engineer/NC Programmer : Determine  tooling
        Process  Planner : Determine  tolerance  of  cuts
     Quality  Assurance : First - article  inspection
             Operator : Production  machining
  Operator/Quality Assurance : In - process  inspection
    Material  Handling : Move  parts  to  next  operation
     Plant  Maintenance : Machine  maintenance
 Operator/Production Liaison : Report  design  problems
```

Fig. 2-2. In a larger shop or manufacturing operation, the responsibilities of a machinist are distributed among different departments and individuals.

ple are responsible for the tasks of conceptual design, layout, design evaluation, prototype testing, material selection, process planning, numerical control programming, and tool design.

As a company becomes larger, the Design and Manufacturing organizations begin to lose their overall understanding of the business. Each person can do his or her job very well, but loses sight of overall company objectives. There is no effective way to communicate with other departments. Phone messages connect two people, but others have little chance of gaining benefit. Memos and other written information will be read, filed, and forgotten, with no reasonable way to ever find the information again, unless someone happens to remember it.

The same is true for solutions to production or design problems. A part that is nearly identical to an existing part can be inadvertently designed from scratch because a different designer was given the task. This is unlike the "old days," when the small group of designers talked and collectively remembered all the parts the company had ever designed. In the same way, a single manager can no longer control product and process definition on every part, because the organization is simply too large. Company structures change in response to increasing size; the old way of doing business is no longer valid or possible.

TECHNOLOGY GROWTH

The evolution of technology has been responsible for creating the need for CIM almost to the same extent as has organizational change and growth within companies. While providing many powerful solutions, technology also creates its own unique set of problems.

AVAILABILITY OF COMPUTERS

Computers were virtually unknown when many of today's larger companies started doing business. The first computers filled entire rooms, and didn't have enough memory to hold more than a small program and an even smaller amount of data. A good hand-held calculator today offers more power than was ever imagined in the early days of computing. Many personal computers currently provide greater processor speed, memory, and storage capacity than the large *mainframe systems* of the early sixties. (See Fig. 2-3.)

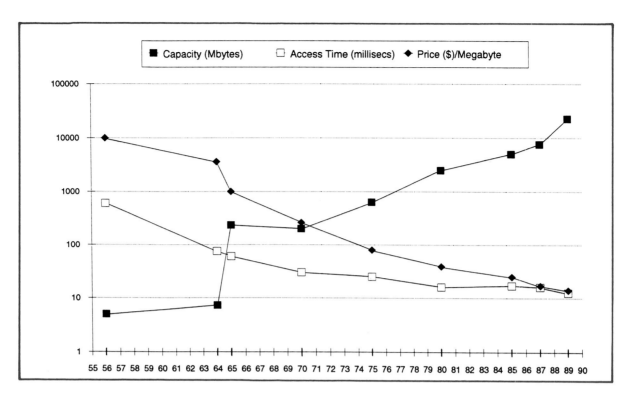

Fig. 2-3. An example of the rapid development of computer technology is the great growth in disk capacity and steep decline in price per megabyte, as shown on this graph. Information for the graph was provided by Bill "Dr. DASD" Donnelly of IBM Corp.

Once computers gained acceptance, they were viewed by most companies as a tool to take data from one place and put it in another, possibly while sorting or modifying the data or summing totals en route. Most applications were for accounting, data collection and report generation functions; most users were aware of the computer only as a mystical "black box" run by powerful magicians to create their treasured reports. Other companies (or people within them) saw computers as tools for speeding up numeric calculations, improving accuracy in the process. The very idea of asking a computer to relate data in ways unknown when the data was collected is still new to many data processing groups.

Computers brought with them obvious problems in the areas of cooling and electrical isolation, data storage, programming, data backups, and archive creation. The more serious problem of *shared data definitions* across an organization was never acknowledged. As computing responsibilities grew, so did the size of the programming groups. These groups often split into smaller ones, each aligned to help a particular set of users such as Accounting, Design, Analysis, or Scheduling. There was no perceived need to share information among these disparate groups; applications and databases were built and optimized to support the needs of small, independent parts of the organization.

NC PROGRAMMING

Numerically controlled machines and robots provide greater accuracy and repeatability, at more consistent rates, than manually operated machines. The use of NC equipment has improved product quality and scheduling accuracy. It has also forced the creation of support organizations of programmers and technicians to keep the machines running.

NC programming has added to process design lead time. A design change can no longer be rushed to the shop floor and immediately implemented by a skilled operator. Instead, NC programs must be written or revised to reflect the design change. The programs should be tested before production use. Thus, the extremely flexible automation made possible by NC programming is not without its costs: it has changed the way companies do business. Through CIM, re-use of product geometry and shop floor information can speed NC program development and improve its quality.

CAD SYSTEMS

Computer-Aided Design (CAD) systems have evolved from simple programs that automated a designer's pencil, eraser, and file cabinet to powerful and flexible design tools. CAD systems are capable of modeling parts in three dimensions and displaying shaded, photo-realistic pictures of parts, Fig. 2-4. They help designers visualize and analyze their creations before committing them to production.

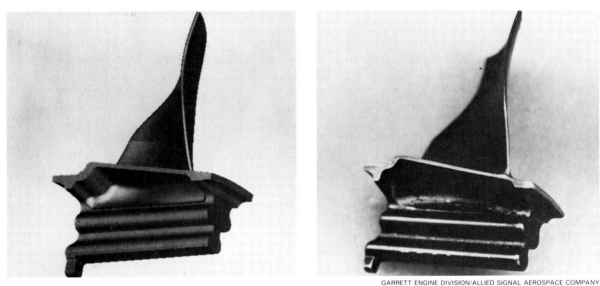

GARRETT ENGINE DIVISION/ALLIED SIGNAL AEROSPACE COMPANY

Fig. 2-4. CAD systems can provide photo-realistic visualization of a part like a turbine blade (left). This permits designers to better visualize and analyze a product before actually manufacturing it (right).

Parametric design tools allow creation of a generic model that can be automatically redrawn to scale after dimensions are entered. *Design libraries* provide quick access to catalog parts (such as bolts and springs) that can be included in a model with a few keystrokes. *Analysis tools* can work with CAD models to evaluate stress, magnetic, thermal and kinematic properties of a design before prototypes are built, Fig. 2-5.

CAD systems, like other computer tools, have some negative aspects. Each CAD system and hardware platform defines geometry and storage structures differently, making data sharing between dissimilar systems difficult at best. CAD systems and models require large amounts of memory, storage space, processing time, and network transmission time. Also, there is some tendency (since electronic designs are so hard to validate and electronic changes are so easy to make) to release faulty designs in a hurry to meet schedule deadlines, with the intent to "fix them later."

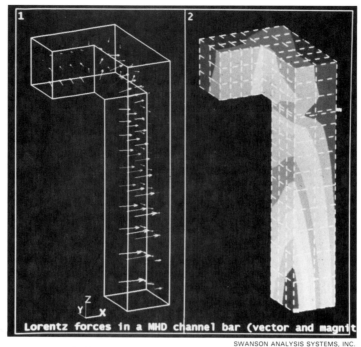

Fig. 2-5. A computer-generated electromagnetic analysis of a conductor for a magnetohydrodynamic generator. The different shadings indicate various magnetic intensities.

DATABASES

A *database system* is at the heart of any CIM architecture. Such systems have been in use for decades, but were largely developed without concern for consistent data definitions across all applications in a company to allow information-sharing. This lack of foresight was not anyone's fault: applications were developed with specific needs and users in mind; only a limited set of data was planned for computer storage. In early systems, *ad-hoc queries* (which let users combine data to create reports and search for answers beyond what available applications provide) were limited by database technology and often restricted to a small number of highly trained data processing professionals.

As a company's information requirements grew, so did its database. Programming staffs could not find time, whenever data requirements changed, to restructure the database and the increasing number of applications that depended on it. The programmers found it much easier to add on to the existing database and write data translation programs to move data between applications. Over time, database and application maintenance became a major part of the workload in many data processing departments.

Need for data sharing

In a CIM environment, *data sharing* is essential. Database technology has advanced to the point where speed, structures, and query languages (to support ad-hoc queries as well as access by applications) are capable of handling many CIM requirements. *Management Information Systems* (MIS) or Information Systems groups (usually referred to, in the past, as *data processing groups*) still have a long road ahead to provide CIM capabilities throughout most companies. Because of the very real and significant manpower cost of modifying existing applications or databases to meet new CIM requirements, most MIS groups are slow to react.

Information users in other departments have found alternatives in personal computer networks and minicomputers outside the immediate control of MIS. Users often find it very easy to create stand-alone databases on personal computers. These databases evolve into important parts of a company's overall processes by giving their users new ways to view information.

There are some dangers involved that many such users never realize, however. Users often neglect to properly back up personal databases, leaving themselves open to data loss if a disk drive should malfunction. More importantly, personal databases usually contain at least some information in common with mainframe databases under MIS control. Since there is no way to ensure the data is identical in both places, decisions based on one set of data or the other may have different outcomes. In addition, other users who could benefit from the information in a given department's minicomputer or a single user's personal computer have no way to access it.

DATA INTEGRITY

Problems of *data integrity* also plague MIS departments. Separate databases, supporting different applications, often will contain duplicate information. Some of this duplicated information is necessary to create relationships between data in different databases. Called "keys" or "key fields" (a *field* holds one piece of data), they provide a common value in two different storage locations or *records* (groups of related fields within a database).

For example, a product definition database may have records that contain a part number, a raw material specification, and an amount of raw material needed per part. Another database, used for production scheduling, might have records that include a part number, due date, and quantity required, along with other information. The common value between the two databases, *part number*, allows someone in Purchasing to determine when to order which raw materials, and how much to order. Such key fields are necessary to link separate databases, Fig. 2-6.

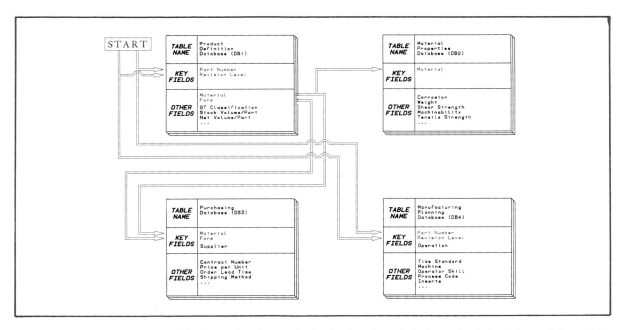

Fig. 2-6. Data that is stored independently can be logically related, and can be linked through key fields common to several databases. Note, for example that "Part Number" and "Revision Level" are common to the databases at upper left and lower right.

Problems arise when key fields have different definitions across databases for different applications. Data relationship links cannot be made without first translating key field values into a consistent format.

Other data definition inconsistencies may impair data integrity. Two data fields may be given the same name but hold information with different meanings, or two fields holding information with the same meaning may be given different names or storage formats.

Although most applications are programmed to prevent entry of the wrong kind of information (such as letters in a numeric field), it is possible in some cases to check data values to be sure they are correct. A machine called out in a routing can be verified by checking against a list of valid machines. A further step would be to check the type of process ("making a hole," for example) to be sure the selected type of machine can actually carry out that process (make a hole). It might also be possible to verify that the specified machine will not be down for planned maintenance or loaded with another job at the time it will be needed, or that the machine is capable of holding the required process tolerances. If the machine selected fails to meet any of the specified conditions, the program would require the user to enter a new machine number.

These types of *data validation* are possible but costly. Few information users are willing to pay the expense of asking a computer to make all possible checks, even if programmers were smart enough to know all the potential problems to check for. For critical information, the expense of extensive data validation may be justifiable. For other data, there comes a time when it becomes more economical to trust the typing skills of the users. Some bad information will inevitably get through to the database. Unfortunately, data deemed not critical when the database and applications were designed may become critical when used for a different purpose.

Other concepts important to data integrity involve traceability, security, and referential integrity. *Traceability* means that older data can still be retrieved and tied to products by serial number, lot number, date, or other reference. Traceability is important to companies making products that may impact public safety, such as aerospace, automotive manufacturing, and pharmaceutical firms. If an accident should occur, information such as a serial number or date stamp can be used to determine how the product was made and what materials went into it. If properly managed, traceability information can be used to track down other products made the same way or from the same materials, so that they can be recalled before further accidents occur.

Related to traceability is the concept of *revision control*. As drawings, plans and other data change, there must be a system of marking to indicate which information is most current and which revision should be used by each application. For example, a part change may not become effective until a current inventory of older parts is used up. An assembly plan can still point to the older part and related drawing, even though a newer revision has been released.

Data security is used to prevent the making of unauthorized changes in information. As the use of CIM increases, more data is accessed by a larger number of people. The accuracy of each data item becomes more important. For this reason, only those people with the proper responsibility and training for updating a specific data item should be allowed to do so. Security may also be used to prevent unauthorized reading or copying of information in a database.

Referential integrity means that data references and relationships must not be damaged unintentionally by changes in data. It is possible, for example, to remove a routing from a manufacturing database. What happens if tooling or NC programs designed for the routing are not also removed? In a poorly designed system, the routing might provide the only relationship to the tooling or programs.

A database could be designed in such a way that looking at a tool alone, there is no way to know if a routing uses it or not. An operation in the routing calls out the tool. Without the routing, there is no way to know if the tool is still needed, or if it can be scrapped. In a system that has referential integrity designed in, it should be impossible to remove an item (like a routing operation) without also removing any active references to it (such as a tool). Information should not be "orphaned" in the database, Fig. 2-7.

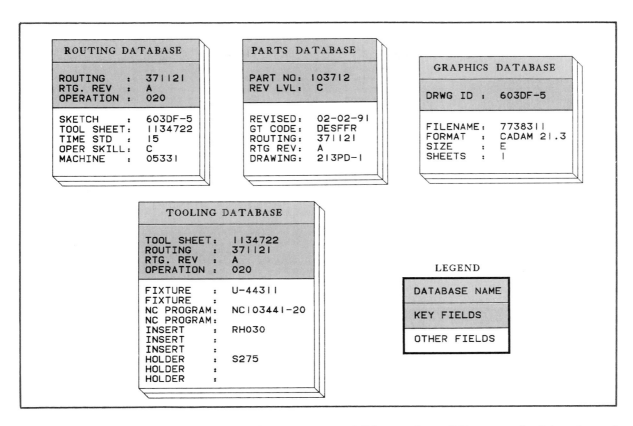

Fig. 2-7. *Referential integrity prevents the occurrence of "data orphans." For example, it is not possible to delete operation 020 of routing 371121, revision A, without also changing or deleting tool sheet 1134722, since these two items refer to each other. In the same way, graphics file 7738311 cannot be erased without also changing operation 020, since that operation refers to a drawing (603DF-5) that is in that graphics file.*

Problems of data integrity reinforce the need for a **CIM data architecture** that defines storage structures, data definitions (both *physical*, such as length and data type, and *descriptive*, including meaning of data as well as possible values), security structures, network structures, and access structures. Access structures govern how users get applications or query languages to access data, including user and system interface standards. Such an architecture, or plan, provides a single, centralized view of a CIM environment. It is perhaps the most effective tool for fighting data integrity problems.

A CIM data architecture is not intended to be a single "best guess" at one point in time that is written down, used once, and forgotten. Instead, it must be continually maintained and managed to ensure the best possible integration of information, applications, and users as information requirements and applications evolve.

CURRENT CAPABILITIES AND APPLICATIONS

Computer technology continues to advance at an accelerating pace. Many of the tools and techniques necessary for CIM implementation already exist. As CIM requirements grow, computer tools will evolve to meet them.

NETWORKS

A number of different *networks* exist to permit interconnection of computers, terminals, data collection devices, and automated machine tools such as NC machines, robots, and coordinate measurement machines. There are several competing network standards, such as MAP (Manufacturing Automation Protocol), DECnet® and TCP/IP (two forms of Ethernet, the former supported by Digital Equipment Corporation and its line of computers, the latter supported mainly on engineering workstations and other minicomputers), SNA® (System Network Architecture, an IBM proprietary protocol), and all sorts of personal computer networks, such as Xenix®. Networks can pass data using various types of wires, fiber optic cables, and even microwave transmission. Selection of the best network for a given situation depends on the types of computers existing and planned within a company, and the network capacity and transmission rates needed, Fig. 2-8. Off-the-shelf solutions or consulting services can be purchased to make nearly any system talk to nearly any other system. Networking is important, but lack of network technology is no longer a block to implementing CIM.

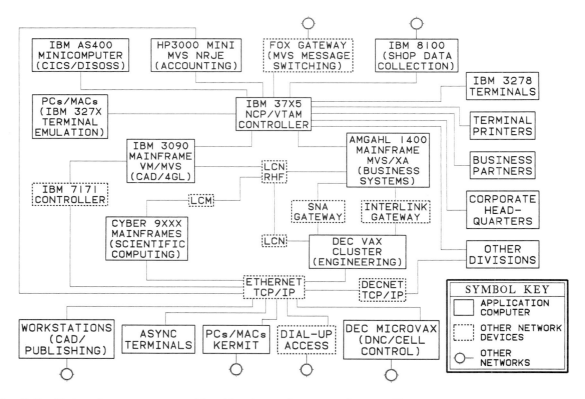

Fig. 2-8. Networks can vary considerably, depending upon the specific needs of a company. This chart shows the network applications of one company.

HARDWARE COMMUNICATIONS

Typical shop floor hardware connected to the rest of a company's system through networking consists of NC machines, coordinate measurement machines, electronic gages and terminals for data collection and display, and printers for paper output. DNC (Distributive Numerical Control) systems are available from a wide range of suppliers to function with the specific computer system in use. Such systems will support NC and CMM program distribution, and may go so far as managing program libraries or providing editing capabilities.

Electronic gages or terminals can be used to collect process information, such as temperature, vibration, tolerances produced, or other performance indicators. Time and attendance information can also be collected electronically.

The use of bar code readers can save time and improve data integrity if part numbers, operator ID numbers, machine numbers, and other information can be scanned from *travelers* (paperwork) accompanying an order through production. An even better solution is to keep the paperwork in the computer system and display it electronically. This helps to keep the paperwork synchronized with the work being performed. Operators should then be able to select their entries of part numbers, operation numbers, and so on, from a computerized list, in order to improve data integrity and reduce typing skill requirements.

EMBEDDED COMPUTERS

Embedded computers provide additional capability in the shop, supplying local processing power or memory to free departmental or mainframe computers for other tasks. Most NC equipment and coordinate measuring machines have computers built in to provide editing capabilities and possibly *adaptive control* (real-time response to changing process conditions). Some NC controllers go so far as providing graphical programming environments.

Embedded computers also support off-line data collection, Fig. 2-9. Small, hand-held bar code readers or *OCR* (Optical Character Recognition) scanners can be used when picking items from inventory, at receiving and shipping, and when moving parts through the shop. A user can scan bar codes or text and enter quantities, condition codes, and other data, without the need for a terminal connected to a cable. At the end of a shift, or at shorter intervals, the device is connected to the network and the data sent to a central database. A variation of this is *buffering* (saving until it is uploaded to the network) of electronic gage data or labor transactions. One drawback to both off-line data collection and buffering is that the new information will not be available to other users until it is uploaded to a central database.

GARRETT ENGINE DIVISION/ALLIED SIGNAL AEROSPACE COMPANY

Fig. 2-9. Data collection and entry can be performed quickly and accurately with devices such as the barcode reader being used here.

SYSTEMS INTEGRATION

There are numerous consulting firms specializing in *systems integration*, the process of interfacing different types of computer and network hardware or in getting software applications to talk to each other. Many hardware vendors, such as IBM and Digital Equipment Corporation, are also able to provide such services. There are two problems with using consultants for systems integration. First, after the consultants depart, there may not be anyone left who understands the system they have designed. Second, the consultants will not understand a company's environment (cultural, software, and hardware) as thoroughly as the company's own employees.

Despite these concerns, consultants have a useful role: they can bring an unbiased and educated perspective to solving CIM problems. Specialists in change psychology can assist company personnel in overcoming the deep-rooted fear of change; consulting firms specializing in systems modeling can help a company to understand its current "as-is" process limitations and develop a "to-be" model representing an ideal business process to work toward. With the combined use of consulting and in-house expertise, most system integration issues can be solved.

PROBLEMS TO OVERCOME IN IMPLEMENTING CIM

Even though most of the technical problems associated with CIM can be solved, and help is available in solving cultural issues, there are several very real obstacles involved in the planning, selling, and implementing of CIM in a company.

INTERDEPARTMENTAL SUPPORT/POLITICS

A company's organizational structure itself often discourages or even prevents interdepartmental projects. Employee performance appraisals seldom provide any incentive for working with other departments on activities ranging beyond a specific job that's been done the same way for years. Large companies, especially those where Department of Defense contracts are in effect, require charge numbers against a specific department budget. Because of the way accounting systems have evolved, it's difficult for different departments to charge against the same project. Another problem in larger companies is that jobs have become highly specialized. This makes it difficult to find a project team that understands enough about interdepartmental processes to carry out a successful CIM project.

A common organizational problem is the NIH ("not invented here") syndrome. One department may be uncooperative in working with another because it feels the other will get the credit, or because it would rather see a solution that meets its own specific needs, without compromise. A department may be unwilling to accept the initial project investigations and resulting justification done by another department, and may insist on recreating work from the beginning rather than validate or accept what has already been accomplished. There is strong competition within a company for resources, and a department that follows another will not believe it has the same chances for more people and funds as a leading department. Management must find incentives to overcome these misconceptions.

CIM JUSTIFICATION

Before management will support a CIM project, there must be sufficient justification to demonstrate that the project is more important to pursue than others competing for the same funds and other resources. Above all, justification needs quantifiable values for costs and savings. No matter how good an idea sounds, businesses exist to make a profit. Even if the real reason for pursuing a project is strategic (for competitive advantage or increased market share, for example), there must be a reasonable belief that the project won't harm the company on cash flow impact alone.

A *cost/savings measurement plan* is a necessity. Such a plan shows what to measure, how to measure it, and how long to measure it. Management must be able to see a project's success or understand its failures before it will approve any similar ventures. If savings projected in a justification cannot be measured, they should not be presented. To do so reduces the presenter's credibility for future projects.

Also, the measurement effort must be in balance with the overall savings plan. If it costs $100,000 to measure a project with $150,000 savings, then the net savings is really only $50,000. This may still be enough to go forward with the project, but savings might be improved with a better measurement plan.

Finally, measurement for a specific project should probably stop sometime shortly after **ROI** (return on investment) or initial savings goals are reached and trends indicate the project is moving in a positive direction, Fig. 2-10. Otherwise, as more and more projects accumulate savings through the years, it may appear as if the company is saving more than it makes or spends, and the numbers become meaningless. Once a project becomes part of "the way of doing business," it should stop being counted as an improvement to the bottom line.

For a variety of reasons, costs and savings may be difficult to measure on a project. First, the true drivers for costs and savings may not be understood; the wrong things may be measured. Second, the appropriate data may not be tracked in existing accounting systems. In measuring the average cost to the company for processing an engineering change, for example, the impact on Manufacturing may be hidden. This could occur because costs due to a change are absorbed into department

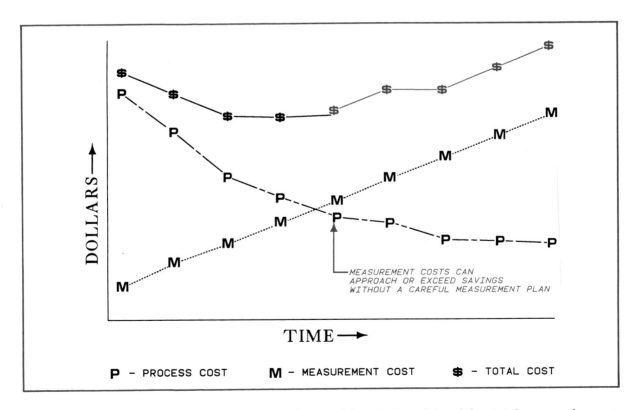

Fig. 2-10. *Process measurement costs must be considered when determining total process improvement benefits. High measurement costs could, at some point, overshadow the cost of the project itself. This would reduce or negate process savings.*

overhead scrap rates and not charged to a specific part, product, or cause. Finally, expected savings may span multiple projects. Two or more projects may both plan to reduce *work-in-process* (WIP) inventory costs by 60%. Each may, if considered independently, be able to achieve that goal. A naive executive might suggest implementing both projects to reduce inventory costs by 120 percent, making money for the company on each item in inventory!

Clearly, the projects must be considered together and a more reasonable savings estimate offered. Project leaders may agree to apportion the savings, with each claiming a lower percentage or one claiming a percentage of what remains after the other project is completed. Care must be taken so that if one project fails and drives the performance measurement the wrong direction, the other project isn't penalized.

There are additional benefits to considering projects together, rather than separately. Common interface or integration problems can be solved once and applied to both. Where similar interfaces need to be done, one team can do the work for both. Training schedules can be coordinated so that users can get a unified understanding of both projects, rather than learning about two sets of changes from seemingly competitive projects at the same time. At first, this combined effort may push back schedules or inflate project costs, because each project is doing more work than was planned. However, the benefits of working together will usually overshadow the apparent increase in implementation costs, Fig. 2-11.

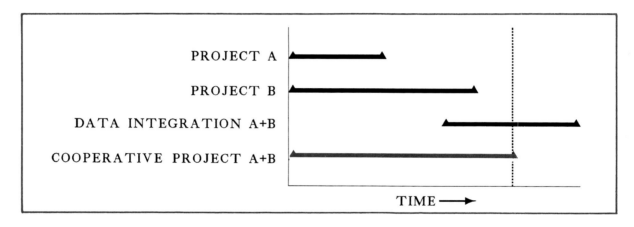

Fig. 2-11. Cooperative project management can reduce the implementation time needed for a project. For example, the cumulative time required to do project A and project B cooperatively is longer than A or B alone, but less than doing both A and B and the necessary after-the-fact data integration.

INTANGIBLE BENEFITS

There are many intangible benefits to most CIM projects; benefits which, for reasons described above, cannot be measured. Improved worker morale, decreased lead times without increasing or reducing hours worked, improved competitive edge, and faster throughput are all examples of intangibles. Some of these (like faster throughput) may seem easy enough to measure, but justifying such savings can prove futile if the intent is to improve worker productivity without reducing the workforce. This scenario inevitably results in statements like, "we'll have more time to do it right" or perhaps, "we can spend more time on the really important issues." These may lead to better quality or other intangibles, but provide no quantifiable (measurable) savings. Still, some executives will settle for the "intuitive goodness" of a project even if benefits cannot be measured.

ADDITIONAL ASPECTS OF CIM

One book cannot hope to fully cover every issue related to CIM. To introduce other aspects of CIM not included in later chapters, several additional topics are briefly presented below. These topics should be considered in CIM planning activities.

SIMULATION

Software simulations of processes, networks, and business plans have been possible for years. Advances in CIM tools now make available other forms of simulation. Three-dimensional solid modeling graphics systems make possible *soft mockups*, which are on-screen images based on part models, Fig. 2-12. This software lets designers check for interference between components and their surroundings. Some systems, such as CATIA® from Dassault, let a designer move parts around on the screen, to be sure they can be installed or repaired. Without such a tool, companies must invest heavily in hard mockups made of wood, plastics, and other materials. These mockups are used to test physical part shapes before committing them to real tooling and machining processes.

Another form of physical part simulation is provided by a process called *stereo lithography*, Fig. 2-13. This process uses a three-dimensional CAD model to guide a laser through a polymer solution. The laser hardens the solution within the boundaries of the part, one thin layer at a time, until a completed part can be removed. The part is a real, three-dimensional mockup of the part described by the model, and is invaluable for visualization.

GARRETT ENGINE DIVISION/ALLIED SIGNAL AEROSPACE COMPANY

Fig. 2-12. Solid modeling can provide an electronic ''mockup'' of a product, such as this turbine engine. The model can be used for visualization, for validating component assemblies, and for determining appropriate wiring and plumbing needed to connect the turbine to the airframe.

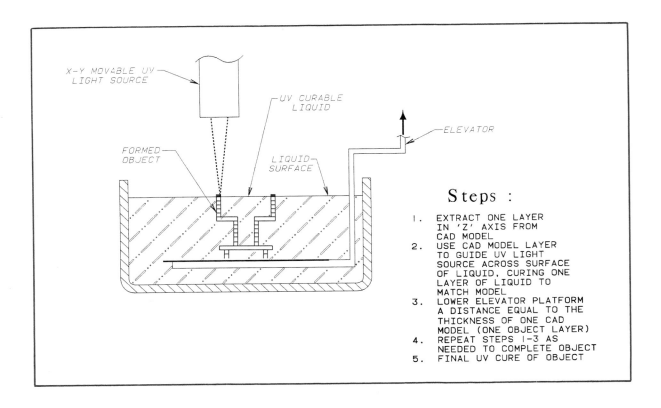

Steps :

1. EXTRACT ONE LAYER IN 'Z' AXIS FROM CAD MODEL
2. USE CAD MODEL LAYER TO GUIDE UV LIGHT SOURCE ACROSS SURFACE OF LIQUID, CURING ONE LAYER OF LIQUID TO MATCH MODEL
3. LOWER ELEVATOR PLATFORM A DISTANCE EQUAL TO THE THICKNESS OF ONE CAD MODEL (ONE OBJECT LAYER)
4. REPEAT STEPS 1-3 AS NEEDED TO COMPLETE OBJECT
5. FINAL UV CURE OF OBJECT

3D SYSTEMS, INC.

Fig. 2-13. The stereolithographic process produces three-dimensional physical mockups of a part, without the need for tooling or machining. A—How the process works. B—Several part mockups made using stereolithography.

Depending on tolerance requirements, it may be possible to use the plastic part in making mold patterns, testing machining processes, and in aerodynamic analysis. Such parts can also be used in physical mockups of the product and for marketing purposes.

Another simulation made possible by CIM tools is in the field of electronics, where software "components" can be used to construct circuit boards or chip logic. The resulting model can be validated before any real parts are made.

On-screen simulation of program steps can be used for NC program verification. Various levels of sophistication are possible, from watching a circle representing a cutting tool move through a two-dimensional *wireframe view* (where all lines in a model are visible, as if it were made from glass), to watching shaded solids be "milled" on-screen by a solid model of the cutting tool. Many robotics programming languages can also graphically simulate the motions of a robot within a model of its workspace.

SYSTEM DESIGN (CASE) TOOLS

System design tools continue to improve. Of special significance for CIM are *Computer-Aided Software Engineering* (CASE) tools, Fig. 2-14. These are programs to help design components of an application and the information flows between them. CASE tools include programming aids (sometimes called "software building blocks"), debugging tools, software design, and configuration management utilities for a wide variety of computers and programming languages. CASE tools are also available for designing and maintaining data definitions in a central data dictionary accessible by other applications, providing a backbone for CIM integration efforts.

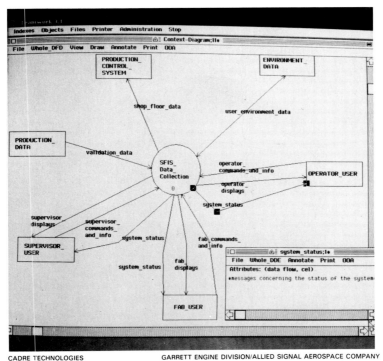

Fig. 2-14. Computer-Aided Software Engineering (CASE) tools can model data definitions and relationships, provide visibility of program and data relationships, and manage model integrity from concept through code maintenance.

ORGANIZATIONAL AWARENESS

Although "awareness" is not a CIM tool per se, every person in an organization must be aware of how his or her job relates to the overall company, where information used in the job comes from, and how information created or changed in the job is used in other business processes. (See Fig. 2-15.) People need to be aware of the company's long-term goals and short-term objectives. They also need to understand how change can be good for a company. Without this understanding, they will not accept changes brought about by CIM projects and may, in fact, prevent such projects from succeeding. There are numerous consulting and training firms that can provide help in designing and administering education plans. This is one aspect of CIM implementation that must be addressed by any company.

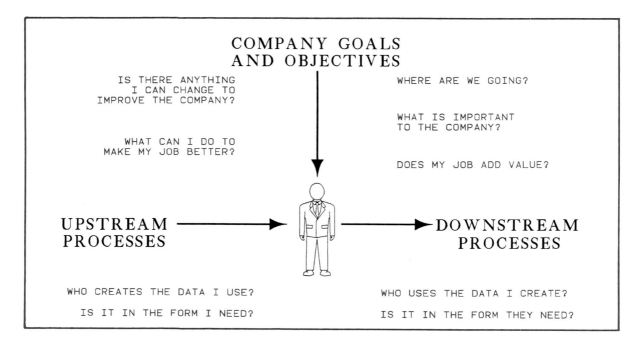

Fig. 2-15. Organizational awareness can improve a worker's self-esteem and improve productivity and effectiveness. It also can help improve the worker's understanding of the company's goals and concern for the company's future.

FILE MANAGEMENT SYSTEMS

Not all CIM data is easily stored in a database. A database is designed to accept data in the form of records with specified fields, such as part number, order quantity, due date, and buyer's ID code. Such considerations as which data fits in which columns of a record, how data is indexed for faster searches, and so on, are all neatly defined. This works well for a lot of CIM information, but some data is stored in large files with structures too complex to describe to a database designed to handle more discrete types of information. A CAD model, scanned image, NC program, or product documentation are all examples of information best stored as a file and not as discrete data records.

The problem with storing data in files is that a file can be read back only by the application that created it, or must be translated to another application's format. Either way, files limit data sharing.

Files are also difficult to track through revisions, backups, and *archiving* (off-line storage for less-frequently accessed files). Archived files are still accessible after some time delay. Taking them out of on-line storage will make room for new information needing quicker on-line access.

File management systems, Fig. 2-16, provide a solution to all these problems. They store any kind of file, without worrying about what it contains, and provide tools for making backups, archiving, and revising. In addition, they store information about the file, such as a description, what application created it and when, who created it, links to similar files, and other information needed by a company. For applications other than the one that created (and can read back) the file, it may be sufficient to know that a file exists and to access the information about the file, without regard for its actual contents. A number of file management systems are commercially available.

THE "PAPERLESS FACTORY"

A goal of many companies is to move toward what has been termed the "paperless factory," where all production information and instructions are accessed electronically. Paper is viewed as bad because it takes so much longer than electrons to move through a system. If justification can be found, the move toward fully electronic information distribution can be supported by existing technology that is still improving at a rapid rate.

Of course, machinists may still want to work with printed instructions and drawings, so they can flip through a large stack of drawings or process steps, pencil in notes, and check off completed steps. Paper may continue to be a viable medium in a production or assembly shop; the conception that paper is inherently "bad" may be wrong.

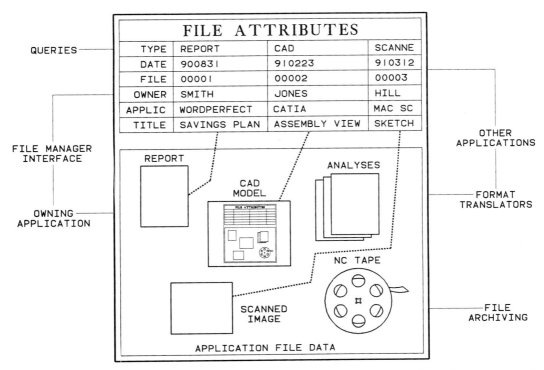

Fig. 2-16. *A file management system maintains descriptive records for each file in a common format. This permits users and applications to find and access data stored in various formats.*

The "bad" aspect of paper systems is that paper is normally used as a control mechanism. Process steps are delayed while waiting for paper document distribution. Signoffs on paper require a serial distribution through all reviewers, because the same document must be signed by each. If the controlling mechanism can be made electronic, and paper used only as a reference, process lead times can be reduced along with most other problems associated with paper systems. It may prove more cost-effective, for example, to print operator instructions at a tool crib for each operation, rather than provide an electronic display at each operator station. Paper should be used as a reference, not as a control.

FEATURES-BASED DESIGN SYSTEMS

As described earlier, most CAD systems store information in files that are not easily understood by other applications. An evolving trend in CAD systems is the move toward *features-based design systems* that store geometric information as lists of features with parameters to define details of each feature type and how it connects to the overall part. This is in contrast to the conventional method of storing geometric information as lists of points, lines, curves, edges, and surfaces. These systems offer great productivity gains, since a designer can, for example, resize a hole and automatically reconnect it to the surrounding surfaces by merely changing a diameter. In traditional CAD systems, the same operation might require erasing all line segments of the hole, drawing a new hole, and intersecting it with the surfaces it penetrates.

A features-based design system also offers new potential for process design as well. The features database for the model can be queried by process planning software to see what features are required on a part, their dimensions, tolerances and placement. Such a system can improve accuracy and reduce the manual effort needed to describe parts to most current planning systems.

EVOLVING STANDARDS (IGES, PDES, CALS)

The *Product Definition Exchange Specification* (PDES) blurs the distinction between CAD data and other non-geometric data needed to design, schedule and process parts. PDES grew out of efforts to support the concept of exchanging all information describing a product between dissimilar applications. One of those efforts is the *Initial Graphics Exchange Specification* (IGES), which lets CAD systems exchange some geometric information by translating to and from an intermediate but common IGES file format, Fig. 2-17.

CALS (*Computer-Aided Acquisition and Logistics Support*) is a United States Department of Defense initiative aimed at improving access to CIM data between defense contractors and the Department. CALS is a logical extension of CIM data sharing, but presents new problems in security and raises questions of ownership. In the past, the Defense Department settled for ownership of a design, but not the processes to create production hardware. Through CALS, it hopes to gain access to process plans in addition to designs, so that other companies (although they are potential competitors) can be given the designs and plans to make the same products. CALS compliance will definitely benefit large corporations with a number of sites by providing them the capability to exchange information within the corporation. CALS may also pave the way for easier exchange of CIM data between companies working together on the same contract. CALS will also help the Department of Defense monitor contract work, potentially saving money for taxpayers.

CONCURRENT ENGINEERING

Concurrent Engineering (CE), also called *Simultaneous Engineering*, is a *cultural* tool as opposed to a *computer* tool. It is a (largely undefined) methodology for getting more departments in a com-

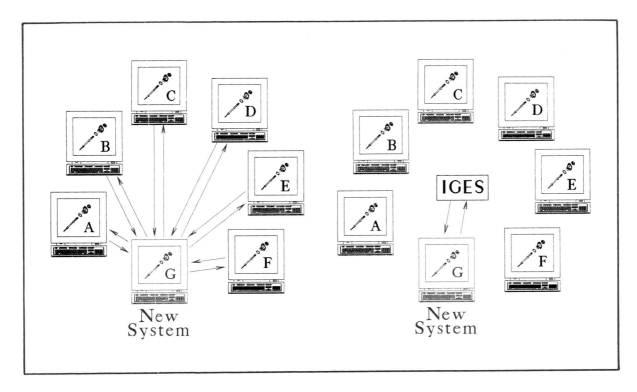

Fig. 2-17. The Initial Graphics Exchange Specification (IGES) concept greatly simplifies communications between different CAD systems. Without IGES (as shown at left), translators must be provided to connect a given system with all other types of CAD systems. In this example, System G would require a total of 12 translators to communicate with systems A-F. For complete system-to-system communication (allowing any system to communicate with any other system), a total of 84 translators — 12 x 7 — would be needed. With IGES, as shown at right, only two translators (to and from IGES format) are needed for each system. Each system can communicate with any other system through IGES. In this example, only 14 translators — 2 x 7 — would be needed, instead of 84.

pany, and even close suppliers, to work together on a design. The group will assess the design (beyond the traditional parameters of cost and performance) for its ability to be manufactured, assembled, repaired, replaced, and inspected. The goal is to be reasonably certain the design will meet the needs of all departments in the company as well as those of customers, contractors and suppliers, all before the design is released for production.

SUMMARY

As manufacturing companies' operations grow and become more complex, the need for Computer-Integrated Manufacturing also increases. The availability of computers, the widespread use in industry of CAD systems, databases, and numerically controlled production machines, have made essential the efficient sharing of information.

Data communication is made possible by a wide variety of networks that link computers, terminals, data collection devices, and automated machine devices such as DNC networks and robots. The integration of these systems has become very difficult, from both hardware and software perspectives. Integration consulting firms are often contracted to assist in this effort. The company's organization structure is a major barrier and must be changed. This requires cultural changes from the management of the business.

Several methods are available to quantify values for costs and savings when justifying expenditures to management for implementing CIM systems. There are both tangible and intangible benefits. Many of the tangible savings may be difficult to document since the appropriate data may not be tracked in existing accounting systems. Much of the justification of CIM systems must be based upon numerous intangible benefits. These cannot be measured accurately, but are a necessity to advance a company to the status of a world-class manufacturer. Intangible benefits include such things as worker morale, competitive edge, faster throughput, and data integrity.

Topics that might prove important to CIM implementation include software simulation, Computer-Aided Software Engineering (CASE), file management systems, Electronic Document Distribution (EDD), feature-based design systems, Product Definition Exchange Specifications (PDES), Computer-Aided Acquisition and Logistics Support (CALS), and Concurrent Engineering (CE).

IMPORTANT TERMS

adaptive control
ad-hoc queries
analysis tools
archiving
buffering
CIM data architecture
Computer-Aided Acquisition and Logistics Support
Computer-Aided Design
Computer-Aided Software Engineering
Concurrent Engineering
cost/savings measurement plan
data integrity
data security
data sharing
data validation
database system
design libraries
embedded computers
features-based design systems
field

file management systems
Initial Graphics Exchange Specification
mainframe systems
Management Information Systems
networks
OCR (Optical Character Recognition)
parametric design tools
Product Definition Exchange Specification
records
referential integrity
revision control
ROI (return on investment)
shared data definitions
Simultaneous Engineering
soft mockups
stereo lithography
systems integration
traceability
traveler
wireframe view
work-in-process

QUESTIONS FOR REVIEW AND DISCUSSION

1. Match each of the acronyms below to the best description:

   a. JIT   _____ Network for machine instructions.

   b. CMM   _____ Information exchange with external companies.

   c. DNC   _____ Ordering strategy to minimize inventory.

   d. IGES   _____ Electronic drawing or modeling tool.

   e. CALS   _____ Translates drawing files between systems.

   f. MRP   _____ Programmable inspection tool.

   g. CAD   _____ Back-schedules orders from due dates.

2. What problems would you expect to encounter in a project to pass a three-dimensional model of a new part from Design Engineering to NC Programming? Consider both technical and cultural issues. What aspects of justification can be quantified in a project like this?

3. How can lead time be decreased if actual process time spent is not reduced? Consider product definition lead time (the time needed to design a product and its associated processes). Assuming actual design and process engineering time is fixed, is it possible to make a design production ready sooner?

4. Which of the following functions in an organization would benefit from knowledge of time standards (the predicted time needed to complete each operation in a process plan)? Briefly describe how the function could use this information:
   a. Conceptual Design
   b. Thermodynamic Analysis
   c. Accounting
   d. Customer Support
   e. Facilities Planning
   f. NC Programming
   g. Shop Scheduling
   h. Production Planning
   i. Contracts Administration
   j. Purchasing
   k. Quality Assurance

5. Divide the following data types into four separate databases supporting the functions of Product Design, Process Design, Production Scheduling, and Production Control. A data item can be included in more than one database if necessary (for possible use as a key field relating data in different databases, or if the same type of data can be used for more than one application):

Part Number	Tool Number
Operation Number	Part Name
Operator Number	Cost Per Part
Time Standards	Material Handling Specification
Drawing Number	Maximum Operating Temperature
Machine Number	Dimensional Tolerances
Operator Skill Code	Due Date
Material Specification	Operation Sequence
Quantity	Revision History
Scrap Rate	Maximum Revolutions Per Minute
Machine Burden Rate	Effectivity Date Range
Process Specification	Where-Used? (what assembly or product uses
Signoff Status	this part?)

6. Describe how the development of the need for CIM as a company's operations grow more complex might apply to software development practices rather than the more traditional *manufacturing* practices. For example, consider reuse of code in a new application as a parallel to reuse of proven design techniques, or the division of labor in a large programming department by application type (business, design analysis, manufacturing, accounting, graphics, scheduling, etc.) or by programmer responsibility (system analysis/design, coding, testing, integrating, maintaining, documenting, and so on).

7. In a company with limited funding, upper management believes in the sequential strategy of "Understand, Simplify, and Automate" for improving business processes, but would rather stop current CIM projects after the Simplification step and before Automation. They claim that up to 50% of the savings can be realized for a small fraction of the investment needed to pursue full automation. Answer each of the following questions:

a. What aspects of CIM can the company apply without computers?

b. What benefits will be lost if automation is not carried out?

c. What are the risks involved in proceeding through the steps of understanding, manually simplifying, and planning for automation of business processes with the intention of implementing automated systems when funds are more readily available?

d. What are the risks involved in proceeding through the understanding and manually simplifying steps without considering automation and aspects of CIM?

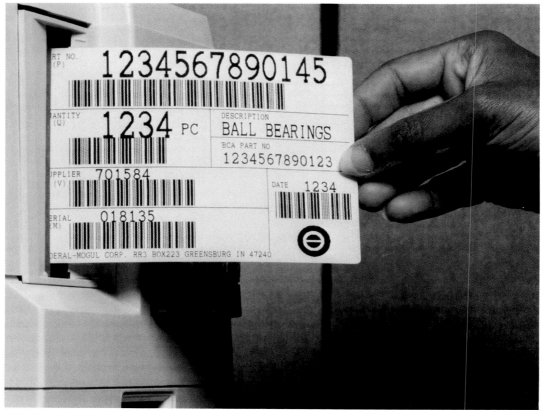

MONARCH MARKING SYSTEMS

Bar code labeling is widely used for part and lot identification, enabling either off-line or on-line data collection. Small thermal printers can be used to generate bar code labels as needed.

3

Manufacturing Product Planning

by W. Van Twelves, Jr.

Overview

Product design for manufacturing, or *manufacturing product planning,* is a multidisciplinary endeavor. To be effective, it must be performed in an *integrated* (blended into a single unit) manner. In this chapter, an integrated approach to product planning in the CIM environment is discussed. It is essential to understand that business objectives, marketing environments, products, and available facilities vary widely from one firm to another. Therefore, it is impossible to present a detailed product planning method that will fit the needs of every manufacturing enterprise. Instead, this chapter presents a framework that may be adapted and tailored to specific situations.

W. Van Twelves, Jr. is a Lead Design Engineer for Pratt & Whitney Division of United Technologies Corporation in West Palm Beach, Florida.

PRODUCT PLANNING

Within the confines of the manufacturing arena, the question, "What is manufacturing product planning?" can be answered in a fairly concise manner. *Manufacturing* is a series of interrelated activities and operations that involve product design, and the planning, producing, materials control, quality assurance, management, and marketing of that product. *Product planning* can be defined as a systematic program intended to identify products that will generate maximum profits for a manufacturing enterprise.

Thus, *manufacturing product planning* requires input from both the business and the technical functions in a company. Activities that fall within the scope of manufacturing product planning include selecting the correct products, managing company resources, engineering products for economical production, and organizing the production operations for efficient manufacturing.

Product planning in the CIM environment is potentially more effective than conventional planning methods in developing a fundamentally correct product specification. This is possible because planners can use relevant computerized knowledge bases to gain access to essential information. A *product specification* is a clear, complete statement of the technical requirements of a product, and of the procedure for determining if those requirements are met.

An intelligently developed product specification involves more than merely function, geometry, and material. A product that will be successful must meet the needs of the customer in ways beyond function alone. The specification can be affected by the issues of cost, performance, ease of maintenance, availability of spare parts, reliability, safety of operation, skill levels, and product appearance.

ISSUES AFFECTING THE PLAN

The manufacturing plan can be affected by the issues of *producibility* (whether the product can be made), *procurability* (whether parts or materials can be obtained), in-house processing capabilities, scheduling, tooling requirements, and assembly methods. Material handling, storage, warehousing, packaging, and shipping may also impact the manufacturing plan. Other requirements, such as security, liability, or other legal considerations, and even social acceptability, may have to be addressed in the planning stage.

The key to addressing these considerations is having an appropriate computer-accessible knowledge base coupled with a well-defined procedure for product development. There are general criteria that should be part of the check list in any product planning endeavor. The potential for success of a new product improves when that product is one:

- that fits logically into the company's existing lines of business or that is compatible with management's projected thinking.
- for which there is an anticipated market demand.
- for which raw materials can be obtained.
- that can be produced with existing plant facilities. If a new plant is required, the product should be one for which capital investment and resulting depreciation charges are not excessive in relation to the anticipated sales revenue.
- for which the company possesses the financial capability, the manpower, and facilities to develop, manufacture, and introduce.
- that can be manufactured and sold profitably over at least a normal payout period.
- for which the development and introduction costs can be amortized in at least a normal payout.
- upon which the return on investment can be estimated with reasonable accuracy.

Another aspect of successful product planning is establishing appropriate project or program management. Specific tasks, activity sequences, resource allocations, schedules, personnel assignments, and budgets for putting a new product into production all must be defined. Regular project status reporting is then essential to maintain schedules and budgets.

MARKET RESEARCH AND FORECASTING

There are numerous approaches to the subjects of market research and forecasting. Regardless of the research method used, accurate information is essential if an analysis is to be valid. In the manufacturing world, precise forecasting of the rate and volume needed for a production run is very important. An accurate forecast is necessary to schedule equipment, determine the proper type of tooling, order raw materials, budget money and resources, and carry out a host of other activities.

The primary goal of market research and forecasting is to identify products that have the potential to make a profit for the company. The products with sales potential that have been uncovered by market research must be evaluated in terms of both producibility and cost.

One market research method that has withstood the test of time can be summarized in ten basic steps:

1. Lay out your company's objectives as your reference point. Number each objective.
2. Segment your industry into its major components.
3. Segment each of the major components of Step 2 into its major sub-segments.
4. Segment the major sub-segments of Step 3 into their minor sub-segments.
5. Eliminate those areas that do not meet your objectives (as outlined in Step 1).
6. On the left side of a sheet of paper, list the problems in each area of Step 4. These problems represent opportunities to provide solutions by offering a product or service.
7. Eliminate those problem areas whose solutions do not conform to the objectives you listed in Step 1.
8. Make a *matrix* (grid) to use in analyzing your results. On the sheet of paper listing the problems down the left-hand side, write the numbers of your objectives in a row along the top. Draw vertical lines to make columns. Place a check mark in each column where the solution to the problem matches your objective. Select the most promising areas for future action.
9. Along the far right-hand side of your matrix, list probable solutions for each of the problems. Do not hesitate to list many probable solutions to each problem.
10. Select the best market gap (or gaps) and begin verifying them through testing.

Market research doesn't have to be extremely complex. First, break the problem down into appropriate categories, then secure accurate information on market size, product pricing, and manufacturing costs. Information on market size can be obtained from a Chamber of Commerce or similar business organization. Pricing and manufacturing costs can often be identified from historical data or similar products. A basic market profile for a medical instrument manufacturing company is shown in Fig. 3-1. This simple chart illustrates the type of information that can be obtained without a major expenditure of time and effort.

Formatting information in the manner shown in Fig. 3-1 makes it fairly easy to decide which market segments should be given the first priority. If sales history, pricing information, and market segment statistics are available from an updated, on-line computer database, product planning and market strategy tasks can be accomplished quickly and accurately.

Computer programs structured like the one shown in Fig. 3-2 can assist in *forecasting* (predicting) the required production rate and volume. The ability to quickly evaluate various strategies can improve a firm's competitive edge. There are, of course, other factors that enter in to a complete market research analysis. A computerized market knowledge base for a firm's specific product line will reduce the time required to accurately evaluate the profitmaking potential of a new product.

PRODUCT DESIGN

Product design is the function that creates the marching orders for virtually all manufacturing operations. Until the specifications and the product drawings (or CAD drawing files) for a new product are released, no action can be taken by procurement, tooling, manufacturing engineering, production engineering, quality assurance, scheduling, safety, or any of the other factory support functions.

For a variety of reasons, communication between the persons performing the design function and

Medical Segment	Probable Sales Price	Market Size	Distribution Machinery
Single-doctor offices	$2,500	82,000	(1) Medical equipment reps (2) Factory agents (3) Catalog houses (4) Medical distributors (5) Factory direct offices
Multi-doctor clinics	$6,500	22,000	(1) thru (5) above (6) Hospital equipment reps (7) Hospital distributors
Private hospitals	$8,500	1,300	(1) thru (7) above
Government hospitals	$9,100	450	(1) thru (7) above
Universities/colleges	$8,300	690	(1) thru (7) above
Single-dentist offices	$1.250	49,500	(8) Dental equipment reps (9) Dental distributors (2), (3), and (5) above
Multi-dentist clinics	$2,300	10,350	(2), (3), (5), (8), and (9) above
Homes for the aged	$1,000	15,000	(2), (3), and (5) above

Fig. 3-1. Basic market profile for a medical equipment manufacturer. The profile provides information on market size and methods that could be used to reach each market segment.

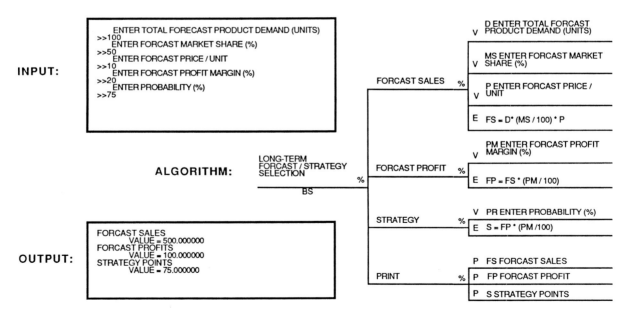

Fig. 3-2. A forecasting algorithm is a computer program that can be used to predict the expected market share and other factors — such as pricing and required production volume — that affect the profit potential of a product.

those in charge of the manufacturing function in many firms is often inadequate. This frequently results in product designs that are unduly expensive to manufacture. The traditional means of addressing this problem is the design review. A ***design review*** is a meeting in which representatives from the manufacturing areas can provide input to the product design. This permits needed changes to be made before the drawings are released for production. See Fig. 3-3.

The effectiveness of the design review depends on the commitment of the individuals involved, the commitment of management to making it work, and the expediency of the schedule.

The CIM environment has the potential to alleviate the problems created by knowledge bases that are isolated from each other. Currently, communication between CAD and CAM is less than complete, due in part to system incompatibilities.

The problem of data exchange between the product definition database (generated by the product design) and downstream activities, such as manufacturing or quality assurance, seem likely to be resolved in the near term. Adherence to standard data formats, such as ***IGES*** (Initial Graphics Exchange Specification) will be part of the solution. Standard data handling protocols will also be useful. More powerful hardware and software capable of simultaneously operating multiple systems and handling data exchange with neutral databases will solve many compatibility problems.

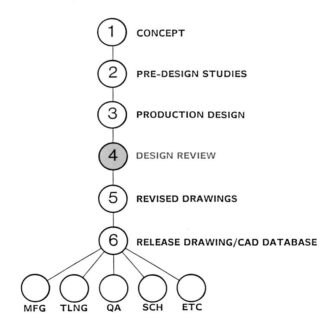

Fig. 3-3. Major milestones in the process of product development. The design review *is an important point in the process, since it allows needed changes to be made before drawings are released for production.*

EXPERT SYSTEMS

As these problems are satisfactorily resolved, the opportunity for further expansion of CAD capabilities will emerge. Perhaps one of the most significant areas of growth for CAD lies in the development of expert systems that will allow the designer to address issues beyond the function and stress analysis of parts.

Expert systems are computer programs that represent a particular domain of human knowledge. They permit a computer to gather information from a database and use it to draw a conclusion or select a course of action. Fig. 3-4 compares the current method of product development with a potential method, making use of expert systems, that can offer critically needed benefits. The important difference between these two sequences is *when* the information input about producibility of the product is introduced. The shortcomings of the traditional method (product recalls, engineering changes, schedule slippage, quality problems, and cost overruns) should provide ample motivation to look for ways to create product designs that are fundamentally correct before production begins.

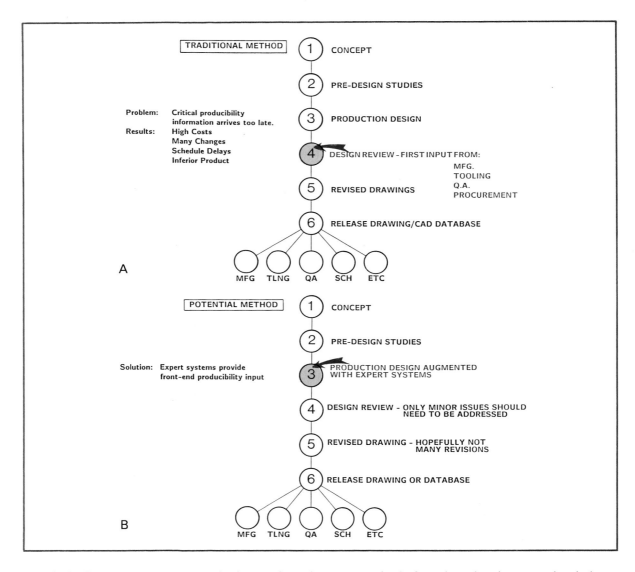

Fig. 3-4. Expert systems are at the heart of an alternate method of product development that helps to produce a product design that is fundamentally correct before being released to production. A — The traditional method of developing a product. B — The potentially improved method involving use of expert systems.

Concern for producibility issues should, in theory, be shared by program management, project engineers, lead design engineers, and individual designers. In many cases, however, the individuals in these positions lack the necessary training or experience to deal with producibility questions. Since the product designer ultimately determines what actually goes into the engineering drawing, he or she should have the assistance of on-line expert systems. An expert system programmed to analyze producibility could eliminate much of the frustration presently found in production operations.

Product definition occurs in several exploratory stages, ranging from concept to predesign studies to production design. For worthwhile innovation to occur, the brainstorming stages must be free from constraints. Producibility issues should be addressed at the point in the process when the detailed production drawings are being created, Fig. 3-5.

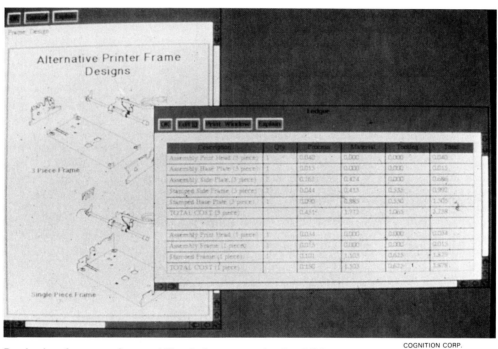

COGNITION CORP.

Fig. 3-5. Designing for manufacturability is important in the CIM context. This screen is generated by an expert system that helps to identify high manufacturing cost areas and provides methods for reducing them. It can be used to aid in process selection and in make-or-buy decisions, as well.

Expert systems should also be developed and incorporated into the knowledge base to address product requirements for interchangeable parts, timely procurement, maintainability, logistics support, ergonomics, and safety.

An important element that must be associated with these premanufacturing expert systems is a means of validation. The design engineer must be able to measure the impact of various product design enhancements against measurable criteria. Lack of a testing capability and an optimization logic could lead to misapplication of logic by the expert system. Such a misapplication could result in faulty design specifications.

With more intense concentration placed on the planning of a product at the front end of the development cycle, the succeeding activities will become more efficient. If the design of the product is thoroughly analyzed and frozen before being committed to production, the disruption created by any later engineering changes will be minimized.

DESIGN CONSIDERATIONS

As the design engineering environment becomes more sophisticated, production efficiency will undergo significant improvement. Whether the environment is a thoroughly developed CIM corporate system or a small manually operated shop, there are a few fundamental considerations that should never be forgotten by the product designer. If a product is to have inherent and long-term sales appeal it must:
- have a sound functional design.
- have eye appeal.
- exhibit quality in both material and workmanship.
- be convenient to maintain.
- be priced competitively.
- be available to the customer when needed or desired. See Fig. 3-6.

To fulfill each of these requirements in a manual engineering system, the product engineer will obviously have to consult with a number of other specialists in the company. In a CIM environment, the use of appropriate expert systems can allow the design engineer to do the job right the first time, with a minimum of outside consultation. A fundamentally correct product design is achieved most easily when as much knowledge and thought as possible is applied at the front end of the engineering effort.

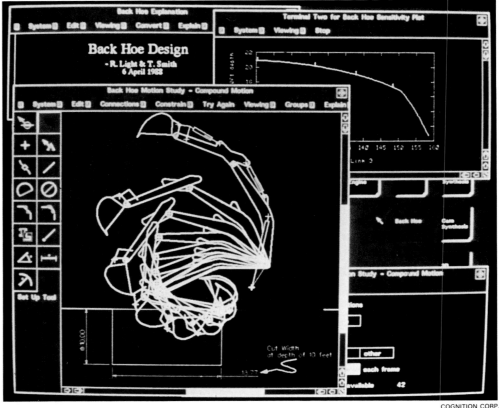

COGNITION CORP.

Fig. 3-6. A well-designed product must be functional. Performance modeling and analysis software allows the designer to perform "what if" analyses of alternative materials or configurations. Such computer software both speeds up and improves the design process.

PRODUCT ENGINEERING

Product engineering is, of necessity, closely related to product design activities. Product design creates the primary geometry of the product, documenting it in the form of drawings or CAD databases. Product Engineering must perform the technical analysis to make sure the design meets the performance criteria spelled out in the product specification document. The design also must meet producibility guidelines.

While organizations vary from company to company, the activities that usually fall into the Product Engineering domain include stress analysis, shock and vibration testing, thermodynamics, material selection, corrosion control, fatigue analysis, chemical analysis, and other product-related analysis functions.

In the CIM environment, with computerized analysis tools such as finite element modeling and knowledge-based material selection systems, Product Engineering can perform tasks more accurately and quickly than manual methods currently allow.

STANDARDIZATION

Within the scope of manufacturing product planning, some truly significant benefits can be realized by applying standardization to company operations. For any company planning to develop a CIM capability, one of the most urgent needs for standardization is in the information management system. Effective integration of a company's design and manufacturing operations demands communication between:
- *CAD* (Computer-Aided Design).
- *CAM* (Computer-Aided Manufacturing).
- *CAE* (Computer-Aided Engineering).
- *CAPP* (Computer-Aided Process Planning).
- *CAI* (Computer-Aided Inspection).
- the *GT* (Group Technology) database.
- the *MIS* (Management Information System).

Communication between these activities requires a common database format and a common data dictionary, Fig. 3-7. The best time to address these issues is at the beginning of the integration effort, before hardware and software are procured.

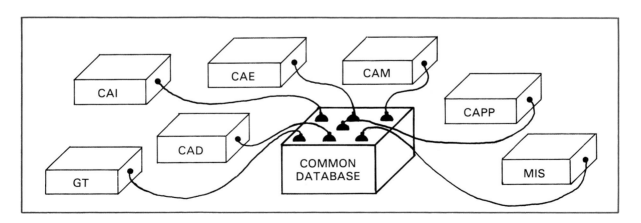

Fig. 3-7. A common database format permits communication between various computer applications involved in the manufacturing enterprise.

Beyond the computer communication issue, there are a number of other areas where significant improvement could be achieved through the application of appropriate standards. Specific applications include:

- *Front end design rules.* Major benefits have been seen where efforts were made to incorporate appropriate design rules and standards. Some of these are **DFM** (Design for Manufacture), **DFA** (Design for Assembly), **VA** (Value Analysis), and **MPD** (Modular Product Design).
- *Development of a standard component manual for design.* Expense, lead time, assembly, and maintenance requirements can be reduced by using standard components in designs.
- *Tooling standardization.* Many opportunities for reducing inventory and variety exist in the tooling area. Coordinating tooling and design functions can result in substantial savings. Standardization opportunities may be found in perishable tooling, jigs, fixtures, and machine tool software routines.
- *Production layout.* By applying group technology, it should be possible to set up machine tool cells for specific families of parts. Standardizing tooling, setups, and routing frequently results in savings of up to 50% in production costs.

- *Material handling.* Coordinating the requirements of the material handling system with design and production engineering can reduce the variety of carts, trucks, basket conveyors, and pallets needed on the shop floor.
- *Packaging and shipping.* Defining standard containers for packaging and shipping can result in consistent loading, simplify tooling, and reduce the variety of packaging materials needed to support a product line.

Standardization can reduce unnecessary variety, simplify logistics within a company's operations, and provide opportunities to obtain price breaks by ordering standard components and materials in large quantity.

PERFORMANCE/PRODUCIBILITY/EVALUATION

Manufacturing product planning requires that a proposed product be carefully evaluated in terms of both performance and producibility.

Product performance criteria may vary widely from one product to another. For example, concerns that must be addressed for a petroleum distillate, such as paint thinner, are much different than the considerations which are involved in an aircraft landing gear trunnion. No matter how widely products may vary in nature, the key to intelligent performance analysis is a detailed specification document.

 A performance specification document should include, at minimum, a general product description, a function description, a statement of the environmental envelope (stress, thermal, chemical, electrical, shock, vibration, pressure, etc.) in which the product must operate, reliability requirements, life expectancy, and maintenance requirements.

The CIM environment provides the opportunity to build a computerized knowledge base and appropriate expert analysis systems to ensure that the front-end engineering of a product is performed correctly.

Producibility issues ideally should be (and within CIM, actually *can* be) addressed early in the design effort. Providing manufacturing constraints by means of on-line design rules for CAD operators will effectively eliminate many production problems and expenses. Beyond DFM (Design for Manufacture) practices, a number of other considerations must be examined. The materials the product requires must be evaluated in terms of cost, processibility, and availability.

The types of surface treatments (if applicable), tolerance requirements, degree of resistance to in-process damage, amount of anticipated scrap or rework, labeling and packaging methods, distribution, required processing facilities and skills, quality standards and need for inspection equipment, logistics for field support, and the need for user documentation all must be viewed as parts of a comprehensive and integrated program. The timely accomplishment of these tasks requires development of computer-based algorithms and the establishing of relevant knowledge bases.

After the front-end performance engineering and producibility analyses have been performed, an ongoing evaluation process is required. Overlooked design and production issues always seem to turn up. Statistical reports generated from process monitoring and control systems can provide virtually real-time feedback for ongoing evaluation activities.

GROUP TECHNOLOGY

Group Technology, in the fundamental manufacturing sense, is a method of organizing information about manufactured products, materials, and processes. The intent of organizing such descriptive information is to make possible intelligent decision-making about basic design, production, and marketing issues.

Group technology is based on the concepts of classification and coding. *Classification* involves grouping similar things according to general likeness, then discriminating within those groups according to specific differences. Industrial and business activities have been classified and segmented, for purposes of the U.S. Bureau of Census, by means of the *Standard Industrial Classification* (SIC) system, as shown in Fig. 3-8. This representation of the SIC system shows major components within the industrial and business sectors of the economy.

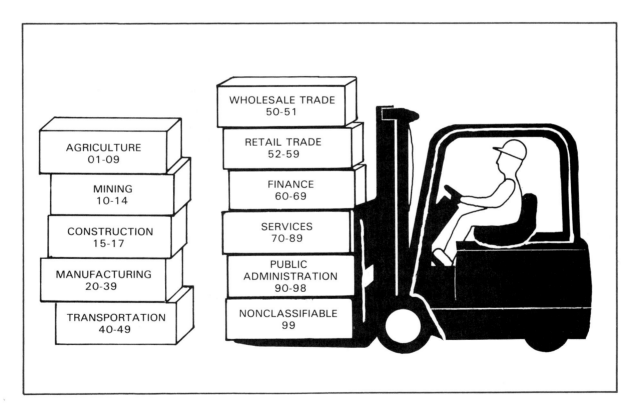

Fig. 3-8. The SIC (Standard Industrial Classification) system divides the economy into major divisions, such as agriculture, manufacturing, communications, and transportation. Each division is assigned a range of two-digit numbers, which are used in turn to represent major groups within that division.

SIC SYSTEM SEGMENTS

Major divisions of the SIC classification include agriculture, forestry, and fishing; mining; construction; manufacturing; transportation, communications, electric, gas, and sanitary services; wholesale trade; retail trade; finance, insurance, and real estate; services, and public administration. Each major SIC division is identified by a letter: Division C, for example, is Construction; Division G is Retail Trade.

By following the market research procedure described earlier, each of the major industry or business divisions then may be categorized into its major and minor subsegments. For example, in Fig. 3-9, the manufacturing division has been subdivided into major manufactured product groups.

Manufactured product groups and codes are:

20. Food and kindred products.
21. Tobacco products.
22. Textiles.
23. Apparel and furnishings.
24. Wood products.
25. Furniture and fixtures.
26. Paper products.
27. Printing and publishing.
28. Chemicals.
29. Petroleum refining.
30. Rubber and plastic products.
31. Leather products.
32. Stone, clay, glass, concrete.
33. Primary metal products.
34. Fabricated metal products.
35. Machinery and computers.
36. Electrical equipment.
37. Transportation equipment.
38. Instruments, photo goods.
39. Miscellaneous goods.

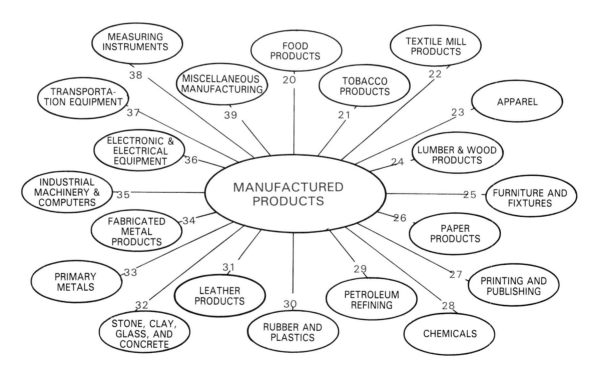

Fig. 3-9. Major groups within the manufacturing division are represented by the numbers 20 through 39. For example, transportation equipment manufacturers use the number 37 as the first two digits of their SIC code.

Each of these major product subgroups may be further refined and classified as finely as desired into similar product groups and eventually into sub-assemblies and components. For example, the general SIC code for transportation equipment is 37; for aircraft manufacturing, it is 372. Fig. 3-10 shows a very small sampling of the many types of personal, domestic, and military aircraft that fit the SIC code 372. A specific type of military aircraft, represented by the complete six-digit SIC code 372111, is the F-16 fighter plane shown in Fig. 3-11.

In conjunction with the product grouping method just described, there are various classification and coding systems available to capture detailed information about individual components. The fundamentals of these component classification systems, as they relate to design retrieval and component manufacturing, are discussed elsewhere in this book.

Within the scope of manufacturing product planning, appropriate application of group technology can help to develop a cost-effective product. Group technology is often underutilized because it is thought of as nothing more than a classification and coding methodology for creating part families and machine tool cells. These functions are among its primary benefits, but there are many more that can be obtained by sophisticated use of the group technology database, particularly at the product level.

Using database statistics from product coding systems, (such as the SIC code) permits the extracting of important product cost information, product performance information, customer information, and even strategic design and production information. This knowledge base, if used properly, can substantially reduce product design and development costs during the new product development cycle.

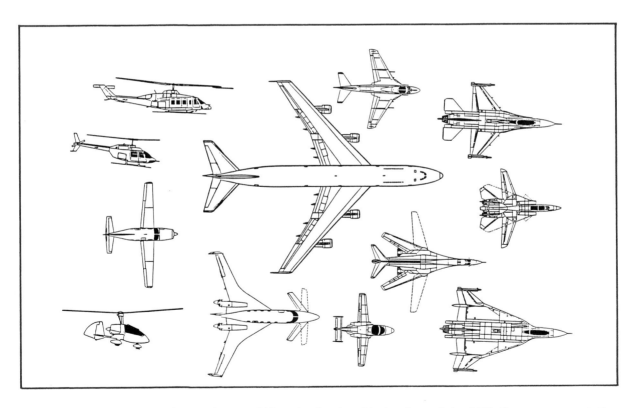

Fig. 3-10. Air transportation equipment SIC codes begin with the three digits 372. Some representative examples of commercial and military aircraft are shown.

Fig. 3-11. A specific type of aircraft can be identified by a complete, unique six-digit SIC code. The code for an F-16 fighter plane is 372111. This aircraft has been modified for aeronautical research.

The product knowledge base can also be used to effectively support computer-aided product planning and market planning. For this task, algorithms can be developed to incorporate expert system design rules and use common data stored in the product database. One important example of using the product database is the linking of CAD and CAM by effectively using the bill of materials generated during the product design stages to provide information for downstream production activities. This linkage between the product planning and design database and the production technology database is easily accomplished by using standard codes, formats, and variables. A group technology database that has been organized and coded adequately can be a key component to systems integration in factory operations.

PRODUCTION DOCUMENTATION

The CIM environment presents some special challenges to traditional methods of production documentation. New procedures must be adopted to take full advantage of computer databases. Production documentation can be divided into two broad categories, the CAD drawing database and production support documentation.

The nature of the CAD drawing makes it necessary to establish and enforce some new policies and procedures in the normal process of the drawing release cycle. Since use of the signature block on a drawing is not practical in an electronic database, alternate means of authorizing the drawings for release must be established. Also, since the drawing database is open to anyone with a CAD terminal, Fig. 3-12, access to the drawings (in anything other than a "read-only" mode) must be carefully controlled. Unauthorized redlining or changes can ruin the integrity of the drawing database. Definite

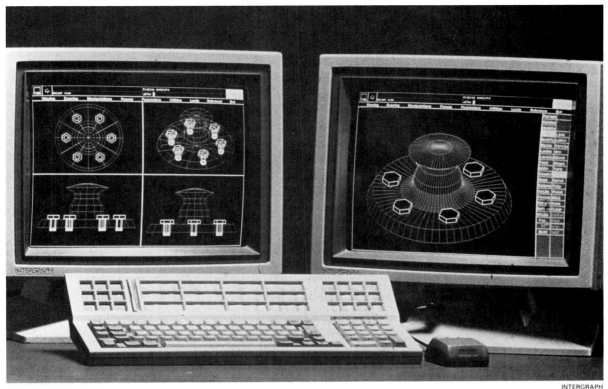

Fig. 3-12. Access to drawing databases must be carefully controlled to prevent unauthorized or undocumented changes that would compromise the integrity of the database.

procedures for incorporating engineering changes and revisions into the database must be established and strictly adhered to.

The versatility of the CAD database can also be a liability, if it is not managed properly. For example, it is possible to edit the number in a dimension callout without actually changing the drawing to conform to the revised dimension. If this is done, the integrity of the database is violated. A *hard copy* (print) of the drawing will appear to comply with the most current revised specifications, but when an NC cutter path is generated from the line geometry, it will be sized according to the unrevised version. Such incidents have made many CAD system users painfully aware of the need to carefully control procedures and prevent unauthorized access to production drawing databases. Production support documentation includes the parts lists, bills of materials, procurement documents, process plans, process specifications, routing sheets, inspection plans, tool drawings, bills of tools, manufacturing standards, and schedules. As these documents are computerized, the same issues that affect CAD databases apply. Access to the database for updating and editing must be carefully controlled, and procedures for release and revision must be established. Checking for compliance with these procedures is vital to insure the integrity of the documentation.

TOOLING CONSIDERATIONS

Tooling can virtually make or break a manufacturing operation. Conventional wisdom regarding tooling states that the objective is to produce a required quantity of acceptable parts, within the specified schedule, at a minimum overall cost. A commonly heard axiom further states that the amount spent on tooling should never exceed the savings it will provide in the production run.

Nothing in the CIM environment changes these basic truths. So far as tooling is concerned, the primary difference between the CIM environment and a manual operation lies in the opportunity to better coordinate information between tooling, manufacturing, and design. A CAD library of standard tooling components, jigs, and fixtures that support the company product line can considerably speed the effort of tool design.

The variety, cost, and set-up time of tooling needed for production can be reduced by designing generic tooling in standard sizes with standard locating points and adapter interfaces for part families based on group technology. Reduction in the variety of perishable tooling can be realized through adopting design rules based on producibility for fillet radii, hole diameters, thread standards, sheet metal bend radii, and other parameters.

A computerized tool management system can improve productivity by making certain that tools are available when and where they are needed. A general reduction in tool inventory can be realized by having the management system track inventory and identify reorder points. Continual tool status updating eliminates the need for expensive buffers of inventory. For complex parts, qualifying the tooling and/or the NC software can reduce inspection requirements.

Of special interest in CIM operations that involve automatic workpiece transfer is the issue of moving the part from one workstation to another without losing orientation and positioning, Fig. 3-13. Tooling must be designed to accommodate this requirement if the full potential of CIM is to be realized.

TOYODA

Fig. 3-13. Maintaining accurate positioning and orientation of the workpiece is important in automated operations. This component, being machined on a Flexible Machining Center, is firmly attached to a pallet for accurate orientation to tool movement axes. The pallet is being held by a tilting trunnion table that permits precise positioning.

LINKAGES TO MANUFACTURING SYSTEMS

In a sophisticated manufacturing facility, efficient operation requires a great deal of planning, scheduling, and coordination. For a number of years, computers have been used in activities involving process monitoring and control, inventory control, scheduling, tool management, facility layout, production simulation, NC operations, material handling, automatic inspection, tool design, estimating, and management information processing. There have been, as a result of computer applications, worthwhile improvements in the performance of each task. Generally speaking, however, each of these applications has been conducted in isolation; the synergistic benefits of integration have not been fully realized.

The challenge that lies ahead is to bridge the gap between CAD and CAM. Developing a fully integrated operation is not a trivial undertaking. Tools, both software and hardware, are becoming available that make the effort easier. Much of the effort must be expended by individual companies, because each product has unique characteristics and requirements. The knowledge bases that are needed to support specific product lines can be developed only by those with the relevant skills and experience.

For manufacturing product planning, an integrated system with a valid knowledge base, appropriate algorithms, and linkages with all manufacturing activities will mean that much greater intelligence can be applied to the planning, evaluating, and decision-making process.

SUMMARY

The Manufacturing Product Planning process and its relationship in the entire CIM strategy involves two basic concepts: *Manufacturing* and *Product Planning*. *Manufacturing* may be defined as a series of interrelated activities consisting of product design, planning, production, materials control, quality assurance, management, and marketing of consumer goods. *Product Planning* can be identified as using the intelligence of a systematic program to identify products that will maximize profits of a manufacturing enterprise. The product design specifications must meet the needs of the customer involving cost, performance, ease of maintenance, reliability, safety, assembly, and aesthetics. The product planning phase must also encompass market research and forecasting for new product development.

In a CIM structure, product designs are developed using CAD (Computer-Aided Design) exclusively. No action can take place in procurement, tool engineering, manufacturing engineering, production engineering, numerical control programming, quality assurance, production scheduling, and factory planning until product designs are released. In the traditional approach, these functions were completed in a serial fashion; the entire cycle was rather long. In a CIM system with good integration and communication, these functions can be done in a simultaneous mode. Thus, Simultaneous (Concurrent) Engineering has become the modern method, in which product design engineering, manufacturing engineering, tooling engineering, production scheduling, and factory planning are done simultaneously. This method reduces the total cycle time by 30%–40% and produces a more profitable product at less cost to the company.

The concepts of *Product Standardization* and *Group Technology* require a computer information management system. The integration modules of CAD (Computer-Aided Design), CAM (Computer-Aided Manufacturing), CAE (Computer-Aided Engineering), CAPP (Computed-Aided Process Planning), CAI (Computer-Aided Inspection), GT (Group Technology), and MIS (Management Information Systems) are absolutely necessary for product standardization. Also, the following functions benefit from a standardization methodology resulting from the adoption of CIM:

- Design rules (value management).
- Standard component designs.
- Tooling standardization.
- Production layout.
- Material handling.
- Packaging and shipping.

Group Technology may be defined as the method of organizing information about manufactured products, materials, and processes. It is intended to facilitate intelligent decision making about product design, production, and marketing issues for increased profitability. Group Technology uses concepts of computer classification and coding to develop families of parts for manufacturing.

IMPORTANT TERMS

CAD
CAE
CAI
CAM
CAPP
classification
design review
DFA (Design for Assembly)
DFM (Design for Manufacture)
expert systems
forecasting
GT
hard copy

IGES
integrated
manufacturing
matrix
MIS
MPD (Modular Product Design)
procurability
producibility
product planning
product specification
Standard Industrial Classification
VA (Value Analysis)

QUESTIONS FOR REVIEW AND DISCUSSION

1. Why is product planning in a CIM environment potentially more effective than conventional planning methods? What would be the desired outcome?

2. There are a number of criteria that can be applied to assess the potential for success of a new product. List at least four of them.

3. Discuss validity and implications of the statement "Product design is the function that creates the marching orders for virtually all manufacturing operations."

4. If your company plans to develop CIM capability, standardization in a number of areas is highly desirable. What should be standardized first? Why? What are some other areas in which standardization will be beneficial?

5. What items should (at minimum) be included in a performance specification document?

6. Why is it important to establish and strictly define procedures for incorporating engineering changes and revisions into a database?

7. What are some of the benefits of a computerized tool management system?

Manufacturing Production Engineering

by David Berling

Key Concepts

- ❏ Planning requires familiarity with production processes and knowledge of the capabilities of specific equipment available.
- ❏ The types of computers used by production engineers differ in the amount of memory they have and in the speed with which they process information.
- ❏ In distributed processing, individual machines and processes are controlled from the lowest level in the network, closest to the action being performed. DNC (Distributive Numerical Control) is an example of distributing information (NC programs) to machine tools, but where all processing to direct the tool is handled by the individual machine's controller.
- ❏ The geometric database created by a CAD system is the nucleus for total integration of the design and manufacturing processes.
- ❏ Interactive graphics permit programming to be created or modified by the machine operator.
- ❏ Computer simulation is an effective support tool for both facilities planning and process evaluation.

Overview

It is the goal of the integrated CIM factory to build a common database and link together all aspects of the operation from customer order to final shipping. In a CIM factory, design engineers create the product on a computer-aided design and computer-aided engineering (CAD/CAE) graphical system. Production engineers then access the CAD database to generate manufacturing documents, computer-aided process plans (CAPP), and programs to direct the functioning of the computer-aided manufacturing (CAM) equipment that will actually produce the product. In this chapter, we will focus on Production Engineering, and how engineers use computer systems in a CIM factory environment.

First, the chapter will briefly describe the major job functions of Production Engineering. Second, it will review the basic forms of computer technology that are commonly employed by production engineers. Third, it will focus on the four major computer technologies used by Production Engineering in a CIM environment: CAD (Computer-Aided Design), GT (Group Technology), CAPP (Computer-Aided Process Planning), and CAM (Computer-Aided Manufacturing).

David Berling is a Project Leader for Garrett Engine Division, Allied Signal Aerospace Company in Phoenix, Arizona.

PRODUCTION ENGINEERING

Production Engineering is a general term that is often used to refer to several engineering specialties that are concerned with planning and controlling manufacturing processes. The two primary branches of production engineering are manufacturing engineering and industrial engineering. There are multiple aspects to each of these engineering disciplines. As manufacturing technologies evolve, the job descriptions of these two disciplines tend to increasingly overlap.

MANUFACTURING ENGINEERING

Manufacturing engineering is a specialty that requires the education and experience to understand and apply modern engineering procedures to manufacturing processes. A manufacturing engineer normally selects the methods that will be used to make a product, and produces process plans for manufacturing. The manufacturing engineer is also responsible for developing all supporting material, such as tooling, NC programs, and special process information. The engineer must know the capabilities of factory machines and equipment. In a CIM environment, he or she will also work with integrated systems to produce quality products more economically, Fig. 4-1. There are multiple aspects to manufacturing engineering. Three major specializations within this field are:
- Process planning engineering
- Tool design engineering
- NC programming engineering

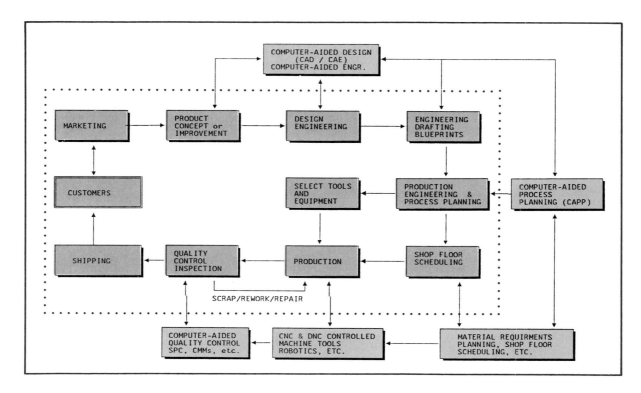

Fig. 4-1. The basic cycle of product design and manufacture is depicted inside the dotted line. CIM and CAD/CAM applications are located outside the dotted line at the points where they apply.

Process planning engineering

Process planning is a complex task that requires a high level of manufacturing and engineering expertise. An effective process planner usually has the education of a manufacturing engineer and the experience of a skilled machinist. Good process planning requires familiarity with a wide variety of production processes. It also requires knowledge of the specific capabilities of various equipment on the factory floor, Fig. 4-2.

In addition to education and manufacturing experience, successful process planning requires access to the company knowledge base. One goal of CIM technology is to provide the process planner with the means to access this collective manufacturing knowledge base of the company. For example, a process planner may recognize similarities of a given product to other products. Access to the company knowledge base will permit the planner to rapidly construct the appropriate process plan.

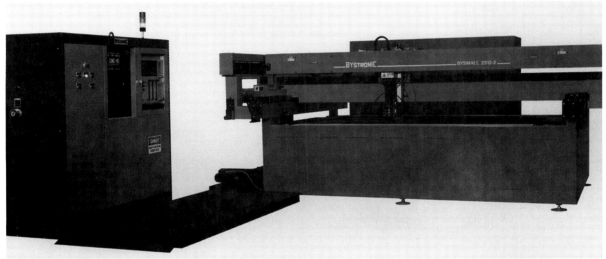

BYSTRONIC INC.

Fig. 4-2. A process planner must have a solid working knowledge of both production processes and the operating characteristics of specific machines, such as this laser cutting system.

The planning process. The task of process planning usually begins with an engineering blueprint and related data. The process planner or manufacturing engineer studies the blueprint drawing for essential manufacturing information, and outlines a general production (process) plan. Next, the engineer evaluates all aspects of manufacturing, such as geometric tolerancing, tooling, fixtures, gages, equipment, etc. Basically, the engineer formulates the sequence of manufacturing operations needed, from beginning to end, to produce the product. Once the *sequence* (flow) of operations has been determined, the written details of *what* and *how* are prepared. These details may include highly specific instructions, procedures, graphic illustrations, routing procedures, specifications, and any other information needed to produce the part. This detailed process plan will manage manufacturing hardware through the production cycle in the factory.

A good process plan will not only satisfy the requirements of Design Engineering, but will also target such issues as costs, quality, and delivery schedules. A process planner should evaluate alternative routings and select the best plan.

Manual process planning presents several problems that CIM and computer technology can address and solve. A process planner must deal with a vast amount of information while developing a plan. The planner needs to be well informed about machines available on the shop floor, including the performance and tolerance capabilities of each tool.

Process planner qualifications. The best process planners tend to be persons who have spent several years working for a company and have an acquired knowledge of part types that can be produced in that company's facilities. Someone new to process planning or to a particular company usually learns the art of process planning by working alongside an experienced planner. New process planners can also learn by reviewing process plans developed by veteran planners. Process planning knowledge is frequently lost when an employee retires or moves on to a new position. In such situations, the novices must work hard to preserve some portion of their predecessor's wealth of experience and knowledge.

Another problem associated with manual process planning is that it is very time-consuming, with little or no guarantee the routing chosen will be the most efficient. With manual process planning, it is quite common to revise the routing several times in a trial-and-error effort.

Automation of process planning. Rapid developments in computer technology and CIM have now made it possible to automate many of the tasks of process planning. When a computerized CAD system is used for process planning, a process plan for a new part can be developed by retrieving existing process plans for a similar part and quickly modifying them.

Although this is effective, the problem with this method lies in knowing *which* parts might be similar. Without this knowledge, the process planner must search a large number of files to account for every feature of the part until a similar part and its routing are located. A much more effective use of computers to aid in the preparation of process plans, called computer-aided process planning (CAPP) will be covered in more detail later in this chapter.

Geometric tolerance stacking. Process plans are normally developed around a series of individual manufacturing operations that take the part from initial roughing operations through final finishing operations in a step-by-step fashion to achieve dimensions and tolerances specified on design engineering drawings or blueprints. See Fig. 4-3.

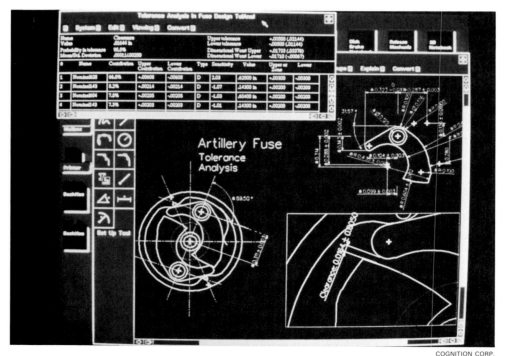

COGNITION CORP.

Fig. 4-3. Tolerance analysis software allows a manufacturing engineer to identify all contributors to tolerance variation and take them into account in process planning.

Since engineering drawing dimensions and tolerances cannot normally be portrayed until the final steps in the machining of the part, there must be some method to determine the intermediate dimensions and tolerances. A tolerance chart or **tolerance stack sheet** must be developed for each part. This is a mathematical method used to determine how the engineering drawing dimensions and tolerances can be distributed throughout the machining operations to obtain the desired final results.

To develop a sequence for the chart, an outline of the planned operations must be prepared. The chart is then filled out to list the routing or manufacturing operations. The dimensions and tolerances arrived at through the use of the chart, are representative figures for producing that part. The outline can then be developed into a process plan, using these values to control each step in the machining processes.

Previously, each tolerance stack or stack sheet of a process plan for a part had to be produced manually. These stack sheets were time-consuming and prone to math errors that could seriously impact manufacturing. Today, the ready availability of computers permits production engineers to quickly and precisely develop a geometric tolerance stack for a particular routing.

Tool design engineering

Another aspect or job function of manufacturing engineering is tool design engineering. The tool design engineer is responsible for the design and coordination of tooling and gages used for production. He or she normally designs the necessary tooling, estimates material and time costs, then coordinates the building of the tools and gages. In the past, tool designs were usually prepared manually on drafting boards, but now it is quite common to develop new tool designs by using a computer graphics terminal and related computer data systems. In a CIM factory environment, the tool designer can easily access the common database to capture the graphical information from design engineering and process planning that is needed to quickly design tooling.

NC programming engineering

A third major specialization within manufacturing engineering is NC programming engineering. Numerical control (NC) machine tools, introduced in the early 1960s, have become a vital part of most metalworking manufacturing facilities. By definition, numerical control is the operation of a machine by an electronic control that gets its instructions in number form. The numbers, of course, must be in a format that can be used to direct the operations of the machine. The numbers are actually coded instructions, such as XYZ coordinates, that refer to a specific distance, position, motion, or function that the machine tool must perform to machine a part.

When numerical control machining methods are used, the part drawing is studied by the NC programmer, who then converts the information on the drawing to the necessary code numbers. The code numbers represent every movement or action the machine tool must perform to properly machine the part described by the engineering drawing. The complete series of codes necessary to produce the part is called an **NC part program**. The programmer's function is to write programs that efficiently produce the parts called for by the process plan. NC programming will be briefly discussed later in this chapter as one aspect of computer systems used by production engineers to support CAM (computer-aided manufacturing). NC programming will also be addressed in depth in a later chapter of this book.

INDUSTRIAL ENGINEERING

Industrial engineering is another specialty area of professional engineering. An industrial engineer must have the education and experience necessary to understand, apply, and control modern engineering procedures in manufacturing processes. Typically, an industrial engineer is concerned with assigning time standards for each process planning operation, determining details of operator methods, and selecting the appropriate feeds, speeds, and perishable tooling. The industrial engineer's job may also require the ability to plan factory layouts, establish capacity, run simulations for scheduling and throughput analysis, and administer shop performance systems.

Although establishing time standards sometimes requires actual stopwatch observations, computerized systems have become quite effective in establishing methods and time measurement values directly from the CIM database.

Manufacturing engineering and industrial engineering are the two primary (but not the *only*) engineering disciplines involved with production in a manufacturing environment. In a small company, one engineer may perform multiple functions. In larger companies, each job function tends to be performed by a different person, group, or department. To avoid job title confusion, the inclusive term of *production engineering* will be used from this point onward.

COMPUTER FUNDAMENTALS

The fundamental technology that links all the elements of a CIM environment is the computer. Without today's computer technology, CIM would not be possible. The rapid development of ever-faster computers, vast memory/storage, and easy-to-use software has made CIM an economic reality.

Computers used for CIM by production engineers can be classified in three major groups. From smallest to largest, they are: microcomputers, minicomputers, and mainframe computers. The boundaries between the major computer types are not clearly defined, since computer technologies develop so rapidly. As a result of this rapid development, the capabilities of yesterday's mainframe have become the capabilities of the mini or micro of today.

Fundamentally, all computers operate in the same way; so when we talk about such categories as micro, mini, or mainframe, we are really talking about differences in the amount of **memory** (storage) they have, and how *fast* they can process information.

The unit of measure for memory size is the **byte**, which is roughly equivalent to one *character* (letter, number, symbol). Thus, one *kilobyte* (1000 bytes) of memory can store 1000 characters, or the equivalent of 200 five-character words. Microcomputers typically have 640 kilobytes of working program memory, extendable to megabytes; minicomputers and mainframe have memories that are far larger. When speaking of memory, most often what is meant is RAM (**random access memory**). As a computer operates, data is written to, and read from, RAM. The data in RAM is *volatile*, which means that it is lost when the computer is shut off. The other form of memory used by a computer is ROM (**read-only memory**). ROM contains programming instructions and constants that the user cannot change. Since RAM contents are lost when the computer is shut off, a more *permanent* means of storage is needed. External storage devices, such as hard disks, floppy disks, CD-ROMs, and magnetic tapes provide non-volatile read/write memory capability.

The *speed* of a computer is determined by a number of factors. One main factor measured is how fast the synchronous internal clock is ticking. Another key element is the number of instructions per second a computer can execute. This is often expressed in **MIPS** (millions of instructions per second). The size of binary words (8-bit, 16-bit, 32-bit) that the computer's architecture can process also has a major effect on speed and computing power. Some computers also offer an architecture that allows multiple processors, permitting parallel processing.

MICROCOMPUTERS

The smallest type of computer normally used by production engineers is the microcomputer, Fig. 4-4. Usually, a microcomputer will be contained on a single printed circuit board. The microcomputer is commonly referred to as a personal computer, or by such names as desktop computer, home computer, portable computer, laptop computer, or small business computer. Microcomputers are used extensively by production engineers in today's CIM settings, but this is a relatively new develop-

Fig. 4-4. The desktop microcomputer has achieved almost universal use by production engineering personnel. The individual units in an office or plant are frequently linked in a local area network (LAN) to permit communication and sharing of files.

ment. Microcomputers became available on a large scale in the early 1980s. Faster and more powerful processors have continued to be released each year since that time.

A microcomputer is normally a single-user computer. Typical applications of microcomputers in production engineering include engineering desktop software, word processing, spreadsheets, databases, process planning, and use as single-user CAD/CAM workstations. Microcomputers are also used for local area networking (LAN) and as remote terminals for larger minicomputers or mainframe systems.

MINICOMPUTERS

The minicomputer lies somewhere between the microcomputer and the mainframe in both size and computing power. There is really no easy way to define where a microcomputer stops and a minicomputer starts. It's equally hard to define where a minicomputer stops and a mainframe computer starts. A minicomputer is essentially a scaled-down version of a mainframe computer. However, in exterior size, the unit may resemble a microcomputer or it might look like a small mainframe. The *interior architecture* of a minicomputer might look like the interior of a microcomputer or a mainframe. Typically, a minicomputer will have an electronic architecture much closer to that of the mainframe.

Normally a minicomputer will be capable of simultaneous shared access by multiple users. Minicomputers are also commonly used as very powerful, single-user workstations. Typical applications of minicomputers related to production engineering include: use as single or multi-user CAD/CAM workstations, process planning, factory-wide networking, DNC systems, cell or FMS (Flexible Manufacturing System) controllers, factory scheduling systems, shop floor information systems, and MRP.

MAINFRAME COMPUTERS

Full-scale *mainframe* computers are generally quite expensive, have vast amounts of memory, and have extremely fast processing speeds. The extreme processing speed, however, may not be readily apparent, due to the large number of simultaneous users on the system.

Mainframe computers frequently handle hundreds and hundreds of users, all accessing the system simultaneously. Typical applications of mainframe computers related to production engineering include: general business applications, large multi-user CAD/CAM systems, process planning, corporation-wide networking, and MRP.

DISTRIBUTED PROCESSING

Many companies, due in part to developments in CIM technology, have developed interconnected computer systems, or *computer networks*, that form a pyramid-like structure. Computers at the shop floor or office level are linked to larger, more powerful systems, all the way up to the corporate mainframe. This type of configuration, Fig. 4-5, is labeled a *hierarchical structure*. It is important to point out that information flows both up and down the pyramid structure.

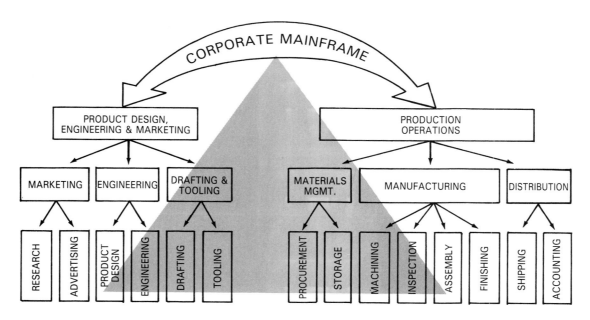

Fig. 4-5. The computer network of a manufacturing enterprise is organized into various levels or hierarchies. In effect, the network takes a pyramid-like form with a broad base at the shop-floor level and the corporate mainframe at the apex.

At each level the information is processed, summarized, and passed upward, if required. Factory operation data, parts status information, and other operational data are passed upward through each level of the computer system.

Commands and schedules are passed down from the corporate computer throughout the rest of the processing structure, as required, to direct the activities of the total factory. Whenever possible, computing and directions come from the *lowest* level, closest to the action being performed. This frees the upper-level computers to do overall planning, controlling, and reporting, and doesn't burden them with controlling and directing individual factory machines or processes. This type of activity is often referred to as ***distributed processing***.

MAJOR COMPUTER APPLICATIONS IN PRODUCTION ENGINEERING

Computers are useful in a variety of ways in production engineering. They serve as the nucleus of the increasingly sophisticated CAD graphic systems utilized for design and drafting. They also serve as CAM workstations to prepare and direct manufacturing.

The four major applications of computer technology in production engineering include:
- **CAD** Computer-Aided Design systems.
- **GT** Group Technology (family parts coding).
- **CAPP** Computer-Aided Process Planning.
- **CAM** Computer-Aided Manufacturing applications.

CAD SYSTEMS

Computer-Aided Design (CAD) is the most widely used application of computer technology in design and production engineering, Fig. 4-6. CAD is an interactive computer graphics system used to automate the design and drafting aspects of engineering. (CAD is also referred to as CADD, which stands for Computer-Aided Design and *Drafting*. For simplicity, we will use CAD as a term to cover both design and drafting).

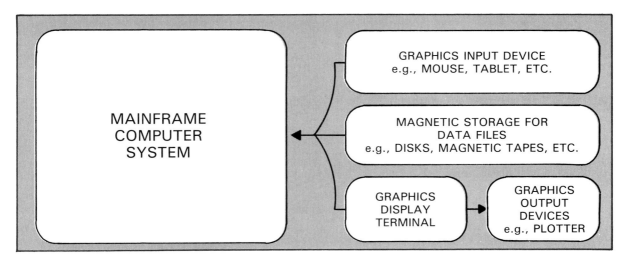

Fig. 4-6. A basic CAD system configuration that might be used in a large manufacturing business. A smaller company's CAD system might be based on a minicomputer acting as a file server, or even on a network of microcomputers.

To perform as an effective computer-aided design and drafting system, the computer must be used in an interactive or "give-and-take" manner. This means that CAD is designed to be operated in a conversational mode between user and machine. The user makes known his or her intentions and other information through various input devices. The computer prompts the user along, acknowledges input, complains about mistakes, and displays progress as a graphical image. An engineer usually describes the part or design to the computer, performs analysis on it, and changes the design on the basis of that analysis until specifications are met.

Geometric *modeling capability* is the cornerstone of the CAD system. The graphical images shown on a CAD display screen are based on mathematical coordinates and equations, just like the description of points, lines, and circles taught in geometry. This descriptive information is calculated and stored in the computer as digital electronic data. The computer makes it possible to retrieve, process, transmit, and store this data quickly and accurately. The computer display screen makes it possible to instantly construct and show the data on the screen as an electronic drawing.

The geometric database created by the CAD system serves as a nucleus for total integration of the design and manufacturing process from initial design concept to final finished product.

There are numerous benefits to CAD:
- it significantly reduces design and drafting time.
- it ensures that mating parts will fit together.
- it keeps track of hundreds of details in the database.
- it speeds the process of generating working drawings.
- it simplifies making drawings for similar products.
- it facilitates group technology.

Although CAD initially required the capabilities of a mainframe or minicomputer system, powerful microcomputer-based CAD systems are now available that put the benefits of CAD within the financial reach of small engineering and production engineering departments. Some of today's CAD systems are even popular for private home use.

Equipment for CAD

From a hardware standpoint, a basic CAD system consists of three major components that work together:
- the graphics terminal workstation.
- the CPU (Central Processor Unit).
- the output devices.

Graphics terminal (workstation). The graphics terminal or graphics workstation must permit the user to enter, retrieve, and manipulate drawing data (both pictorial and alphanumeric), and to initiate calculations, analysis procedures, and hard copy output. See Fig. 4-7. This unit serves as the human/machine interface for the CAD system. It provides input/output devices for the acceptance and display of graphics and text, the presentation of operating instructions, and the retrieval of previously stored data.

There are several methods for entering data. The graphics display screen is essentially a vector-generating cathode ray tube (CRT) used to display lines, points, and alphanumeric characters in response to user input. This input will normally come from one or more of the input devices described below. To permit input/output transactions, the graphics terminal is connected by a bi-directional link to the Central Processing Unit (CPU) and can utilize several types of input devices:

A *light pen* is an object, shaped like a pen wired to the computer, and can be used to locate points on the screen. See Fig. 4-8.

A *mouse*, Fig. 4-9, is a small hand-held device that's moved across a smooth special reflective surface or simply across a desk or tabletop. It provides precise and rapid cursor control.

A *digitizing tablet* uses a pencil-like device (stylus) that the CAD operator moves around on a digitizing surface. It may also use a mouse-like device. See Fig. 4-10.

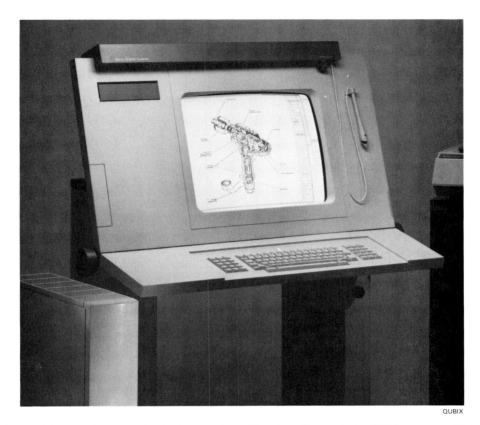

QUBIX

Fig. 4-7. A graphics workstation is the human/machine interface of the CAD system. This unit is set up to use both a keyboard and a light pen as input devices.

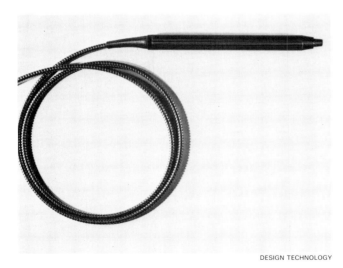

DESIGN TECHNOLOGY

Fig. 4-8. A light pen is an input device that can be used to locate and alter objects on a graphic display.

NUMONICS

Fig. 4-9. Cursor movement and object selection for movement or other manipulation is easily and quickly accomplished through use of a mouse like this one.

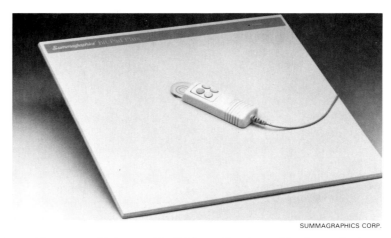

Fig. 4-10. Digitizing tablets are widely used in CAD applications. The input device may be a mouse-like unit like the one shown, or a stylus that resembles a pencil.

Another device the CAD operator can use is a *keyboard*. The alphanumeric keyboard is essentially a standard typewriter keyboard. Sometimes, special function keys are added to simplify data input and selection of software functions. The keyboard is normally located just in front of the display screen and is used by an operator to input all required alphanumeric data.

Some systems have a separate *special function keyboard* to simplify operation or to activate various software functions. Each key calls a specific function which, in turn, may call or allow the user to invoke other special choices.

Central processing unit. At the heart of any CAD system is the CPU or central processing unit. On larger multi-station systems, the CPU is frequently a mainframe or minicomputer networked to various workstations. At the workstation level, a lesser computer can be employed for local processing and storage. Workstations tend to be either minicomputers or microcomputers.

The local workstation computer accepts input from the graphics terminal input devices, performs calculations, stores/retrieves data, and serves as a communications link to the larger central computer. The central computing facility of the system normally includes the main computer and all its peripheral equipment, memory, on-line disk storage, off-line magnetic tape storage, and system software.

One trend particularly worth noting is that, as microcomputers become more powerful and disk storage becomes larger, more and more systems are using inexpensive personal-type computers.

Output devices. CAD output devices include plotters, magnetic tape decks, perforated tape punches, microfilm systems, and any other equipment required to produce output in forms useful to the particular installation.

Two general classes of devices are used for hard copy output: **electrostatic printers** that are much like an office copying machine, and pen plotters, which use colored pens to draw on paper of various sizes. See Fig. 4-11.

Computer graphics software

CAD is an *interactive* computer system, providing the operator with an instant visual display of what he or she is doing. Because a CAD system allows an operator to manipulate the images quickly and effectively, it can accomplish, in moments, tasks that would take hours or days with paper and pencils.

The screen of a video display is essentially a space with two dimensions. While a number of schemes exist for dealing with two-dimensional spaces, the most common is the use of *Cartesian coordinates*.

Each point in the space is represented by a pair of numbers that correspond to its distance from the intersection of two axes at right angles to each other. On a video display, for example, this pair of numbers can correspond to the scan line number and picture element within the scan line. This can be referred to as *display space*.

For many problems in computer graphics, a second space is used. This is called the *problem space*. It corresponds to the description of the problem or part geometry, as opposed to the screen itself. The units of problem space may be miles, feet, or anything that relates to the item being described.

CAD software normally lets the operator work in problem space coordinates. The system software will then map this onto the display space to create a graphics display image. The complex mathematics required, involving vectors and matrices, is beyond the scope of this chapter. However, it is important to understand that *internally* CAD is building a database of points, equations, and mathematical descriptions. This is normally transparent to the CAD user, who is working more on a descriptive geometry level, identifying points, lines, circles, intersections, etc.

For the reasons just discussed, the graphics screen itself is actually a *window* through which the user views the "electronic drawing board" in the computer. Although the graphics screen limits the window area (usually 12 to 19 inches square), the "electronic drawing board" is virtually unlimited in size. This means that the part geometry can be drawn to any size and at any distance from the origin. However, since mathematical precision decreases as distance increases, the area available is effectively limited to a given number of units in each direction (determined partly by the size of the computer system being utilized).

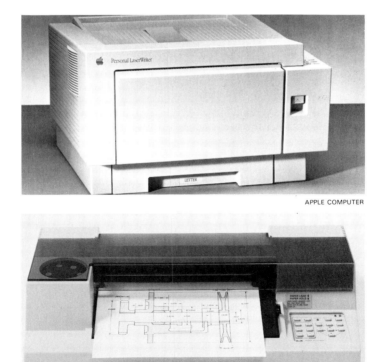

APPLE COMPUTER

HEWLETT-PACKARD

Fig. 4-11. Typical hard copy output devices used for CAD systems. A—An electrostatic (laser) printer can produce both text and graphics output. B—This desktop pen plotter can use up to six pens for different line widths or colors.

The best way to appreciate the window concept is to think of a camera focused on the center of a huge piece of drawing paper. The camera could be moved around the drawing to focus on different portions of the paper. The camera could also "zoom-in" for a close-up or "zoom-out" to show a larger portion of the drawing at any given time. See Fig. 4-12. The camera could even be tilted or rotated, which would effectively make the drawing appear to rotate to a different viewing angle.

Menus, icons, symbols, prompts, system messages, and status information are also displayed in given positions on the display screen.

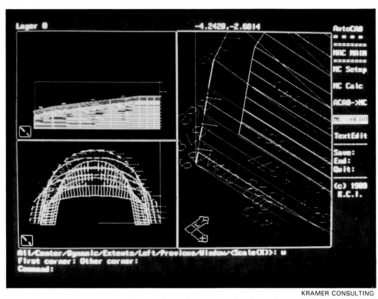

KRAMER CONSULTING

Fig. 4-12. The graphics screen can be thought of as a window (or series of windows) that will allow the user to show an overall view of a drawing and also to "zoom in" to a closeup view, as at the right side of this screen.

CAD workstations are equipped with a select/indicate device to allow the user to *point to* selected areas on the screen. This device may be a mouse, a light pen, or other digitizing device. Whatever the device, its purpose in the CAD system is the same: it is used to *select* an element on the screen or to *indicate* an approximate location for various purposes. The select function identifies an existing element on the screen (a point, line, symbol, note), so the computer can perform some operation with it. The indicate function identifies a position on the screen where no geometry currently exists.

In addition to such useful features as "windowing," "zoom-in," and "zoom-out," CAD enhances productivity of designers and operators because of four basic software functions: replication, translation, scaling, and rotation. When a part or design has features that are repeated, CAD's ability to *replicate* means that the CAD operator can copy part of the image and use it in several other areas of the drawing without having to redraw it each time. *Translation* means the CAD operator can move portions of the drawing around from one location to another. *Scaling* means that CAD can be used to change the proportions or size of one part of the image in relation to the others. Finally, CAD's ability to *rotate* lets the CAD operator move the design around to see it from different angles or perspectives.

CAD systems have seen rapid and continuous improvement since their introduction. The first generation of CAD, still in wide use, is 2-D CAD. It can best be described as a computerized drafting system. The CAD user can construct a drawing, add dimensions, insert text, and manipulate the image in ways impossible with pencil and paper. Another major benefit of the system is that it is interactive. The displayed drawing will respond instantly to every addition, change, or instruction from the CAD operator.

The second generation of CAD, also widely available, is 3-D CAD. With a 3-D CAD system, the operator can produce an image of a part, using either wireframe models or *surfacing*, a technique that displays the surface of the objects. This gives a three-dimensional perspective to the part, as shown in Fig. 4-13.

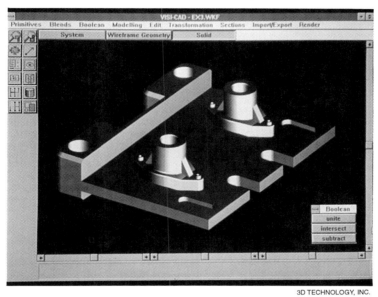

3D TECHNOLOGY, INC.

Fig. 4-13. Some CAD systems permit construction of a three-dimensional model of a part. This software automatically develops the 3-D model from the 2-D drawings.

A third generation of CAD, also available but not as widely used, allows the operator not only to draw the object in three dimensions, but also to view the part realistically. Users can rotate, move, and view the part from any angle. The images produced can seem quite realistic.

CAD is the primary computer application used in production engineering. It allows the design engineer and the production engineer to share a common database to describe a part or product. The production engineer quite often will select views from the overall product drawings to produce illustrations for manufacturing process plans.

GROUP TECHNOLOGY

Many production engineering CIM systems are the result of applying group technology, or could benefit greatly from the application of group technology. *Group Technology* (GT) is a philosophy in which similar items are identified and grouped into families. This permits you to take advantage of the items or similarities in manufacturing and design. This is a very important concept, and one that must be in place before CIM can be fully implemented.

Group technology's primary application in production engineering is to group parts into families. It is common to speak of similar part characteristics as *attributes*. Attributes might represent physical shape, detailed design characteristics, or manufacturing processes required to produce a part. Parts are normally grouped into families because they have common or very similar attributes. Any mixture of part attributes might be utilized to establish the pattern of similarities. Once these similarities are identified, the parts are arranged into part *families*; sometimes further subdivided into *subfamilies*.

The first applications of GT were designed to permit quick retrieval of similar designs. Manual coding was used to identify attributes and classify parts into families. While there were definite benefits, manual coding and retrievals were time-consuming. Rapid developments in computer technology have produced a renewed interest in GT. The coding and classification of parts has become a common tool for design and production engineers. It is not unusual for a large company to have over 100,000 active part designs on file. Today, the power of the computer is available to quickly code parts and to perform the necessary lengthy database searches.

GT can significantly improve the productivity of design and production engineering personnel. It does so by decreasing the amount of work and time involved in designing new parts and process plans. Quite often, a new part will be very similar to one that is already designed and in the database. With CAD systems, it's easy to retrieve the similar design and process plans and make slight modifications for the new part. Modifications require less time and work than starting from scratch.

GT is a fundamental aspect of CAPP (Computer-Aided Process Planning), which we will cover later in this chapter. Group Technology has also been popular for several years to improve **batch manufacturing**. Grouping parts into families allows Manufacturing to run similar parts in more efficient fashion. Often, a group of machines can be organized into a **manufacturing cell** to produce one family, or several similar families of parts. This concept of cellular manufacturing will also be covered later in this chapter.

Part families

There are three common methods used to group parts into families:
- sight inspection.
- routing sheet inspection.
- parts classification and coding.

The simplest method is **sight inspection** of parts, but this is also the least sophisticated and the least accurate method. Sight inspection involves looking at actual parts, photos of parts, or blueprints of parts. It is strictly a manual, subjective evaluation.

The **routing sheet inspection** method involves reviewing the process plans or routings to establish similar manufacturing processes. This method can be a quite effective way to group parts into families for cellular manufacturing. The major drawback is that this process depends upon the current routings, which may not be the best way to route the parts for efficient part family or cellular type manufacturing. The routing sheet method is sometimes referred to as the *production flow analysis* method. This evaluation can be done manually or through the use of a computer.

The **parts classification and coding** method normally involves the use of special computer systems and software. To code the parts by this method, various design and/or manufacturing characteristics of a part are identified, listed, and assigned a code number. Significant characteristics are referred to as **attributes**, and are assigned alphanumeric symbols. Typically, a part is evaluated through a series of questions. The answers to the questions result in codes and select the branches to follow for additional questions. The process continues until the part is completely coded and classified. There are three types of software packages that exist to support parts classification and coding:
- design attribute coding.
- manufacturing attribute coding.
- combined attribute coding.

Typical *design* attributes include dimensions, tolerances, shape, finish, and material. Typical *manufacturing* attributes include manufacturing processes, sequence of operation, perishable tooling required, fixtures required, and material handling requirements. The most effective parts classification and coding systems work with a *combination* of design and manufacturing attributes to establish parts coding and families.

Manufacturing cells/FMS

Group Technology is also the cornerstone of cellular organization and *flexible manufacturing systems* (FMS). Once parts have been coded into families, many companies are reorganizing their factories around manufacturing cells.

The concept of cellular manufacturing involves producing a family of parts, from start to finish, in a single area. Instead of the traditional arrangement of machines according to *function* (milling machines all together, lathes all together, drilling equipment all together, etc.), each cell has a mixture of machines and processes designed to complete the parts within the cell. (If special outside processes are required, parts leave the cell as little as possible.) A family of parts is produced in a cell, manufacturing personnel can reduce the leadtime needed to make those parts. They can reduce wasted time that parts normally spend traveling between departments and waiting to be processed. Setup times, material handling times, direct labor times, and the amount of in-process inventories are all reduced. Quality improvements are also significant when parts are manufactured utilizing GT and cellular manufacturing concepts. Production engineering systems that apply GT both enhance engineering and manufacturing productivity.

The FMS, Fig. 4-14, differs from a manufacturing cell in several respects. A primary difference is the level of automation: in a manufacturing cell, quite often parts are manually moved from machine

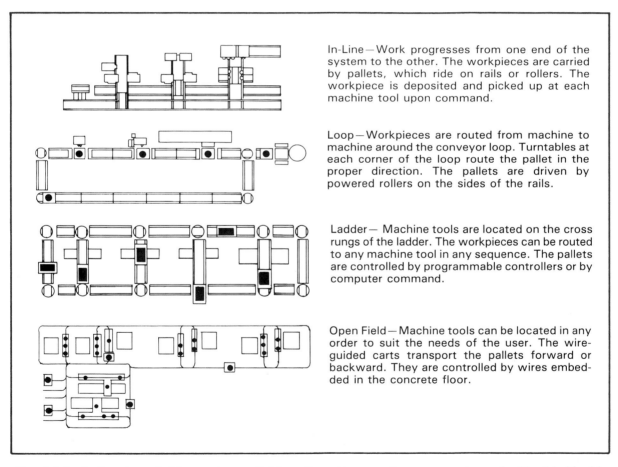

In-Line—Work progresses from one end of the system to the other. The workpieces are carried by pallets, which ride on rails or rollers. The workpiece is deposited and picked up at each machine tool upon command.

Loop—Workpieces are routed from machine to machine around the conveyor loop. Turntables at each corner of the loop route the pallet in the proper direction. The pallets are driven by powered rollers on the sides of the rails.

Ladder— Machine tools are located on the cross rungs of the ladder. The workpieces can be routed to any machine tool in any sequence. The pallets are controlled by programmable controllers or by computer command.

Open Field—Machine tools can be located in any order to suit the needs of the user. The wire-guided carts transport the pallets forward or backward. They are controlled by wires embedded in the concrete floor.

Fig. 4-14. In-line, loop, ladder, and open-field configurations are the common types for Flexible Machining Systems.

to machine; in an FMS, parts usually move under computer control on an automated material handling system. In a cell, a worker or robot loads and unloads each machine, while loading and movement in an FMS is automated (using computer-controlled conveyors or automated guided vehicle systems). Cells are usually smaller and limited to producing only a few families of parts; FMS are usually larger and tend to handle a much wider variety of parts and families.

CAPP SYSTEMS

Process planning is the complicated task of determining and specifying the step-by-step sequence of manufacturing operations needed to produce a given part or product. The process plan, or *route sheet*, will normally list the production operations, manufacturing equipment, and all associated tooling required.

Until recently, process planning has been a task requiring a very high level of expert intuition. However, to produce a process plan was often a manual and rather clerical task. Good process planning has been very dependent upon the experience and judgment of the production engineer. Each production engineer could have a different opinion about what constituted the best routing. Very similar parts might have quite different routings, simply because different engineers prepared them. GT and cellular manufacturing concepts encourage standardization of process plans.

In recent years, computer software has been developed to capture the analysis steps, decision process, and experience required for this important function. Automatic generation of documentation has also been a goal of CAPP (*Computer-Aided Process Planning*) systems. Today, computerized expert decisions and automated generation of documentation are being used in some companies. See Fig. 4-15. In fact, GT-based CAPP systems have been used successfully in a variety of industries for many years. Once the basic characteristics of a given part are described to the computer, the

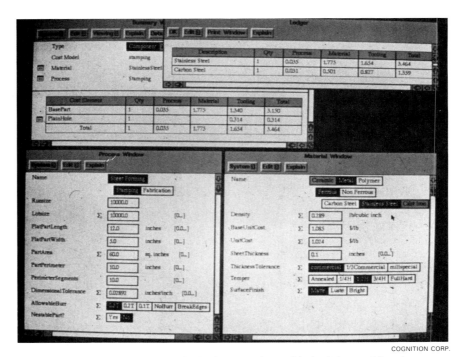

COGNITION CORP.

Fig. 4-15. Expert systems are increasingly being used to aid decisionmaking in process planning for manufacturing operations.

CAPP program will automatically generate the sequence of manufacturing operations and develop manufacturing documentation. CAPP systems can significantly improve the consistency and quality of routings while reducing the skill level and clerical time required to produce them.

Basically, there are two systematic approaches to computer-aided process planning:
- variant-type CAPP systems.
- generative CAPP systems.

Variant process planning systems

A *variant-type CAPP system* is also commonly referred to as a *retrieval-type* CAPP system. The key to a variant-type system lies in being able to identify whether a part is similar to some previous part or family of parts. The part is identified by GT to place it in a particular part family (and possibly a specific subfamily).

To develop plans for a new product (part or workpiece) using the variant approach, existing process plans for similar parts are retrieved from a database and modified to suit the processing requirements of the new part. A routing for the new part is produced on the basis of the similar workpieces.

Often, the family of parts will have a predetermined *standard routing*. All parts in the family follow the same standardized sequence of operations. Variations may occur within any specific operation (certain operations may be skipped as unnecessary), but the general outline of operations is very standardized. Computerizing variant-type process planning is a very practical first step toward generative process planning.

Generative process planning systems

One goal of CIM is the automation of process planning. This is often referred to as *generative process planning*. With the generative approach, a detailed description of the workpiece and its features is used by the CAPP system to automatically generate a process plan.

The concept of generative process planning has attracted a great deal of attention for some time. The idea of describing a part in simple terms through a computer terminal, then pushing a button for the computer to generate a detailed process plan, is very exciting. There is no question that such a system would increase planning productivity.

There are several other advantages of a generative process planning system over manual or even computerized variant-type planning methods. If it worked properly, a generative system would increase the quality of the planning department output. Subjective decisions and opinions would be replaced by firm rules derived from experts. Process plans developed using the generative approach would also be more adaptive to new technology and changes in factory conditions.

Although a great deal of interest has been expressed in generative process planning, no universal generative process planning system exists, as yet. The key word here is *universal*. Limited success has been achieved in systems tailored for a given type or family of parts. Developments in this area of CIM are moving rapidly, with some companies claiming each year to have achieved generative process planning.

Most systems marketed as *generative* use the term rather loosely. This has made the line of separation between variant and generative very fuzzy. A system labeled as generative is most likely to be a variant-type system with *some* generative features.

To accomplish true generative CAPP in a CIM environment may require the application of *Artificial Intelligence* (AI) or *expert systems* software. AI and expert systems software could have significant advantages over conventionally structured computer programs to develop a generative CAPP system.

One major advantage is that expert systems offer a modular architecture for building large programs. Knowledge (in the form of production rules) may be added, deleted, or modified in the program's knowledge base, without any alteration in the main shell program. The sequence of rules and their relationships can be input in natural language to build the program logic. Later, the program would be able to express and explain its logic in the form of natural language output.

Finally, the ability of an expert system to manipulate symbols permits users to choose a variety of representation schemes for components, equipment, and tooling. It would also permit translation of CAD data directly into the system.

CAM SYSTEMS

Computer-Aided Manufacturing (CAM) is often referred to as the final link in the CIM factory. CAM has been described as the effective use of computer technology in the management, control, and operations of the factory floor. It involves a computer interface, either direct or indirect, with the manufacturing equipment and shop personnel. As the level of computer control increases in all areas of manufacturing, it becomes evident that the level of production engineering support for CAM must expand.

In this final section, we will discuss three common CAM applications involving production engineering:

- NC (numerical control) part programming.
- DNC (distributive numerical control).
- simulation software.

NC part programming

Most computer graphic systems that support CAD also support various CAM applications. In fact, this computerized link of CAD and CAM and the sharing of a common database is the heart of CIM philosophy.

NC (numerical control) part programming is a common CAM system used by production engineering. Production engineers can often make use of the CAD graphics already created by design engineering or process planning to produce NC (numerical control programs). Tool paths and machine cycle functions are added to produce a source program, Fig. 4-16. This source program must then pass through a post-processor to produce a machine-readable part program.

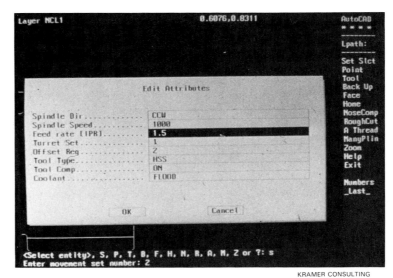

KRAMER CONSULTING

Fig. 4-16. Programming for CNC equipment can be generated by computer software. This screen shows tool parameter input for such a program.

92 *Computer-Integrated Manufacturing*

Computer-aided programming of NC machine tools began in the aerospace industry, where the main emphasis lay in finding the solution to very complicated geometric problems to allow production of complex aircraft parts. Since that time, NC has become a very popular form of CAM. It provides flexible automation that is best suited to batch-type manufacturing. NC programming will be covered in detail later in this book.

Group Technology (GT) is also important to NC programming. Many NC part programming departments have created substantial libraries of part programs. Many part programs can be produced more efficiently by modifying existing programs, if they can be readily identified. However, programmers often elect to develop a new part program, rather than spend hours trying to retrieve an existing one.

The inability to quickly identify previously written part programs that are similar to the new part program has always been a problem. Quick retrieval, if it could be achieved, would permit rapid production of the new NC program. Often, the old program can be copied, modified, and released. Group Technology coding and classification systems now provide a workable method of identifying existing part programs so they can be retrieved and edited to produce new part programs. This coding and classification can also be used to create standards for tools, fixtures, and gages.

One exciting new feature of CNC machine tools is the addition of controls that feature *interactive graphics*, Fig. 4-17. These NC/CNC controls are capable of on-the-machine graphical programming by the operator or setup person. Automatic selection of cutting sequence, selection of tools, and calculation of feeds and speeds are all accomplished through a graphical description of the part, material, and tooling library input at the machine. This provides the option of programming either off-line in the office, or on-line at the machine tool directly. CNC controls with graphical input/automatic programming capability often split the control into foreground and background operation. This permits machining of one part while the next program is being prepared.

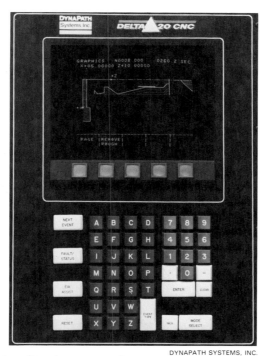

DYNAPATH SYSTEMS, INC.

Fig. 4-17. Interactive graphics, like the part profile display on this two-axis turning center control unit, allow the operator to edit the part program as necessary.

DNC

The same three-letter combination, DNC, is used to describe two different but related methods of control. They are known as *Direct* Numerical Control and *Distributive* Numerical Control. In both cases, a powerful central (host) computer is interfaced with one or more CNC machine tools, forms the basis for a centralized, tapeless NC operation. The essential difference is that, in *direct* numerical control, the central computer is connected directly to each CNC machine controller; in *distributive* numerical control, there is an intermediate layer of control, consisting of (usually) a minicomputer connected to a group of CNC machine tools. Responsibility for operation of the machine tools is thus "distributed" among the minicomputers and machine controllers, rather than being totally a function of the central computer. A feature of both direct an distributive numerical control systems is two-way communication with the DNC computer. A large library of part programs, stored in mass storage units, is the essential feature. Part program information is transmitted electronically from the DNC unit to the machine controllers (in *direct* numerical control) or to the minicomputers which then pass it on to machine controllers (in *distributive* numerical control). Information can also be sent back to the host computer and used for factory management information reporting purposes.

Simulation software

Another increasingly important CAM tool for production engineering is the use of **computer simulation**. The application of computer simulation techniques to modeling manufacturing facilities is recognized as an effective decisionmaking support tool. Simulation has been successfully used in many manufacturing industries to help size and design new facilities, evaluate new equipment, and establish optimum part flow. Before relatively inexpensive minicomputers and microcomputers became available, attempts to use computer modeling for day-to-day operation of the factory were largely unsuccessful.

Today, however, several trends indicate modeling will be a vital element of the advanced CIM factory. Applications have been developed in all types of manufacturing industries in the following areas:
- determining the limiting capacity of the existing or proposed facility and identifying the processes that are the *bottlenecks* or *pacing* items.
- determining the impact of bringing additional worker into a facility or subcontracting work out of the shop.
- sizing and designing a proposed new facility.
- evaluating the addition of new equipment or processes to an existing facility to determine how best to maintain balance among the various elements of the production process.
- formulating management policy for both existing and proposed facilities.

In these applications, a model of the system is developed and simulated on the computer. Alternative configurations of the system are also evaluated. System performance can be observed with output that is either statistical and/or graphical. Many systems support color graphics to simulate actual factory operations. Graphic animation's of product flow are improving the understanding of complex manufacturing processes and shortening the time for model validation.

The key advantage of factory simulation is its ability to analyze numerous alternatives. Each alternative can show varying affects on throughput time, bottlenecks, and outputs. Computer simulation modeling seems to be the best alternative because of the inherit advantages of low capital investment and quick answers. Most manufacturing plants have used the method of *trial and error* to arrive at some conclusion or decision. As many managers have found out, this technique tends to tie up too much machinery and resources and is often cost prohibitive. These experiments are also carried out in real time and therefore prolong the decision-making process. Computer simulation modeling is a technique that allows building of and experimenting with a model of a real system on a computer. A production simulation of a validated factory model for a period of one year can now be completed in a few hours.

Much of the success of the simulation model is dependent on the methodology used by the simulation team. A discrete event simulation project should consist of eight phases:

Phase l: Define the problem

Phase 2: Design the Study
Phase 3: Design the Conceptual Model
Phase 4: Formulate Inputs, Assumptions, and Process Definition
Phase 5: Build, Verify, and Validate the Simulation Model
Phase 6: Experiment with the Model and Look for Opportunities for Design of Experiments
Phase 7: Document and Present Results
Phase 8: Define the Model Life Cycle

These phases are listed in a sequential fashion, but one might need to go back to a previous phase because the scope and objectives may change throughout the duration of the project.

Today's simulator software tools for a single license can range in price from $1000.00 to $l00,000.00.

COMPUTER-INTEGRATED MANUFACTURING

Computer-Integrated Manufacturing is a very young technology. Future production engineers will have at their disposal much more advanced tools. These tools will include advanced solid modeling CAD/CAM systems, AI/expert systems, and other very powerful software. Relatively inexpensive microprocessor and communication technologies have provided the opportunity to make productivity gains in production engineering and manufacturing. Continuous training and advanced on-line educational programs will be needed to maintain a knowledgeable production engineering workforce in a CIM factory.

Systems in the future will be primarily computer-based and integrated with all factory CIM systems, including more powerful computer hardware, completely linked databases, greater choice of software and expert systems, and advanced manufacturing systems.

SUMMARY

Production engineering refers to several engineering specialties concerned with planning and controlling manufacturing processes. *Manufacturing engineering* and *industrial engineering* are the two primary branches of production engineering. Manufacturing engineering includes process planning, tool design, and NC programming engineering. The industrial engineer typically assigns time standards for each process planning operation, determines details of operator methods, and selects the appropriate feeds, speeds, and perishable tooling.

With the rapid development of computers and software, CIM has become an economic reality. With faster and more powerful *microcomputers*, *minicomputers*, and *mainframes*, many companies can quickly and efficiently advance into CIM.

Microcomputers are commonly referred to as desktop, portable, personal, or small business computers. *Minicomputers* are more powerful than microcomputers and are often, in a manufacturing setting, linked to a mainframe. They are very practical for Production Engineering uses related to DNC, process planning, cell networking, and FMS systems. *Mainframe computers* are powerful computers with vast amounts of memory and can handle hundreds of users simultaneously.

CADD represents the predominant application of computer technology in design and production engineering. As an interactive computer graphics system, this technology automates the design and drafting aspects of engineering. It also allows the design engineer and the production engineer to share a common database to describe the product.

Group Technology (GT) applications in production engineering group parts into families. Similar part characteristics are known as *attributes*. They represent physical shapes, design characteristics, or manufacturing processes. Three common methods used to group part families are: sight inspection, routing sheet, and part classification and coding.

The use of some form of Group Technology is the cornerstone to cellular and flexible manufacturing systems (FMS). Parts must be put in families, then standard process routings must be generated before the reorganization of machines in the factory can be accomplished. This type of manufacturing can reduce lead time, set-ups, material handling, direct labor, and many other costs.

Computer-Assisted Process Planning (CAPP) can assist in the complicated task of determining and specifying the step-by-step sequence of operations needed to manufacture a part.

Two basic types of CAPP systems are *variant* and *generative*. A *variant* system is commonly called a retrieval-type system. It relies on the retrieval of a similar process plan for a part to create a new part process plan. The existing plan is slightly modified to fit special needs of the new part, and then the modified (variant) process plan is used. The *generative* process plan refers to an automated system used to create new process plans. The process plan is generated automatically from the part attributes coded into the database.

CAM (Computer-Aided Manufacturing) systems complete the CIM network in a factory. The CAM system includes the full Computer Numerical Control Programming system, the Direct Numerical Control (DNC) system to the factory floor and a full Computer Simulation Software system. The simulation software has been applied across all types of manufacturing industries to size and design new facilities, evaluate new equipment needs, and formulate a CIM management factory strategy.

IMPORTANT TERMS

Artificial intelligence
attributes
batch manufacturing
byte
Cartesian coordinates
computer simulation
Computer-Aided Manufacturing
Computer-Aided Process Planning
digitizing tablet
distributed processing
expert systems
flexible manufacturing systems
generative process planning
Group Technology
hierarchical structure
interactive
interactive graphics

mainframe
manufacturing cell
memory
MIPS (millions of instructions per second)
NC part program
parts classification and coding
random access memory
read-only memory
replicate
rotate
routing sheet inspection
scaling
sight inspection
standard routing
tolerance stack sheet
translation
variant-type CAPP system

QUESTIONS FOR REVIEW AND DISCUSSION

1. Describe the usual responsibilities of a manufacturing engineer. How do these differ from the duties of the industrial engineer?

2. Describe the differences and varying uses of mainframe computers, minicomputers, and microcomputers.

3. What does the phrase, "Whenever possible...directions come from the lowest level" mean when it is applied to a distributed processing system? Why is this important?

4. What are some of the benefits of CAD in terms of a CIM operation?

5. Distinguish between *display space* and *problem space*, as these terms are used in relation to a Computer-Aided Design program.

6. Similar part characteristics are referred to as *attributes*. How are attributes used when implementing the manufacturing philosophy known as Group Technology?

7. What are some of the advantages that Artificial Intelligence or *expert systems* software would offer, when compared to conventionally structured software?

Manufacturing Production Planning

by Michael W. Pelphrey

Key Concepts

☐ Effective planning and scheduling are vital to efficient functioning of a plant in a CIM environment.

☐ The two major manufacturing environments, *make-to-order* and *make-to-stock*, have different planning and monitoring requirements.

☐ Time fences are critical management decision points in the Master Production Schedule, which serves as the driver for Material Requirement Planning activities.

☐ Manufacturing Resource Planning (MRP II) is a method of planning and controlling resources within the total manufacturing company to attain the desired business objectives.

☐ Successfully implementing MRP II and CIM requires full commitment from top management and the building of a grass-roots wave of interest in the process.

Overview

Planning and scheduling in a manufacturing environment is a complex and demanding, but necessary, task. The demands of competition dictate that the business be functionally integrated, as well as being integrated on a planning and informational basis. This means that management must develop both strategic and tactical plans.

In a CIM environment, planning and scheduling become especially critical to daily operations because the entire factory is integrated. If any single operating element is not congruent with the strategic direction of the business, optimum results cannot be obtained. The complexities associated with planning involve technical, environmental, and people factors. If these factors are not taken into account, imbalances in both planning and execution can result.

This chapter describes planning procedures that meet the needs of most industries. Some industries, such as those involved in government contracting, have particular characteristics that make it necessary to amplify or modify the techniques described.

Michael W. Pelphrey is Director, Manufacturing Services, at BDO Seidman, an accounting and consulting firm in Los Angeles, California.

MANUFACTURING ENVIRONMENTS

There are two broad categories of manufacturing systems in use today. They can be described as "make-to-order manufacturing" and "make-to-stock manufacturing." Each of these has variations.

MAKE-TO-ORDER MANUFACTURING

In the *make-to-order manufacturing* system, the customer specifies the "product" before any resources are committed. Make-to-order manufacturing typically involves longer lead times and meets a need for specific, not normally available, product. The planning in a make-to-order system may involve a project orientation that can cover a span of years, such as a nuclear generating plant. It could also apply to shorter-range products, such as machine tools, that follow more traditional planning and monitoring techniques.

MAKE-TO-STOCK MANUFACTURING

Make-to-stock manufacturing is used for products that are essentially identical and that are produced in large quantities.

Most consumer products, which must be available for purchase at the time when a customer wishes to satisfy a need, are produced in a make-to-stock manufacturing system. The planning and control principles involved in make-to-stock manufacturing require customer demands to be anticipated well in advance of actual consumption of the goods.

Configure-to-order manufacturing

A variation of make-to-stock manufacturing is configure-to-order. A *configure-to-order product* is one that consists of a basic item with various options selected by the buyer to meet specific needs.

CHARACTERISTICS OF MANUFACTURING SERVICE ENVIRONMENTS

CONTINUOUS (REPETITIVE) PRODUCTION	DISCRETE (INTERMITTENT) PRODUCTION
1. Limited line of production.	1. Flexible line of production.
2. Plan and schedule by rate; large quantities.	2. Plan and control by lot sizes; small quantities.
3. Large plants; automated fixed flow path of material.	3. Small plants; may have alternate routings.
4. Equipment dedicated to a range of tasks that allow flow.	4. Similar equipment groupings (machine shop, paint shop, welding)
5. Few low-skill workers.	5. Highly skilled operators.
6. Few instructions needed.	6. Many instructions needed.
7. Low raw material inventory.	7. High raw material inventory.
8. Low work-in-progress inventory.	8. High work-in-progress inventory.
9. Conveyors used for material handling.	9. Trucks used for material handling.
10. Narrow aisles, little material storage.	10. Considerable material storage (on shop floor).

Fig. 5-1. Defining the type of manufacturing environment, continuous or discrete, is important when implementing CIM. Each environment has distinctive characteristics.

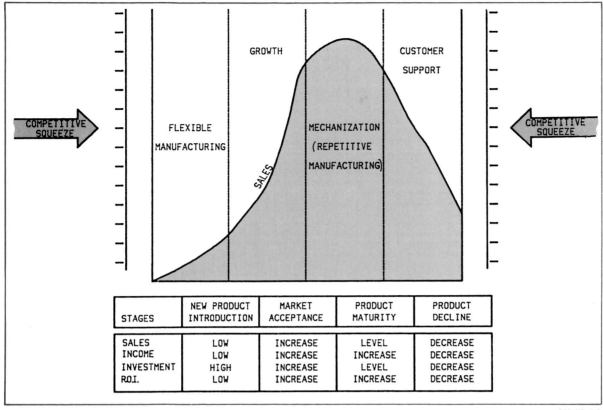

GROWTH

CUSTOMER SUPPORT

COMPETITIVE SQUEEZE

FLEXIBLE MANUFACTURING

MECHANIZATION (REPETITIVE MANUFACTURING)

COMPETITIVE SQUEEZE

SALES

STAGES	NEW PRODUCT INTRODUCTION	MARKET ACCEPTANCE	PRODUCT MATURITY	PRODUCT DECLINE
SALES	LOW	INCREASE	LEVEL	DECREASE
INCOME	LOW	INCREASE	INCREASE	DECREASE
INVESTMENT	HIGH	INCREASE	LEVEL	DECREASE
R.O.I.	LOW	INCREASE	INCREASE	DECREASE

Fig. 5-2. A product has a definite life cycle. During the introduction phase, considerable investment is needed, while sales, income and return on investment are low. During the market acceptance phase, all factors increase. When product maturity is achieved, sales and investment level out, while income and return on investment continue to increase. In the decline phase, all factors decrease.

The options may be standard products themselves, or may be unique designs. The manufacturer has increased flexibility because there is no need to stock completed products; only completed subassemblies must be on the shelf. The challenge for management is to rapidly respond to changing market demands. The configure-to-order business thus aligns itself very closely to the responsive nature of the make-to-stock system.

Manufacturing environments may be viewed as either *continuous* (repetitive), or *discrete* (intermittent) in nature. The characteristics of each environment are shown in Fig. 5-1. It is important for a company to identify the type of manufacturing environment in which it operates, so that the CIM technologies needed for support can be adequately defined. It is not uncommon, however, for a business to have products and demands that incorporate aspects of both environments.

PRODUCT LIFE CYCLE

Another important planning concept is the product life cycle, in which each product follows a cycle from introduction through growth, maturity, and finally, decline. See Fig. 5-2. Planning activities will vary according to the stage at which the product is located in the life cycle. The planning for new products requires heavy involvement of the marketing and engineering functions. Mature products released to the manufacturing floor typically require less marketing and more refinement of the production processes. Products nearing the end of their life cycle require more involvement from the customer support organization through warranty and service part activities. From the standpoint of CIM, early communication and coordination involving the entire organization is vital to effectively develop the processes and technologies necessary to capture the desired market share.

Each functional unit of an organization committed to excellence through CIM must be comfortable with simultaneously acting as a leader and as a follower (servant). For example, during product introduction, Engineering *serves* Marketing while *leading* Manufacturing. In the same manner, Marketing *leads* Engineering while *serving* the marketplace. Each function must be able to operate as the servant or the leader, even though this may be difficult or uncomfortable for its members. The ability to blend these roles is essential to the success of the business.

THE PLANNING PROCESS

The objectives of manufacturing management are to provide service to the customer while maintaining an optimum inventory level. Manufacturers desire productive employees, efficient machine utilization, minimized overhead costs, and increased sales. The best way to achieve these objectives is to establish a formal planning process. See Fig. 5-3.

THE BUSINESS PLAN

The planning process begins by establishing a business plan. The ***business plan*** describes the sales, profit, and capital required to achieve the objectives set for growth, profitability, and return on investment. A business plan is a strategic statement, normally covering a period of about five years. The plan must reflect economic sense, business sense, and common sense.

In the near term, the strategic business plan is normally called the ***annual operating plan*** (AOP). Marketing will develop a sales plan — the expected demand on a product family (product line) basis. This plan is then translated into dollar amounts and called the ***financial plan***.

PRODUCTION PLANNING

The AOP is then converted into a unit ***production plan***, in which rates of production and resource requirements are defined. This conversion process is labeled production planning; the planning document serves as the baseline for master scheduling.

The production plan is stated in broad terms, such as product groupings. It must be further broken down into actual products to be built. As shown in Fig. 5-3, a check of constraining resources is made and resolved at each planning level. If a constraint cannot be resolved, the plan must be adjusted until the constraint is eliminated.

The ***Master Production Schedule*** (MPS) is a further refining of the production plan into item-build priorities. A computer algorithm, ***Materials Requirements Planning*** (MRP), uses the MPS to create a material procurement and shop floor manufacturing plan.

Fig. 5-4 recaps the planning process by correlating it to manufacturing systems.

DEMAND MANAGEMENT

Demand management is a process that identifies, for the factory, the items to be built. It includes the functions of forecasting and customer order servicing. Without demand, supplies are not justified.

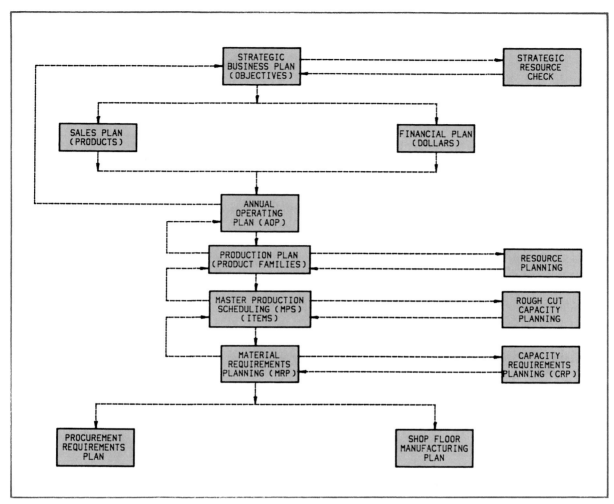

Fig. 5-3. The planning process proceeds from a strategic business plan to an annual operating plan and production plan. The Master Production Schedule and Material Requirements Planning combine to provide information needed for the operative documents: the procurement requirements plan and the shop floor manufacturing plan.

It should be clearly understood that a manufacturing company should not be building any product for which there is no current or expected future demand. The recognition of a demand initiates activity in the factory.

FORECASTING

Ideally, the factory should respond only to known demands expressed as firm customer orders. This is seldom the case in a real manufacturing operation. In the absence of known demand, a *forecast* is required. The main purpose of making a forecast is to gain knowledge about uncertain events that are important to present decisions. True forecasting is a blend of science and art that defies precise definition. A valid forecast, however, should pay off in additional profits.

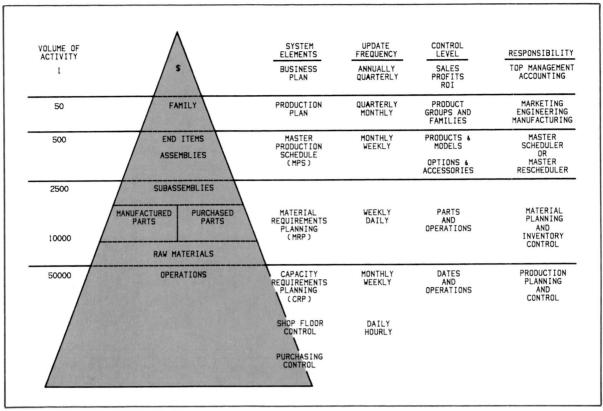

VOLUME OF ACTIVITY		SYSTEM ELEMENTS	UPDATE FREQUENCY	CONTROL LEVEL	RESPONSIBILITY
1	$	BUSINESS PLAN	ANNUALLY QUARTERLY	SALES PROFITS ROI	TOP MANAGEMENT ACCOUNTING
50	FAMILY	PRODUCTION PLAN	QUARTERLY MONTHLY	PRODUCT GROUPS AND FAMILIES	MARKETING ENGINEERING MANUFACTURING
500	END ITEMS ASSEMBLIES	MASTER PRODUCTION SCHEDULE (MPS)	MONTHLY WEEKLY	PRODUCTS & MODELS OPTIONS & ACCESSORIES	MASTER SCHEDULER OR MASTER RESCHEDULER
2500	SUBASSEMBLIES				
10000	MANUFACTURED PARTS / PURCHASED PARTS / RAW MATERIALS	MATERIAL REQUIREMENTS PLANNING (MRP)	WEEKLY DAILY	PARTS AND OPERATIONS	MATERIAL PLANNING AND INVENTORY CONTROL
50000	OPERATIONS	CAPACITY REQUIREMENTS PLANNING (CRP)	MONTHLY WEEKLY	DATES AND OPERATIONS	PRODUCTION PLANNING AND CONTROL
		SHOP FLOOR CONTROL	DAILY HOURLY		
		PURCHASING CONTROL			

Fig. 5-4. *The manufacturing system pyramid, in relation to planning system elements, serves as a recap of the planning process.*

To formulate a reasonable forecast, there must be effective communication between the various functional areas of the business. General business conditions that could influence the forecast must be identified. These may include the desired level of customer service, marketplace trends, the level of competition, technologies available to engineering and manufacturing staffs, and the relative confidence each function has in its input. Forecast characteristics to remain aware of include:
• Forecasts are more accurate for a large group of items than for single units.
• Forecasts are more accurate in the near term than in the long term.
• Forecasts should be tested using historical data.
• Forecasts must include a reasonable estimate of error.
• It should be clearly understood that forecasts can never be totally correct.
• If a *demand* can be calculated, it is not necessary or desirable to forecast.

Techniques used to develop a forecast can be *qualitative* (market research, management decision), *intrinsic* (extrapolation of past into future), or *extrinsic* (external leading indicators). Forecasting is reiterative: gather data, apply the data to a model, temper with judgment, revise, then repeat the process. If the iterations do not occur frequently enough, or adjustments are not made to the forecast periodically, the results should be questioned.

Listed in Fig. 5-5 are various forecasting techniques and their applications. Fig. 5-6 describes how these techniques apply to the product life cycle. Fig. 5-7 shows a method of measuring forecast error by calculating the **Mean Absolute Deviation** (MAD) of a specific forecast trend line.

BASIC FORECASTING TECHNIQUES	ACCURACY IN MONTHS SHORT 0-3	MEDIUM 3-24	LONG 24+	APPLICATION	DEMAND PATTERN	COST DOLLARS	COMPUTER REQUIRED	TIME REQUIRED
QUALITATIVE METHODS								
DELPHI (PANEL OF EXPERTS)	F-VG	F-VG	F-VG	ALL QUALITATIVE ADDRESS LONG–RANGE AND MARGIN FORECASTS, AND FORECASTS OF NEW PRODUCT SALES	UNKNOWN	2000 +	NO	2 MOS +
MARKET RESEARCH	E	G	F-G		UNKNOWN	5000 +	NO	3 MOS +
PANEL CONSENSUS	P-F	P-F	P		UNKNOWN	10,000 +	NO	2 WKS +
VISIONARY FORECAST	P	P	P	''	UNKNOWN	100 +	NO	1 WK +
HISTORICAL ANALYSIS	P	F-G	F-G	''	KNOWN COMPARABLE PRODUCT	1000 +	NO	1 MO +
INTRINSIC METHODS	TIME SERIES ANALYSIS AND PROJECTION							
SIMPLE AVERAGE MOVING WEIGHTED	P-G	P	VP	INVENTORY CONTROL LOW VOLUME	LEVEL	0.005	NO	1 DAY –
EXPONENTIAL SMOOTHING	F-VG	P-G	VP	PRODUCTION AND INVENTORY CONTROL	LEVEL	0.005	NO	1 DAY –
SECOND ORDER SMOOTHING					UPWARD OR DOWNWARD TREND			
ADAPTIVE SMOOTHING					CYCLICAL OR SEASONAL			
BOX-JENKINS	VG-E	P-G	VP	PRO. & INV. CON. LARGE VOLUME	KNOWN	10	NO	1-2 DAYS
X-11 (CENSUS BUREAU)	VG-E	G	VP	TRACKING AND WARNING FORECASTS	CYCLICAL OR SEASONAL	10	YES	1 DAY
TREND PROJECTION	VG	G	G	NEW PRODUCT FORECASTS	KNOWN COMPARABLE PRODUCT	VARIES	NO	1 DAY –
EXTRINSIC METHODS	CAUSAL AND ECONOMETRIC MODELS							
REGRESSION	G-VG	G-VG	P	SALES BY PRODUCT CLASS	UPWARD OR DOWNWARD TREND	100	NO	IDENTIFY RELATION-SHIPS
ECONOMETRIC MODEL	G-VG	VG-E	G	''	KNOWN	5000 +	NO	2 MOS +
INTENT TO BUY SURVEY	P-G	P-G	VP	''	COMPARABLE	5000	NO	2-3 WKS
INPUT-OUTPUT MODEL	NA	G-VG	G-VG	COMPANY'S SALES	KNOWN	50,000 +	YES	6 MOS +
LEADING INDICATORS	P-G	P-G	VP	BY PRODUCT CLASS	KNOWN	1000	NO	1 MO +
LIFE-CYCLE ANALYSIS	P	P-G	P-G	NEW PRODUCT SALES	KNOWN COMPARABLE PRODUCT	1500	NO	1 MO +

KEY: V = VERY P = POOR G = GOOD NA = NOT APPLICABLE
GOOD F = FAIR E = EXCELLENT

Fig. 5-5. *The basic forecasting techniques and their applications. This chart also shows the typical cost in dollars and the time required for the use of each method.*

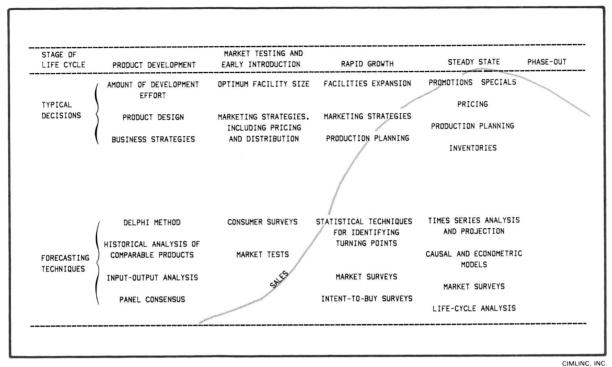

STAGE OF LIFE CYCLE	PRODUCT DEVELOPMENT	MARKET TESTING AND EARLY INTRODUCTION	RAPID GROWTH	STEADY STATE	PHASE-OUT
TYPICAL DECISIONS	AMOUNT OF DEVELOPMENT EFFORT PRODUCT DESIGN BUSINESS STRATEGIES	OPTIMUM FACILITY SIZE MARKETING STRATEGIES. INCLUDING PRICING AND DISTRIBUTION	FACILITIES EXPANSION MARKETING STRATEGIES PRODUCTION PLANNING	PROMOTIONS SPECIALS PRICING PRODUCTION PLANNING INVENTORIES	
FORECASTING TECHNIQUES	DELPHI METHOD HISTORICAL ANALYSIS OF COMPARABLE PRODUCTS INPUT-OUTPUT ANALYSIS PANEL CONSENSUS	CONSUMER SURVEYS MARKET TESTS	STATISTICAL TECHNIQUES FOR IDENTIFYING TURNING POINTS MARKET SURVEYS INTENT-TO-BUY SURVEYS	TIMES SERIES ANALYSIS AND PROJECTION CAUSAL AND ECONOMETRIC MODELS MARKET SURVEYS LIFE-CYCLE ANALYSIS	

CIMLINC, INC.

Fig. 5-6   Different forecasting techniques are used, as aids to reaching different decisions, at various points in a product's life cycle.

CUSTOMER ORDER MANAGEMENT

Another pertinent aspect of demand management is the entry and maintenance of customer orders. Customer orders serve a multitude of functions, in addition to telling shipping what to ship and accounting what to invoice. They also represent current demand patterns, reflect current backlog, and consume the forecast.

Customer order *service functions* include entering the order, negotiating a delivery date using availability information from the Master Production Schedule, managing the order as it is processed, handling inquiries, coordinating shipping, and evaluating performance to schedule. Customer order *service objectives* include prompt response to orders and requests for quotations, rapid order follow-up, accurate communication of delivery dates, reduction of order cycle time, communication of the reasons for order delays, and the development of plans to correct problems causing delays.

To meet customer service objectives, a customer service policy must be defined and performance measured against that policy. For example, in a make-to-stock company, *service level* is determined by measuring availability against requirements. If a line item on a customer order requires shipment of 100 units on May 1 and only 75 units are shipped on that date, the **customer service performance** for that line item is 75%. The performance level must be compared to the policy level desired and corrective action taken as required.

Fig. 5-8 depicts the thought process that an order entry administrator might follow for allocating available inventory. To maintain credibility, each decision point requires accurate information from the manufacturing system. Problems at these decision points must be communicated to the customer or to manufacturing for action or agreement. Frequently, delays are related to supply problems.

THE MEAN ABSOLUTE DEVIATION (MAD) RELATES TO THE NORMAL DISTRIBUTION (BELL-SHAPED OR PROBABILITY) CURVE AS SHOWN BELOW:

THIS SIMPLY STATES THAT ONE MAD WILL NORMALLY COVER 58% OF THE OCCURRENCES TRACKED (IN THIS CASE ORDERS OVER A WEEKLY BASIS).

PERIOD WEEK	FORECAST	DEMAND	ERROR	ABSOLUTE VALUE
1	175	169	−6	6
2	175	180	+5	5
3	175	135	−40	40
4	175	213	+38	38
5	175	181	+6	6
6	175	148	−27	27
7	175	204	+29	29
ETC				
		SUM OF ERRORS:		151

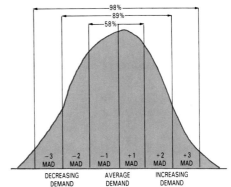

MAD IS CALCULATED BY DIVIDING THE SUM OF THE ABSOLUTE VALUE OF THE ERRORS BY THE NUMBER OF OCCURRENCES: (151÷7) THE MAD IN THIS CASE EQUALS 21.57 OR 22. THEREFORE, 175 WOULD BE THE AVERAGE DEMAND, AND

PROBABILITY	MAD	EXAMPLE
58%	+/−1	+/−22 (153−197)
89%	+/−2	+/−44 (131−219)
98%	+/−3	+/−66 (109−241)

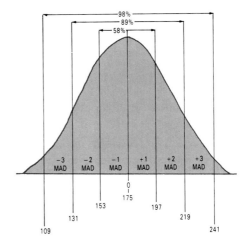

USING THE NORMAL CURVE AND MAD ALLOWS THE FORECASTER TO

• ESTABLISH A PROBABLE RANGE OF ERROR

• IDENTIFY EXTREME DEMANDS

• MONITOR THE QUALITY OF THE FORECAST.

NOTE THE SIMILARITY TO STATISTICAL PROCESS CONTROL (SPC)

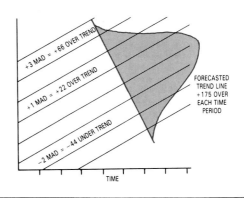

Fig. 5-7   *The Mean Absolute Deviation (MAD) is an accepted statistical method of measuring error in a forecast.*

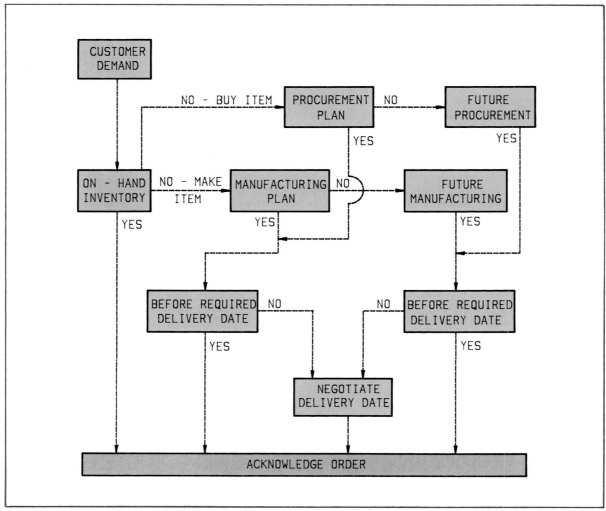

Fig. 5-8. The order-filling process may require a series of decisions that involve the Purchasing and Manufacturing functions. Accurate information is vital to making sound decisions.

SUPPLY MANAGEMENT

As noted earlier, the Master Production Schedule (MPS) maintains the anticipated build schedule for specific configurations with their related quantities and due dates. The MPS is a "game plan" for the company. The purpose of the MPS is to orchestrate manufacturing resources for optimum efficiency and inventory investment while driving costs down.

The MPS serves as the driver for the Material Requirements Planning (MRP) activity. The MPS establishes priorities and functions as a communication medium to ensure continuity between manufacturing, marketing, and finance functions. Changes in priority can be costed to provide management with a tool to evaluate profit margin effect due to changes inside the specified management decision points, or *time fences*.

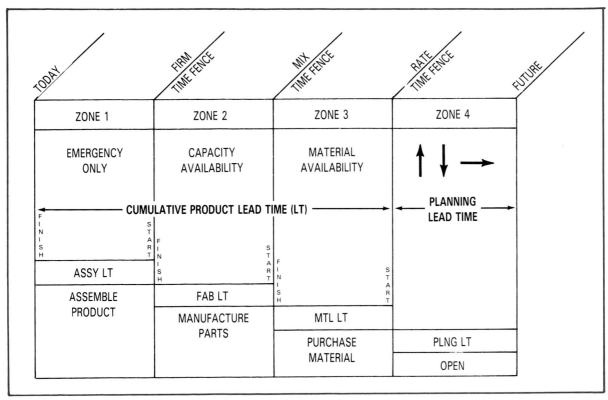

Fig. 5-9. *Time fences represent critical management decision points in the processes of planning and manufacturing a product.*

TIME FENCES

The concept of the MPS time fence is critical to understanding supply management. Fig. 5-9 demonstrates these critical management decision points. One of the fundamentals of master production scheduling is that the MPS horizon must be *equal to* or *greater than* the cumulative lead time of the product being scheduled.

A critical action is identification of the point in the life of an order at which it is not economically feasible to accept further change. This *firm* (no change) *time fence* occurs within the cumulative product lead time. Accepting change inside the firm time fence, particularly if it occurs within the final assembly lead time, is very costly. Making changes in less than cumulative lead time infers that plans have been made for such change, and that extra matched sets of parts and extra capacity have been made available to accommodate the original order.

A master production schedule may be aggregated into the **production plan**. See Fig. 5-10 for a comparison of the production plan and the master schedule. A rough-cut capacity check can be performed by using a **bill of resources** that relates, for example, the capacity of a turning machine center to the product being built in the MPS. If 100 units of product A are to be scheduled, and each unit of production requires 12 minutes on the turning machine center, then 20 hours of capacity will be required. If the available capacity is more than the 20 hours needed, the MPS can be entered. If not, other activities will have to be considered.

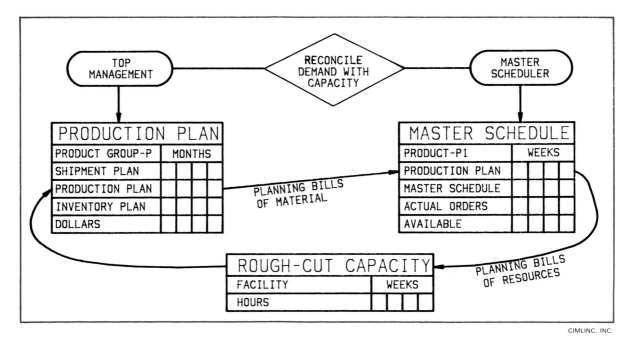

Fig. 5-10. A rough-cut capacity check can be used to reconcile the demand for a product with manufacturing capacity.

FINAL ASSEMBLY SCHEDULE

The process of bridging the MPS completion to the customer orders that consume the MPS is considered the *Final Assembly Schedule* (FAS). Many companies view the FAS as the means to actualize the shipping schedule. Depending on the environment, this may be as simple as issuing finished goods from stock (make-to-stock), or as complex as customizing a product for a specific customer requirement during final assembly (configure-to-order). Fig. 5-11 compares a Master Production Schedule with a Final Assembly Schedule and points out the essential differences.

STABLE MPS

A stable MPS is one of the key requirements of an effective *Manufacturing Resource Planning* (MRP II) system. Changes at the end-item level within a critical time fence will have serious effects on component requirements at lower levels. One effect might be that no-longer-required components are in process and may be completed, creating potential overages. A related effect would be that components not originally scheduled on the MPS will require expediting and management attention. MPS stability is a major management issue; one not easily arrived at in the implementation of MRP II.

Common pitfalls to an effective MPS are:
- The total MPS is overloaded.
- The MPS is front-end loaded.
- The MPS is not stable or managed.
- The MPS is incomplete (for example, service parts are not included).
- The MPS horizon is too short.
- The MPS is nothing more than a "wish list."

See Fig. 5-12 for an example of how an MPS is determined, how it is compared to actual orders, and how an available-to-promise line is derived.

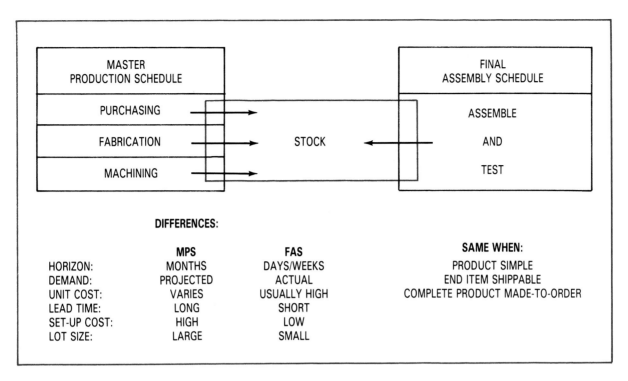

DIFFERENCES:

	MPS	FAS	SAME WHEN:
HORIZON:	MONTHS	DAYS/WEEKS	PRODUCT SIMPLE
DEMAND:	PROJECTED	ACTUAL	END ITEM SHIPPABLE
UNIT COST:	VARIES	USUALLY HIGH	COMPLETE PRODUCT MADE-TO-ORDER
LEAD TIME:	LONG	SHORT	
SET-UP COST:	HIGH	LOW	
LOT SIZE:	LARGE	SMALL	

Fig. 5-11. The Master Production Schedule and the Final Assembly Schedule can differ considerably in such characteristics as lot size and necessary lead time.

BASIC FORMAT - MPS

TIME FENCE	ZONE 1 FROZEN		ZONE 2 FABRICATION LT		ZONE 3 MATERIAL LT			ZONE 4 OPEN	
	QUANTITY BY TIME PERIODS (BUCKETS)							UM: UNITS	
PERIODS	1	2	3	4	5	6	7	8	9
MPS ITEM 1	12		10		8				
MPS ITEM 2	8		10		12				
MPS	20		20		20				
AO ITEM 1	12		6						
AO ITEM 2	8		9						
AO	20		15						
ATP ITEM 1	0		4		8				
ATP ITEM 2	0		1		12				
ATP	0		5		20				

MPS = MASTER PRODUCTION SCHEDULE

AO = ACTUAL ORDERS

ATP = AVAILABLE TO PROMISE

CUSTOMER REQUESTS 8 ADDITIONAL ITEM 1 AND 7 ADDITIONAL ITEM 2

ORDER ADMINISTRATOR CHANGES ACTUAL ORDER COLUMN TO READ

AO ITEM 1	12		10		4					
AO ITEM 2	8		10		6					
AO	20		20		10					
ATP ITEM 1	0		0		4					
ATP ITEM 2	0		0		6					
ATP	0		0		10					

CIMLINC, INC.

Fig. 5-12. The basic format of a Master Production Schedule, showing how an available-to-promise line is derived by comparison of the MPS with actual orders.

MANUFACTURING RESOURCE PLANNING (MRP II)

The growth of computer technology from the early 1960s to the present has allowed continuing evolution of the techniques for manufacturing planning and control. Before the computer came into general use, statistical analysis was employed to identify the point at which an order should be placed to replenish inventory. These early reorder point methods depended heavily on manual manipulation of data. A manufacturing facility processes hundreds of end-items and thousands of unique part numbers. For this reason, a better method was needed to calculate material requirements from a list of items to be sold. That list was called the Master Production Schedule, and the method was initially called Material Requirements Planning (MRP). For a comparison of order-point versus MRP logic, see Fig. 5-13.

	ORDER POINT (INDEPENDENT DEMAND INVENTORY SYSTEM)	vs.	MRP (DEPENDENT DEMAND INVENTORY SYSTEM)
ASSUMPTIONS	• DEMAND IS INDEPENDENT OF OTHER ITEMS • DEMAND IS UNIFORM • DEMAND IS CONTINOUS		• DEMAND IS DEPENDENT UPON PARENT DEMAND • DEMAND IS LUMPY • DEMAND IS DISCRETE AND IRREGULAR
GOAL	• MAINTAIN INVENTORY LEVEL		• INVENTORY TO BE AVAILABLE ON DATE OF NEED
PLANNING TECHNIQUE	• FORECASTING EXTRINSIC INTRINSIC • BASED ON PAST DATA		• COMPONENT DEMAND IS CALCULATED FROM PARENT • BASED ON FUTURE DEMAND
WHAT TO ORDER?	• ITEMS AT REORDER POINT (ROP)		• ITEMS WITH NET REQUIREMENTS
HOW MUCH TO ORDER?	• EOQ—ECONOMIC ORDER QUANTITY • FOQ—FIXED ORDER QUANTITY • FPR—FIXED PERIOD REQUIREMENTS		• EOQ, FOQ, FPR • LFL—LOT FOR LOT (DISCRETE) • POQ—PERIOD ORDER QUANTITY • LUC—LEAST UNIT COST • LTC—LEAST TOTAL COST • PPB—PART PERIOD BALANCING
WHEN TO ORDER?	• DATE THE ROP IS REACHED		• TIME-PHASED • ORDER RELEASE DATE FROM LEAD TIME OFFSET
WHEN TO SCHEDULE ORDER DUE DATE?	• DATE THE ROP IS REACHED (TODAY'S DATE PLUS LEAD TIME)		• TIME-PHASED • PLANNED ORDER RELEASE DATE IS DYNAMICALLY ALIGNED WITH NEED DATE OF PARENT

Fig. 5-13. Material Requirements Planning provides a means of calculating material requirements that is more accurate than using traditional order-point logic.

CLOSED-LOOP MRP

Material Requirements Planning systems initially provided a schedule and priority planning tool that focused on material control. As MRP matured, other functions were integrated, such as shop floor control, master scheduling, and capacity planning. Also added was the function of feeding back performance measurement data. This caused the system to become known as *closed-loop MRP*.

In the late 1970s, it became apparent that links to other business and financial planning areas were needed. A system to plan and control resources within the total manufacturing company was highly desirable. That system, called Manufacturing Resource Planning (MRP II), recognized that engineering, marketing and financial resources, in addition to manufacturing resources, must be planned and coordinated to attain the desired business objectives.

CHARACTERISTICS OF DEMAND

To better appreciate MRP II, it is useful to review the components that typically support it. Demand for a part can come from sources outside the business, or may be derived from the demand on another item. See Fig. 5-14.

INDEPENDENT AND DEPENDENT DEMAND ITEMS

INDEPENDENT DEMAND ITEM	DEPENDENT DEMAND ITEM
DEFINITION:	DEFINITION:
AN ITEM WITH DEMAND NOT RELATED TO DEMAND OF ANY OTHER INVENTORY ITEM	AN ITEM WHOSE DEMAND IS DERIVED FROM THE DEMAND OF OTHER INVENTORY ITEMS
EXAMPLES:	VERTICAL DEPENDENCY:
FINISHED GOODS	PARENT-COMPONENT RELATIONSHIPS THROUGH ALL THE LEVELS
SPARE PARTS	HORIZONTAL DEPENDENCY:
	COMPONENTS AT THE SAME LEVEL FOR A PARENT

NOTE: SOME ITEMS, LIKE SPARE PARTS, MAY HAVE BOTH INDEPENDENT DEMAND AS SPARES AND DEPENDENT DEMAND AS THE COMPONENTS OF A PARENT.

Fig. 5-14. *Demand for a part or a product may be either independent or dependent, as shown. Some items, such as spare parts, may be subject to both types of demand:* independent, *as spare parts, and* dependent, *as components of a parent.*

Independent and dependent demand

Demand derived from an external source is called *independent demand*. Independent demand comes from such sources as the receipt of a customer order or the forecast of an anticipated need. A forecast of independent demand is, at best, a guess. Actual demand patterns and their effects will not be known until the event (a customer order) actually occurs.

In contrast, *dependent demand* derives from a schedule of end products or service parts, the Master Production Schedule. The MPS is processed by *exploding* (breaking down into quantities of needed components) a bill of material that identifies all components and processes to be used in manufacturing. The component part requirements for the end product are then calculated. During the explosion process, calculated demand is called *dependent* because the component demand depends on the supply needs of the parent item, Fig. 5-15.

Horizontal and vertical dependence

As shown in Fig. 5-15, it takes one component B and two component Cs to make one *parent item*, A. *Vertical dependence* is the relationship between the parent and its direct components. If 100 As are needed to satisfy a customer (*independent*) demand, how many Bs and Cs are required? If there are no As on hand (in inventory) and none on order (due in), it is easy to calculate that there is a (*dependent*) requirement for 100 Bs and 200 Cs.

The relationship of availability of components that are common to a given parent is called *horizontal dependence*. As shown in Fig. 5-16, a screw assembly consists of a screw, washer, housing, and spring. It cannot be built until all the components are on hand. Now, multiply parents and components by factors of tens, hundreds, or thousands, and it is easy to understand why a computer is needed to manage the various dependent relationships.

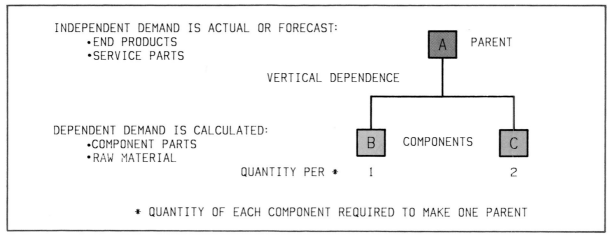

CIMLINC, INC.

Fig. 5-15. Demand for the parent (A) is independent, because it is an end product or is being produced as a spare part. Demand for the components (B and C) is dependent, since the quantity required is determined by the demand for (A). This is also an example of vertical dependence.

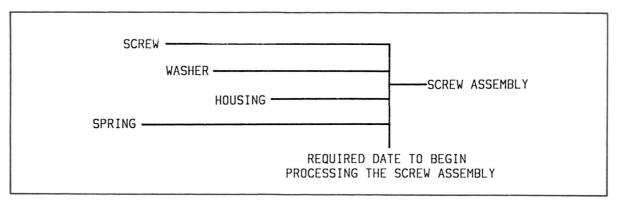

CIMLINC, INC.

Fig. 5-16. In horizontal dependence, all the components (screw, washer, housing, spring) must be available before the screw assembly can be produced.

112   Computer-Integrated Manufacturing

MRP PROCESSING FLOW

The processing flow of a material requirements planning program is shown in Fig. 5-17. The MPS is the driver that initiates the process. From the earlier discussion, it can now be seen that the MPS is a list of significant "parents" to be assembled and shipped to meet the Production Plan and business objectives. The MPS items are exploded through their bills of material into gross component requirements. The *net requirements* are then determined by deducting available inventories (on hand) and supply orders (purchase and manufacturing) from the gross requirements. A quantity (lot size) is then determined. This process is repeated on a level-by-level basis, until all related bill of material items are considered.

Rescheduling

Typical MRP report messages are to increase or decrease order quantities, reschedule orders earlier (*expedite*) or later (*de-expedite*), or cancel a released supply order. This is called *rescheduling*. Fig. 5-18 provides an example of this process.

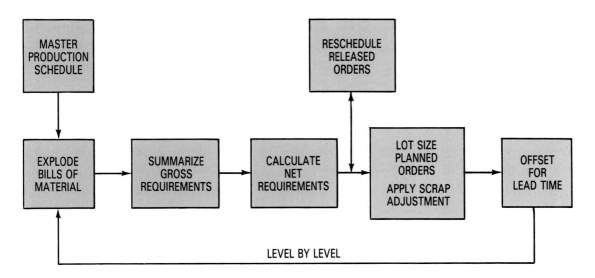

Fig. 5-17. *The typical processing flow of a material requirements planning program, which is initiated by the Master Production Schedule.*

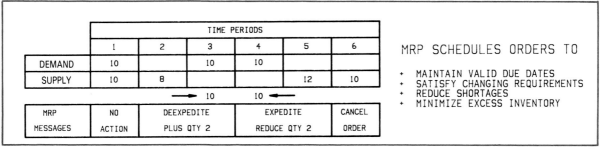

CIMLINC, INC.

Fig. 5-18. *An example of the rescheduling process used by MRP to meet various requirements.*

Time-phased backscheduling

An integral process performed by MRP is the *time-phased backscheduling* of a net requirement. Lead-time elements form the basis for determining the release date of an order. Fig. 5-19 shows how a parent's *due date* (date of completion) is backscheduled into its release date, based on the lead-time offset, then exploded into its component due dates. Note that the process includes all levels of a product's structure; each component of a parent is independently backscheduled, based upon its unique lead-time values.

Supply/demand management

The careful balancing of orders to their requirements reflects the principle of supply/demand management. A *supply* replenishes inventory and includes purchase orders, manufacturing orders, and rework orders. A *demand* reduces inventory. It is represented by requirements (customer orders, service parts, and parent needs). It may be useful to think of a supply as a deposit into a bank account and a demand as a withdrawal from the account.

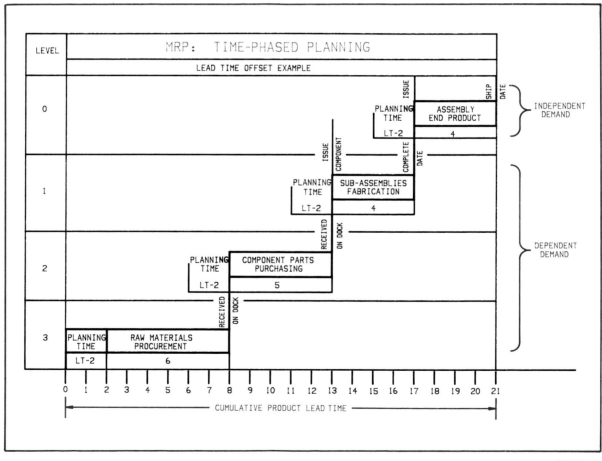

CIMLINC, INC.

Fig. 5-19. Lead-time elements are critical factors in determining the release date of an order. Time-phased planning involves backscheduling from the date of completion to determine release date and component due dates.

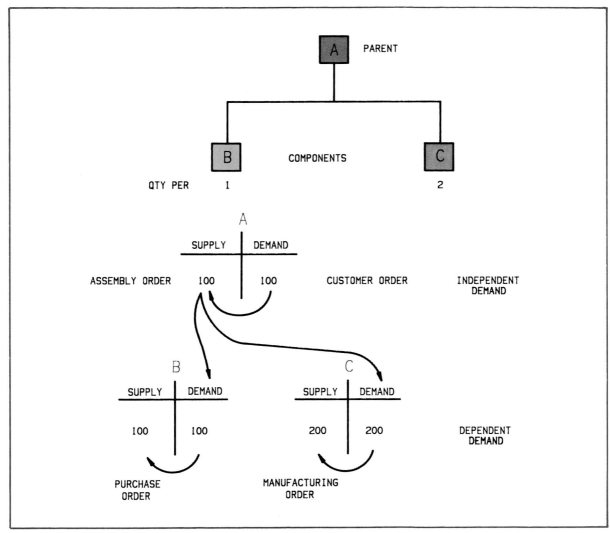

CIMLINC, INC.

Fig. 5-20. Supply and demand are interrelated: supply at each level affects the demand at the next lower level. The process continues until the lowest level is reached.

The interrelationship of supply and demand is shown in Fig. 5-20. As MRP works its way down through the bill of material, demands are placed on the next level. These demands, in turn, generate further supplies and demands until the bottom of the bill of material is reached. At the bottom of a bill of material will usually be either a purchased part or raw material.

Pegging

Another MRP concept that must be understood is "pegging." *Pegging* displays, for a given item, the detailed sources that generate its gross requirements. The net requirement for an item may be composed of a multitude of demands from various parents. Pegging allows you to identify those individual parent items, Fig. 5-21.

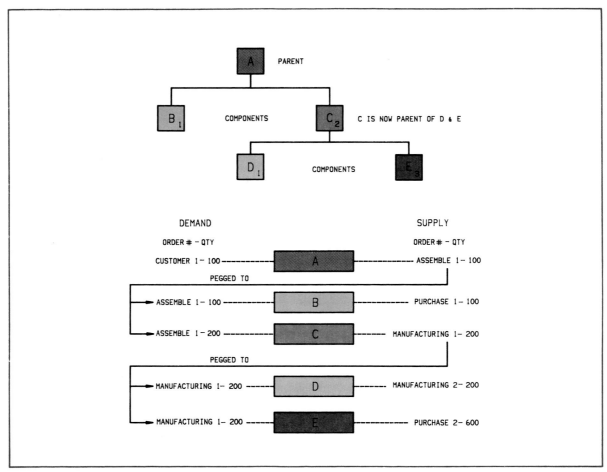

Fig. 5-21. An example of pegging, *in which the net requirement for a given component (such as E) is composed of demands from various parents. In this example, 600 of component E are needed to make 200 of its parent, C, which in turn is a component used to assemble 100 of parent A (the final product).*

USES OF PEGGING

* TO CHECK SOURCE OF REQUIREMENTS.
* TO TRACE IMPACT OF A COMPONENT DELAY:
    OPTIONS:
    —RESCHEDULE ALL OTHER COMPONENTS.
    —RESCHEDULE PLANNED ORDER QUANTITY OR DUE DATE FOR A PARENT ITEM.
    —RESCHEDULE MPS (LAST RESORT).
* TO ANALYZE NEED OF RESCHEDULING AN ORDER FLAGGED BY THE SYSTEM.
* TO TRACE EFFECTS OF ENGINEERING CHANGES.

* TO ANALYZE POSSIBILITY OF PULLING COMPONENTS ALLO-CATED FOR ONE ORDER TO SATISFY AN ORDER WITH HIGHER PRIORITY.
* TO MAINTAIN CUSTOMER IDENTITY FOR LOWER LEVEL COMPONENTS FOR LOT COSTING AND INSPECTION STANDARDS.
* TO OBTAIN PROGRESS PAYMENTS.
* TO PERMIT SHORTAGE REPORTING.
* TO FACILITATE CUSTOMER INQUIRY.
* TO VERIFY ADHERENCE TO CONTRACTUAL SPECIFICATIONS.

Fig. 5-22. Some of the applications for which the technique of pegging is used.

In a commercial environment, it is useful to be able to peg to an end item to make better allocation decisions about a limited component. In a military (government contractor) situation, the ability to separate supplies, demands, and inventories into their respective contracts and priorities may be needed. This requires *full-level* pegging. Fig. 5-22 lists some typical applications for pegging.

Lot-sizing

Determination of the net requirement leads to a process of coverage and lot-sizing. In the past, due to lack of computing power, various algorithms were used to determine order quantities. These algorithms did not calculate a discrete demand; they computed coverage based upon predefined models.

Current computer technology allows you to determine an individual part requirement within a defined time period as short as a single day. The defined time periods are often accumulated into longer increments (**buckets**) of weeks or months. Lot sizes that are greater than discrete (lot-for-lot) needs imply increased inventory. The principle of *Just-In-Time* (JIT) inventory relies on only processing a specific quantity (as close to a lot-size of *one* as possible). See Fig. 5-23.

LOT SIZING TECHNIQUES	LOT SIZING MODIFIERS

• ECONOMIC ORDER QUANITITY (EOQ)
• FIXED ORDER QUANTITY (FOQ)
• FIXED PERIOD REQUIREMENTS (FPR)
• LOT-FOR-LOT, DISCRETE (LFL)
• PERIOD ORDER QUANTITY (POQ)
• LEAST UNIT COST (LUC)
• LEAST TOTAL COST (LTC)
• PART PERIOD BALANCING (PPB)
• WAGNER-WHITIN
	• CUT-OFF DATE
	• AMOUNT OF SUPPLY STATED IN:
	  –DAYS
	  –QUANTITIES
	• MINIMUMS
	• MAXIMUMS
	• MULTIPLES

Fig. 5-23. *There are many different lot-sizing techniques available. The modifiers listed allow the quantity calculated by MRP to be adjusted to better match a supplier's production capabilities.*

Reporting

It is important to note that MRP can generate a large mass of detail for the planning department. A planner could then be overwhelmed by insignificant details and lose sight of the business plan. MRP report output must be tempered by common sense and the use of filtering tools. These *filtering tools* allow the use of ranges (percentages, quantities, and dates), rather than absolute values. For example, a purchase order may not be considered late if it is received within a three-day period of time (rather than on its specific "dock date"). An order might be considered closed if the quantity completed is within a range of five percentage points.

Planned orders

One of the most significant outputs of an MRP system is the generation of "planned orders." These planned orders support the entire MPS *planning horizon* (the time span for which planning has been done). Individual action on these orders by the planner is not required until the order matures and nears its release date. An MRP system is able to handle significant activity (such as engineering changes, customer order reschedules, or manufacturing process changes) without human intervention.

If a planner wants to control a specific order quantity or due date, he or she can intercede through a concept called the *firm planned order*. This type of order may be used to:
- Anticipate a service part requirement that was not entered in the MPS.
- Build up lower level components to reduce the cumulative lead time for a specific end product.
- Deviate from the usual process due to a decision by the planner.

MRP LOGIC

MRP *netting* (process of determining quantity) logic is as follows:

On hand (parts in warehouse/stockroom)
+ In inspection (parts awaiting quality release)
+ Floor stock (parts issued in bulk to shop floor)
= Currently available (as of MRP processing date)
− Demand (requirements)
+ Supply (purchase, manufacturing, rework orders)

If sufficient coverage is present for the period analyzed, MRP moves to the next period and analyzes it, using the remaining projected supply (from the previous period) as the beginning balance. This process occurs across the entire horizon (horizontally) to ensure coverage. If, at any point in analysis, MRP determines that coverage is needed, it will plan a supply order.

With every supply order planned, MRP explodes through the bill of material:

Required supply quantity (from parent)
× Shrinkage factor (calculated from parent)
= Factored supply quantity
× Quantity per assembly (for each component of parent)
× Scrap factor (calculated on component)
= Gross requirements (component)

Shrinkage is the amount of an item that is not acceptable for use. If, each time that a shipment of a particular component is received, it is found that 5 out of every 100 is defective, the shrinkage could be determined by dividing .95 into 1. This would yield a shrinkage factor of 1.053.

Scrap is the quantity of a component that is expected to be discarded (scrapped) while being built into a particular parent. The scrap factor can also be used to express the amount of raw material needed, in excess of the calculated requirement, to build a specified quantity of a component.

This explosion analysis occurs for every item associated with the parent. Net requirements are determined for each component. The pick date of the parent item becomes the due date for each of its components. The *pick date* is the date when components are gathered for assembly of the parent item.

Throughout the analysis process, MRP ensures supply coverage for demands. At the same time, it identifies excess supply. If a planned order is found to be creating excess supply, the order is automatically deleted. Excess supply for *firm planned orders* results in a message to reduce quantity, reschedule later (de-expedite), or cancel the order. Fig. 5-24 demonstrates the process graphically.

Fig. 5-25 incorporates the inputs and outputs of a typical MRP system. As shown in the example, a stable and comprehensive MPS, properly structured and up-to-date bills of material, accurate on hand and on order balances, realistic order due dates, timely transaction processing, efficient disposition of defective or scrap materials, and accurate data elements. The business objectives of MRP are to plan and control inventory, to plan priorities, and to serve as an input to capacity planning.

CAPACITY PLANNING

The function of *capacity planning* is to establish, measure, monitor, and sufficiently adjust levels of capacity to properly execute all manufacturing schedules. Specifically, capacity planning addresses the basic concerns of manufacturers by:

• Establishing operation start dates.
• Summarizing capacity requirements by resource center and time period.
• Analyzing underload and overload conditions.
• Controlling work backlog by work center.
• Reducing manufacturing lead times.

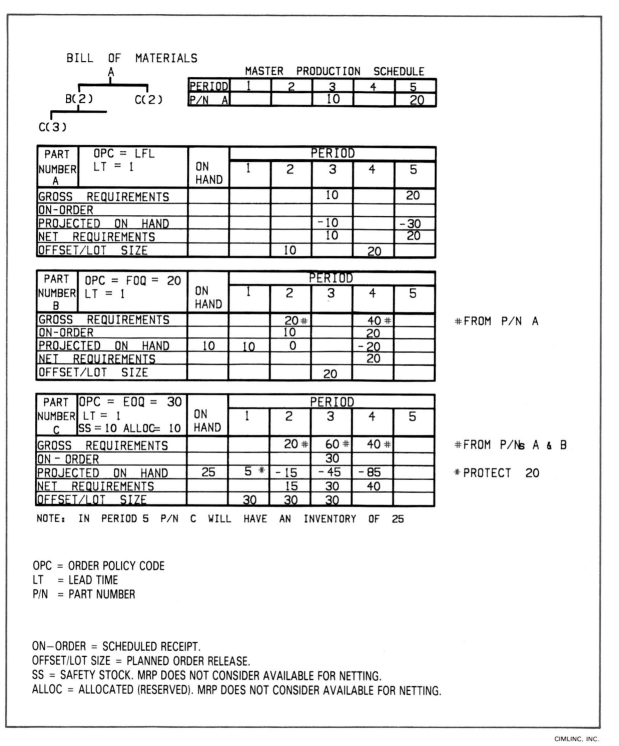

BILL OF MATERIALS

```
           A
      ┌────┴────┐
    B(2)       C(2)
   ┌──┘
 C(3)
```

MASTER PRODUCTION SCHEDULE

PERIOD	1	2	3	4	5
P/N A			10		20

PART NUMBER A	OPC = LFL LT = 1	ON HAND	PERIOD 1	2	3	4	5
GROSS REQUIREMENTS					10		20
ON-ORDER							
PROJECTED ON HAND					-10		-30
NET REQUIREMENTS					10		20
OFFSET/LOT SIZE				10		20	

PART NUMBER B	OPC = FOQ = 20 LT = 1	ON HAND	PERIOD 1	2	3	4	5
GROSS REQUIREMENTS				20 #		40 #		#FROM P/N A
ON-ORDER				10		20	
PROJECTED ON HAND		10	10	0		-20	
NET REQUIREMENTS						20	
OFFSET/LOT SIZE					20		

#FROM P/N A

PART NUMBER C	OPC = EOQ = 30 LT = 1 SS = 10 ALLOC = 10	ON HAND	PERIOD 1	2	3	4	5
GROSS REQUIREMENTS				20 #	60 #	40 #		#FROM P/Ns A & B
ON - ORDER					30		
PROJECTED ON HAND		25	5 *	-15	-45	-85		*PROTECT 20
NET REQUIREMENTS				15	30	40	
OFFSET/LOT SIZE			30	30	30		

#FROM P/Ns A & B

*PROTECT 20

NOTE: IN PERIOD 5 P/N C WILL HAVE AN INVENTORY OF 25

OPC = ORDER POLICY CODE
LT = LEAD TIME
P/N = PART NUMBER

ON-ORDER = SCHEDULED RECEIPT.
OFFSET/LOT SIZE = PLANNED ORDER RELEASE.
SS = SAFETY STOCK. MRP DOES NOT CONSIDER AVAILABLE FOR NETTING.
ALLOC = ALLOCATED (RESERVED). MRP DOES NOT CONSIDER AVAILABLE FOR NETTING.

Fig. 5-24. The process of netting, as used by MRP to balance supply with demand, is shown graphically.

Manufacturing Production Planning   119

- Controlling the level of work-in-process.
- Identifying potential bottlenecks.
- Minimizing resource idle time.

As a result of MRP output, capacity can be time-phased, loads on resources can be calculated, and plans can be made to adjust capacity to changing demands. *Capacity* is the highest sustainable output rate that can be achieved with the current work schedule, product specifications, product mix, equipment, and worker productivity.

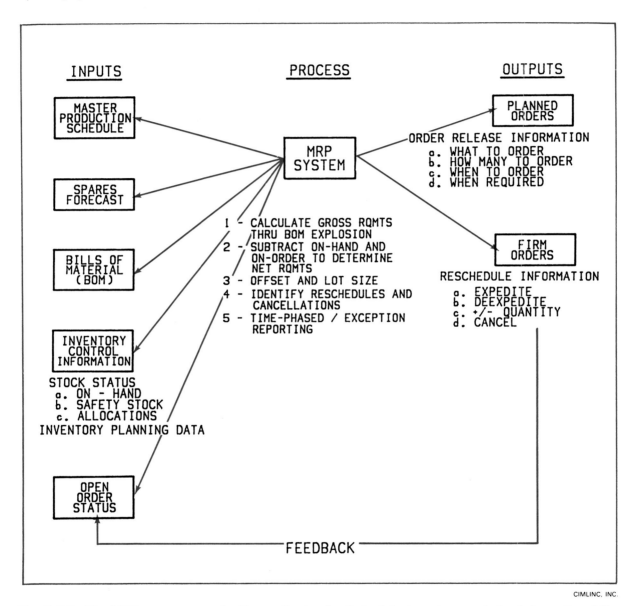

Fig. 5-25. The MRP system uses the Master Production Schedule, inventory, ordering information, bills of material, and other inputs to generate both planned and firm orders as outputs. Feedback of demand information is vital for proper control.

Planning levels

Fig. 5-26 shows the various capacity planning levels, horizons, and methods. Long-range capacity planning or **Resource Requirements Planning** (RRP) considers the load on long-range resources, such as land, facilities, cash flow, capital equipment, and labor skills and availability.

Intermediate-range capacity planning or **Rough-cut Capacity Planning** considers adjustment of resources required for such activities as make/buy, subcontract, hire/lay off, additional tooling, etc. Short-range capacity planning or **Capacity Requirements Planning** (CRP) compares actual to planned output, determines reasons for deviations, and initiates corrective action. Fig. 5-27 relates the capacity-planning periods to the business planning process.

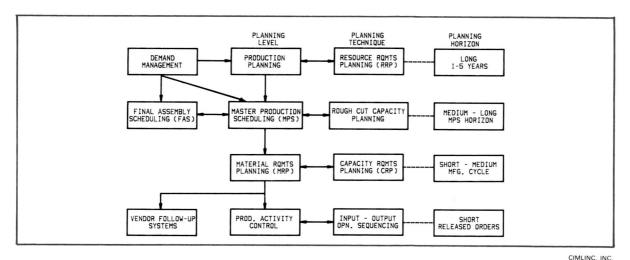

CIMLINC, INC.

Fig. 5-26. Various planning techniques and levels, applied across a range of horizons, are used for capacity management.

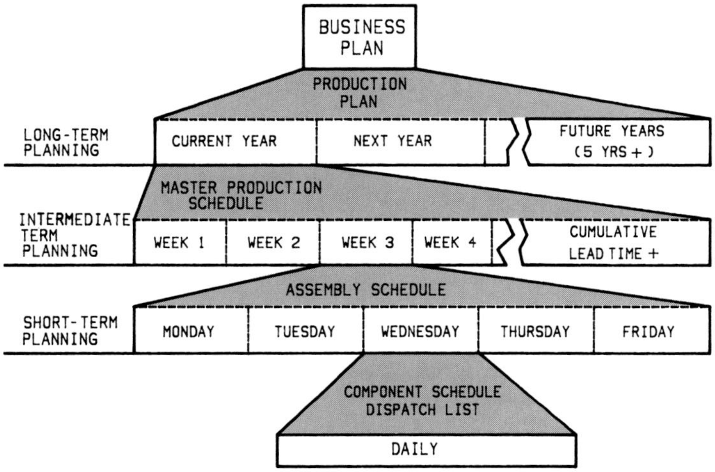

CIMLINC, INC.

Fig. 5-27. Long-term, intermediate-term, and short-term planning techniques all play a role in the overall business plan.

Fig. 5-28 depicts the capacity-planning process. This same process can be applied at any level. The process is to define, calculate, schedule, and balance capacity-sensitive resources. An axiom of capacity management is, "A plan that exceeds capacity will not get produced and will build inventory."

The vicious circle

The problems of lead time, work-in-process, and capacity management are interrelated, as shown in Fig. 5-29. This has been termed "the vicious circle." As business improves, additional orders are released to the shop floor. If the work input is greater than the capacity, queues of work-in-process build up at individual machine centers. This causes jobs to wait longer. As a result, customer orders

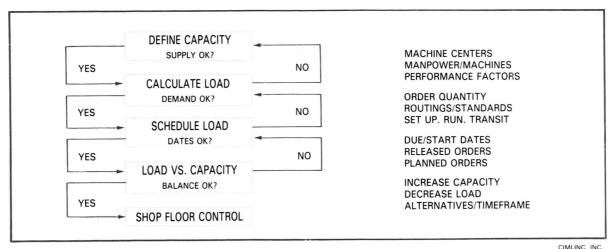

CIMLINC, INC.

Fig. 5-28. The capacity planning process involves a series of decision points that allow the planner to schedule and balance capacity-sensitive resources.

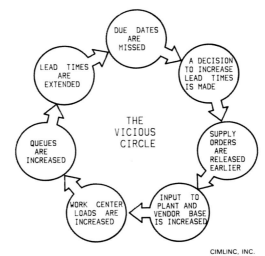

CIMLINC, INC.

Fig. 5-29. If capacity is insufficient, increasing lead times will increase work-in-process, but not improve delivery performance. This is known as the "vicious circle," since further increasing lead times will make the situation worse, not better.

become late. Shop floor personnel plead for more lead time, so the MRP system recalculates supply coverage (start and release dates) with new lead-time values. This further increases work-in-process and queue size.

If lead times are increased still further, shop orders will be released even earlier, creating additional work on the floor. Work-in-process inventory goes up, but (without a corresponding increase in capacity), there is no improvement on delivery performance. The solution to an increasing number of late orders is not to release work to the shop floor earlier, but rather to recognize that it is either a scheduling or capacity-related problem.

Fig. 5-30 illustrates an important principle: MRP is insensitive to capacity and continues to generate workload (orders) to provide sufficient coverage for the MPS. As discussed previously, a necessary condition for MRP is a Master Production Schedule that has been resource-tested (through use of Rough-cut Capacity Planning) to make sure that it is possible to execute. This resource test is reiterative in nature; it occurs any time that the MPS or capacity changes. As illustrated in Fig. 5-31, an imbalance between load and capacity will render any schedule unrealistic.

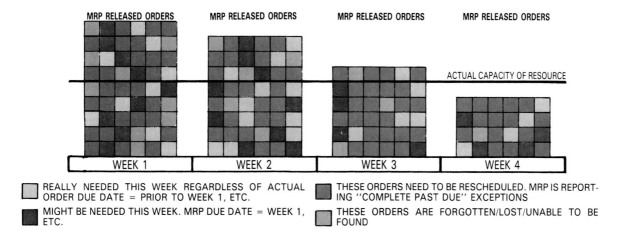

REALLY NEEDED THIS WEEK REGARDLESS OF ACTUAL ORDER DUE DATE = PRIOR TO WEEK 1, ETC.

MIGHT BE NEEDED THIS WEEK. MRP DUE DATE = WEEK 1, ETC.

THESE ORDERS NEED TO BE RESCHEDULED. MRP IS REPORTING "COMPLETE PAST DUE" EXCEPTIONS

THESE ORDERS ARE FORGOTTEN/LOST/UNABLE TO BE FOUND

Fig. 5-30. MRP is insensitive to capacity. It interprets the MPS into item due dates and order release dates, regardless of the capacity of the individual or vendor.

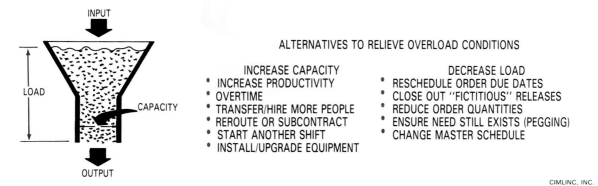

ALTERNATIVES TO RELIEVE OVERLOAD CONDITIONS

INCREASE CAPACITY
* INCREASE PRODUCTIVITY
* OVERTIME
* TRANSFER/HIRE MORE PEOPLE
* REROUTE OR SUBCONTRACT
* START ANOTHER SHIFT
* INSTALL/UPGRADE EQUIPMENT

DECREASE LOAD
* RESCHEDULE ORDER DUE DATES
* CLOSE OUT "FICTITIOUS" RELEASES
* REDUCE ORDER QUANTITIES
* ENSURE NEED STILL EXISTS (PEGGING)
* CHANGE MASTER SCHEDULE

CIMLINC, INC.

Fig. 5-31. An imbalance between load and capacity will result in making any schedule unrealistic. Solutions to such a problem include various methods of increasing capacity or decreasing the load.

MANAGEMENT SYSTEM IMPLEMENTATION

Implementation of a modern manufacturing planning and control system is an awesome task. The business issues facing the project team are both strategic and functional in nature. Fig. 5-32 is a partial list of policy issues by function that must be addressed.

A key question to answer is: "At what point should change no longer be allowed in the MPS?" The relationship between different business functions can no longer be an arms-length or parochial one. This means that, for example, Engineering must talk with Manufacturing to better design products for producibility. Manufacturing must be aware of new materials and processes being designed into products, so the correct resources can be put in place. Computer files must be cleaned up to eliminate obsolete or incorrect records, and so on. In other words, to make the concepts come to life, a major project effort will be needed throughout the company.

POLICY ISSUES BY FUNCTION	
MARKETING	**DESIGN ENGINEERING**

• FORECASTING APPROACH
  —PRODUCT GROUPS
  —END ITEMS
  —NEW PRODUCTS
  —PROMOTIONS
• PRODUCT EVAULATION/INTRODUCTION
• MASTER SCHEDULING/APPROACH
• TIME-FENCE POLICIES
• CUSTOMER SERVICE LEVELS
	• CONFIGURATION MANAGEMENT
	• ENGINEERING BILLS OF MATERIAL
	• NON-SIGNIFICANT PART NUMBERS
	• PARTS STANDARDIZATION
	• MULTIPLE SOURCES
	• ENGINEERING CHANGE CONTROL

	INDUSTRIAL ENGINEERING
MANUFACTURING ENGINEERING	• WORK MEASUREMENT STANDARDS • DATA ACQUISITION METHODS • FACTORY NETWORKS (ALTERNATIVES) • SHOP OUTSIDE PROCESSING (TRADEOFF) • MACHINE LEVEL ACCOUNTING
• MANUFACTURING BILLS OF MATERIAL • PRODUCTION ROUTINGS • WORK CENTER DEFINITION • PROCESS SPECIFICATIONS • TOOLING	
MANUFACTUIRNG	**PURCHASING**
• MASTER SCHEDULE COMMITMENT • DISPATCH LIST ADHERENCE • CAPACITY UTILIZATION/PLANNED LOADS • ACHIEVE PLANNED RATES OF OUTPUT • PRODUCT QUALITY • DATA DISCIPLINE	• ACCURATE LEAD TIMES • ACCURATE OPEN ORDER STATUS • REACT TO REQUIRED DATES • REPORT VENDOR PERFORMANCE • MULTIPLE SCOURCES • ACCURATE COSTS AND PRICE BREAKS

Fig. 5-32. *Implementing a manufacturing and control system raises a number of functional and strategic policy issues. These issues can be divided by function.*

CHALLENGES OF IMPLEMENTATION

The challenges facing a company committed to implementing MRP II and CIM include the magnitude of effort and resources required, and the need to develop a grass-roots wave of interest in the process. Implementing the new system is a very people-intensive procedure — it will involve almost every department, and will require changes in the way that people do their jobs.

CIM and MRP II, to be effective, must involve integration of the company's planning, execution, and performance measurement activities. It requires intensity and enthusiasm, a top-priority orientation, and a willingness to change from the way business is currently practiced. Key elements associated with implementing MRP II and CIM include commitment, education, dedication, and resource allocation.

As described in Fig. 5-33, it is important to make sure that project statements are congruent with management's perception of what is being said. The objective must coincide with the plan, and ultimately, with reality.

PERCEPTION PROBLEM

PROJECT TEAM SAYS:	MANAGEMENT HEARS:	ACTIONS TO RESOLVE
ABC is "complex"	ABC is "too hard"	Setup pilot or model Educate Provide additional assistance
ABC is "very structured" ABC is "formal"	ABC is "inflexible" ABC "constrains actions"	Relate formal environment discipline Educate Relate to performance measurement
ABC requires these "decisions"	ABC IS "controversial" ABC is "risky" ABC makes us "vulnerable"	Demonstrate control that would result Educate Persevere reasonably!
ABC will take "_____ time"	ABC will not address "day- to-day" operations	Persuade Educate Compare old and new
ABC is "difficult to quantify"	ABC might not "measure up"	Indicate ABC will make job easier Educate
ABC is "go." No further analysis is necessary	ABC does not meet "ROI" ABC is an "attitude" I can't "control"	Show example of behavior change Educate Make it easy for top management to "MBWA" (manage by walking around)
Discipline will be the most difficult to implement	Chaos and disruption will prevail	Reward performance Educate

ABC = Feature or function of new system.

Fig. 5-33. A major challenge of implementing MRP II and CIM is avoiding differing perceptions by management and the project team.

Implementation benefits

Why should a company undertake the task of implementation? Some of the benefits are listed below.

- Increased sales resulting from improved customer service.
- Increased labor productivity from valid, attainable schedules.
- Increased visibility, which translates into increased flexibility.
- A disciplined workforce that is more responsive to change.
- A focused management that is more responsive to accountability and cost control issues.
- Reduced purchase costs resulting from use of valid schedules and forward visibility.
- Reduced inventories resulting from the use of valid schedules and decreased shortages.
- Reduced obsolescence as a result of managing engineering changes, forward visibility, and lower inventories.
- Reduced quality assurance costs results from valid schedules and closed-loop information.
- Reduced premium freight cost as a result of shipping on time and reducing expediting charges.

Implementation costs

What are the costs? A good way to view costs associated with implementation is to rank their relative importance using the standard ABC classification: **A** (greater importance), **B** (medium importance), and **C** (least importance).

A. People-related costs
- Project team, full-time project leader.
- Education and training.
- Consulting and other expertise.
- Increases in indirect costs as shortcomings in existing planning and control areas are identified.

B. Information and data costs
- Data acquisition equipment.
- Revised plant layouts.
- Other resources necessary for auditing and timely updating of electronic records.

C. Computer costs
- Hardware.
- Software.
- Technical resources.
- Interfaces with current systems.
- Documentation.
- System maintenance.

KEY ELEMENTS OF IMPLEMENTATION

Key elements in implementing a manufacturing planning and control system are to set up a project team and establish an information center.

The concept of an information center within an organization or as part of a major project is relatively new. Typically, when a manufacturing planning and control system implementation begins, a wealth of information already exists in electronic files. For a number of reasons, the existing data processing programming staff is usually unable to accommodate new system needs. Staffing of the information center usually includes personal computer application programmers and people with fourth-generation report-writing expertise. This allows both old and new files to be manipulated, providing information output aligned with the current needs of management.

Once a project team and an information center are in place, implementation activities usually follow this sequence:

1. Develop a project steering committee.
2. Document existing systems (the "as is" condition).

3. Define project objectives.
4. Develop a project charter that clearly identifies those areas *not* being addressed, as well as those that *are* to be included.
5. Prepare and publish a project plan.
6. Conduct a software definition review.
7. Conduct an information systems audit.
8. Develop functional specifications.
9. Develop and publish a production conversion plan.
10. Plan and execute a conference room pilot.
11. Define and test system interfaces and conversion programs.
12. Conduct and review a production pilot.
13. Prepare and publish an education and training plan.
14. Develop user manuals.
15. Develop a certification program.
16. Define an excellence awards program.

The bottom half of Fig. 5-34 depicts functional relationships that surround the implementation of a comprehensive planning and control system, such as MRP II. Top management must be involved in the funding, support, resource allocation, and results-monitoring aspects of the system implementation. The top half of Fig. 5-34 provides a view of where an MRP II system fits with other possible systems company management might install.

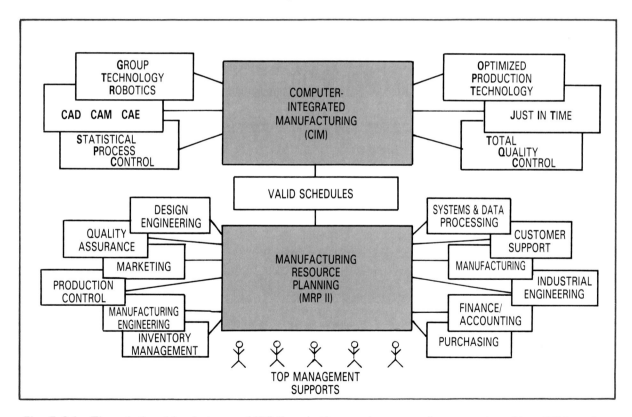

Fig. 5-34. *The relationships between MRP II and other systems usually encountered in a CIM environment. Also shown are the relationships between MRP II and various functions, and the supportive role played by top management.*

INFORMAL AND FORMAL SYSTEMS

A major problem for many companies is lack of information continuity. This lack of continuity arises when the same information emanates from more that one source, or when related information is inconsistent. Such problems result from what is often referred to as the "informal system." The *informal system* consists of methods employees follow in the absence of formal information procedures. Informal systems imply achieving results independently of others. The operation of the informal system has a number of negative ramifications.

First, management's perception of how events occur (formal procedure) is different from reality. Second, deviation from formality in one event tends to lead to a similar deviation in complementary events. Third, as new employees are hired, they are educated more through "street savvy" than the formal (approved) approaches. Fourth, when informality interfaces with formality, inconsistencies occur that give employees the impression that management (and/or the system) is not working. This

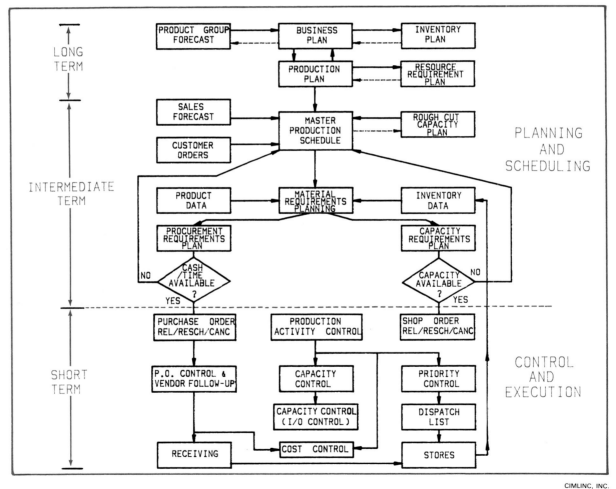

CIMLINC, INC.

Fig. 5-35. A manufacturing planning and control system consists of short-term, intermediate-term, and long-term elements. Long-term and intermediate-term elements are concerned with planning and scheduling; short-term elements are devoted to control and execution.

can lead to an erosion of morale. Fifth, once firmly entrenched, the informal system is extremely difficult to change. The informal system has far-reaching tentacles. By virtue of its influence, the informal system erodes the confidence people may have about whether information is accurate. Related information is contaminated though association or perceived association. The result is insufficient, inadequate, and/or inaccurate communication.

Formal systems imply accountability and measurement. As a manufacturing planning and control system is created, it is important that accountability and measurement be realistic.

As a company moves from manual systems through MRP and into MRP II, its management can lessen the impact of change by allowing tolerances, or filters, to be used. Current MRP II systems are based on the concept of management by exception. If the part or order is in accordance with the MPS and the resulting material plan, no action is necessary, and no report need be created. However, if a condition exists that is contrary to the material plan someone should be alerted. This is the point at which *filters* (tolerances or restrictions on action) can be used intelligently.

For example, when a system is first implemented, the number of orders requiring rescheduling will be large. By introducing a tolerance factor (filter) of "XX" days, the number of messages and exception conditions to be handled can be drastically reduced. As production personnel gain control over the environment, these filters can be tightened. Over a period of time (usually measured in years), they can be gradually eliminated. The business then may be ready for more change, such as Just-In-Time inventory control.

Fig. 5-35 provides an overall view of the planning activities covered in this chapter and identifies some of the control techniques that will be discussed later in this book.

SUMMARY

Under CIM, production planning and scheduling activities become critical to the daily operations of the factory because all systems are entirely integrated. Therefore, any single element of the business not conforming to the chosen strategic direction will produce less than optimum results.

The planning process begins by establishing a business plan. This business plan must include sales, profit margin, capital requirements, and the rate of return on the investment. This is naturally known as the annual operation plan. When this plan is given dollar values, it is known as the financial plan.

The entire business plan is then converted to a production unit plan. This unit plan is broken down into specific production rates that identify the resource plan at each phase of production.

In the late 1970s, Manufacturing Resource Planning (MRP II) began to be implemented in factories to plan, schedule, and forecast production requirements throughout the entire factory.

The backbone to the MRP II system is the Master Production Schedule (MPS). It lists the specific items to be assembled and shipped to meet the production plan as part of the business plan. The MPS items are exploded in the MRP II system as "bills of material." These are, in turn, netted against available inventories, machine capacity, lead times, labor resources, and schedule to determine shortfalls in the production plan. The MRP II system reports possible alternatives: reschedule, expedite, decrease order quantities, or cancel orders.

Capacity Management is the ability to establish, measure, monitor, and adjust levels of capacity to properly execute the production schedule as stated by the MRP II system. The capacity planning process defines, calculates, schedules, and balances capacity sensitive resources. A plan that exceeds capacity will not get produced and will build inventory levels.

The major challenges facing companies committed to implementation of MRP II revolve around the magnitude of effort, the resources required, the development of a grass-roots level of interest, recognition that it is people intensive and that it involves all departments within the company. Management cultures also must change; people have to do their jobs differently. It requires high intensity and enthusiasm and a top priority from management.

IMPORTANT TERMS

annual operating plan
bill of resources
buckets
business plan
capacity
capacity planning
Capacity Requirements Planning
closed-loop MRP
configure-to-order product
customer service performance
de-expedite
demand management
dependent demand
due date
expedite
exploding
filtering tools
filters
Final Assembly Schedule
financial plan
firm planned order
forecast
formal systems
horizontal dependence
independent demand

informal system
Just-In-Time
make-to-order manufacturing
make-to-stock manufacturing
Manufacturing Resource Planning
Master Production Schedule
Material Requirements Planning
Mean Absolute Deviation
net requirements
netting
parent item
pegging
pick date
planning horizon
production plan
rescheduling
Resource Requirements Planning
Rough-cut Capacity Planning
scrap
service level
shrinkage
supply
time fences
time-phased backscheduling
vertical dependence

QUESTIONS FOR REVIEW AND DISCUSSION

1. Explain how different stages of the product life cycle affect production planning activities.

2. What considerations must a planner keep in mind when developing a forecast of production requirements?

3. Why is a stable Master Production Schedule said to be a key requirement for an effective MRP II system? What are some pitfalls that can create MPS problems?

4. Describe how dependent demand is calculated from the Master Production Schedule. Demonstrate this by calculating the dependent demand for components B, C, and D in the following situation.

   A customer order for 200 of the parent item (A) is received. Each parent item consists of one component B, three component Cs, and two component Ds. There are 50 units of component B on hand and 150 on order; 100 component Cs on hand and 200 on order; and no component Ds on hand, but 200 on order.

5. In MRP, how is time-phased backscheduling used to determine the release date of an order?

6. Describe the logic used in the process of determining quantity that is known as *MRP netting*.

7. List five or more ways that a company would benefit from implementing CIM and MRP II.

Manufacturing Control and Execution

by Richard J. Evans

Key Concepts

- ❏ In terms of performance, continuous improvement is preferable to postponed perfection.
- ❏ Control of the inventory asset in a manufacturing company is central to meeting business goals. It requires a constant balancing of the optimism of sales-related functions against the conservatism of finance-related functions.
- ❏ As MRP breaks down the Master Production Schedule into planned orders, the load created on each internal manufacturing resource must be aligned with available capacity.
- ❏ Purchasing Control must secure an efficient flow of materials and components that will permit the end product to be built as scheduled.
- ❏ TQC/JIT is a state of mind; implementation requires a long-term commitment to excellence.

Overview

This chapter will introduce you to the use of three manufacturing control methods: performance measurement, record accuracy, and cycle time analysis. It will describe how these methods enhance the effectiveness of any manufacturing planning and control system. The value of education and training also will be presented. Control concepts in such areas as engineering data, inventory, order release, shop floor operations, tools, purchasing, cost accounting and management, and data integrity are covered in detail.

Richard J. Evans is Manager, Business Systems, for ITT Cannon Components Division in Van Nuys, California.

CONTROL CONCEPTS

In many companies, management periodically begins a "crusade" for performance improvement. Unfortunately, the campaign is often based upon false premises, erroneous or preconceived notions, or unreasonable expectations. Because management is unaware of the current environment or underestimates the magnitude of the task, the performance improvement that results is unlikely to meet expectations. To be effective, the campaign must be used in an environment that includes baselines which can be used to measure performance.

PERFORMANCE MEASUREMENT

Several steps are involved in establishing a ***measurement baseline***:
1. Flowchart the *reality* of events, not the perception.
2. Define the *problem*, not just the symptoms.
3. Analyze the *cause* of the activity that is degrading performance.
4. *Document* the results and establish the baseline.

If measurement baselines are not established, it is difficult to determine how much progress is being made, and to project when goals will be achieved. Fig. 6-1 shows how a typical performance measurement chart could be constructed to improve inventory accuracy. Note that the *current baseline* is shown as a point on the Y axis, and that *time* is incremented in quarters along the X axis. A current target is stated and actual performance is graphed.

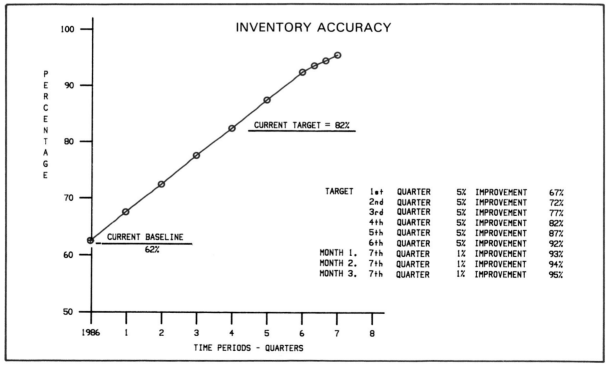

CIMLINC, INC.

Fig. 6-1. A typical performance measurement chart, showing projected improvement in inventory accuracy.

Stating a goal of achieving 95% inventory accuracy in two years, when the current accuracy level is 62%, may be seen as unrealistic. The 95% goal might be better stated as improving performance by 5% per quarter for six quarters, then 1% per month for three months. Each performance measurement should be time-phased, and evaluation criteria clearly defined. Actual performance must be charted, and frequently reviewed by management. The benefits of establishing measurement baselines may be expressed as:

- Measurement, used with a corrective action strategy, will produce the best results.
- Measurement, displayed on a chart visible to all, will enhance performance.

If tolerance factors are to be used in measuring performance, a program should be developed to make known how and when tightening of tolerances will occur.

In attempting to achieve a goal, it is important to *begin* movement (whether positive or negative). From a management standpoint, a continual *positive* trend is desirable. For example, if a system conversion is being planned, the "go/no go" decision should be based on the positive trend toward the goal, not necessarily on final achievement of that goal. *Continuous improvement is better than postponed perfection.* Typical performance measures used in a manufacturing environment are listed in Fig. 6-2.

TYPICAL MANUFACTURING PERFORMANCE MEASUREMENTS

CUSTOMER-RELATED
Daily Unit Production
Daily Shipments
Past-Due Reduction
Reduction in Age of Past-Due Units
Backlog—Monitor to Desired Level
Branch Inventory Level
Customer Support Inventory Level
Customer Order Processing Time

PRODUCTION-RELATED
Comparison of Production to MPS
Level of Productivity
Work Center/Equipment Use
Reduction of Labor Costs
Reduction in Number of Shortages
  (Material, Routings, Tools, Drawings)
Reduction of Scrap/Reject Rates
Reduction of Queues/WIP (Improved Throughput)

RECORD-RELATED
Bill of Material Accuracy
Inventory Record Accuracy
  (Raw Material, Subassemblies, Finished Goods)
Accuracy of Due Dates/On-Order Qty.
  (Purchasing, Manufacturing, Assembly)
Transaction Throughput Performance

SCHEDULE-RELATED
Magnitude of Reschedules
Frequency of Reschedules
Monitor Causes for Reschedules
Actual Performce to Plan
Forecast to Actual Sales Orders
Reduction in Age of Past Due by Requested Date
  and by Promised Date

Fig. 6-2. Performance measurements, in various categories, that are often used in manufacturing.

RECORD ACCURACY

In data processing, a record is composed of *data elements*, or pieces of information about the business. If it were possible to do so to emphasize their importance, each of these pieces should be colored green and handled like money. *Record accuracy* is the value placed on any given data element. It is obvious that, in an integrated system, the optimum decision about a business cannot be made if any single data element (piece of information) is inaccurate.

Efficient and effective use of management time is essential in today's dynamic, fast-paced business environment. Steps should be taken to prevent information overload, and to avoid duplication of information sources for a given piece of data. Current computer technologies are organized around the data element. Manufacturing businesses are organized by function, with each function further broken down into job descriptions.

Manufacturing management

The objective of manufacturing management should be to identify a specific job function and make it accountable for a given piece of information (the source for a data element). In the warehouse, for example, the storekeeper or material handler is responsible for the accuracy of inventory information. In the same way, part number assignment is the responsibility of a specific individual or department (sometimes referred to as "configuration management"). Functions can be organized to manage these pieces of information, and the system programmed to accept the value of a data element only from the designated source. In this way, it should be possible to propagate awareness of the value of a data element (accurate information) throughout an organization.

Accountability

In an organization such as a manufacturing business, it is not uncommon to encounter an endless chain of "finger-pointing" when attempting to establish responsibility for a problem. Individuals must be made aware of the importance of their role in company profitability; the perpetuation of faulty information cannot be tolerated if a company's credibility is to be maintained. Each person in the organization must take the appropriate action to ensure that each piece of information is accurate, and accept responsibility for that accuracy.

People

Companies often state that their most important asset is people; yet the actions of these same companies sometimes show insensitivity to the needs, desires, and objectives of the people who work for them. For example, the company may talk about the importance of maintaining record accuracy, yet fail to assign accountability, provide resources, or expect performance. Such contradictions often appear to be the norm, rather than the exception, and can be traced to a lack of positive attention to the details by management. It is important that managers recognize the need for continually participating in and monitoring the "people system." It is also important to recognize that the actual *opportunity* to enter an item of information correctly, or otherwise perform a job well or poorly, rests with the *employee*, not management or supervisory personnel. Managers and supervisors are the facilitators; individual employees are the performers.

CYCLE TIME

The period of time from recognition of a customer need to shipment of actual product to meet that need is called *cycle time*. All the events that occur in that cycle, from recognition of a need to shipment of the product, form a manufacturing company's reason for being. One of the benefits of a manufacturing planning and control system is that all of the increments that make up the total cycle time of a particular factory are input requirements for the system to execute. These increments are called *lead-time elements*. Typical lead-time elements for both manufactured and purchased items are shown in Fig. 6-3.

What is the cycle time of a particular manufacturing operation? An easy way to check on what an *ideal* cycle time would be is to track an order that has the company president's stamp of urgency, following it through the entire operation. See Fig. 6-4. Any cycle time greater than that ideal should be challenged for possible reduction by the manufacturing organization.

In the face of demands for shipping product, time is too often wasted by expediting orders, extending lead times, and increasing quantities and inflating "planning factors," such as shrinkage, scrap, and safety stock. These activities will increase inventory, release orders more frequently, increase work-in-process, and ultimately, affect lead time to the detriment of the total factory throughput

```
┌─────────────────────────────────────────────────────────────────────────┐
│                         LEAD-TIME ELEMENTS                                │
│                                                                           │
│      MANUFACTURED ITEM                    PURCHASED ITEM                   │
│      Recognition of Need                  Recognition of Need             │
│      Order Processing                     Order Processing                │
│      Pick                                 Order Communication             │
│      Move                                 Vendor Lead Time                │
│      Queue                                Transportation                  │
│      Set-up                               Receiving                       │
│      Run                                  Inspection                      │
│      Wait                                 Put-Away                        │
│      Inspection                                                           │
│      Put-Away                                                             │
│                                                                           │
└─────────────────────────────────────────────────────────────────────────┘
```

Fig. 6-3. Lead time is the estimated elapsed time between recognition of a need and the receipt of the needed item by a customer (external or internal). Lead-time elements like those listed are used by MRP to establish the order in which parents and components are to be released and completed.

```
┌─────────────────────────────────────────────────────────────────────────┐
│                          LEAD-TIME AXIOM                                  │
│               LEAD TIMES ARE WHAT YOU SAY THEY ARE                        │
│                                                                           │
│     • If "YOU" are the person responsible for the maintenance of that     │
│       lead-time increment, the lead time is probably realistic.           │
│                                                                           │
│     • If "YOU" are the person responsible for the performance of a        │
│       department, the lead time is probably pessimistic.                  │
│                                                                           │
│     • If "YOU" are the Sales Manager of the company, the lead time is      │
│       probably optimistic.                                                │
│                                                                           │
│     • If "YOU" are the President of the company, the lead time is totally  │
│       accurate.                                                           │
│                                                                           │
└─────────────────────────────────────────────────────────────────────────┘
```

Fig. 6-4. The accuracy of lead times may depend upon who is establishing them.

(lead time). A benefit of current manufacturing planning and control systems is that exception messages (reduce an order quantity or deexpedite or cancel an order) are a normal expectation for material planners.

Fig. 6-5 depicts the five increments that make up manufacturing lead time. In a CIM environment, these elements apply to administrative and manufacturing support tasks, as well as to actual manufacturing operations. Thus, a manufacturing engineer documenting a process, an industrial engineer developing a standard, or a purchasing agent placing an order are faced with these elements:
- QUEUE (The in-basket)
- SET-UP (File review)
- RUN (Preparation of appropriate paperwork)
- WAIT (Process through the functional bureaucracy)
- MOVE (Process through the organizational bureaucracy)

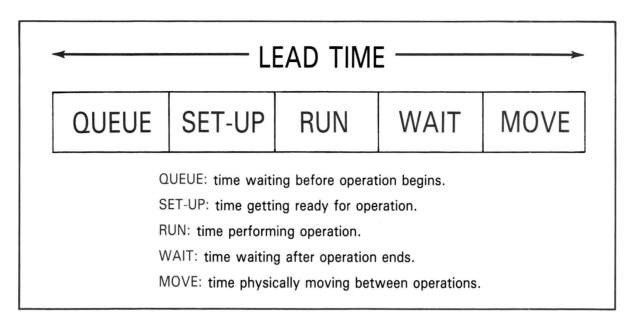

Fig. 6-5. *The five elements (time increments) that make up lead time apply to administrative and manufacturing support tasks as fully as they do to manufacturing operations themselves.*

EDUCATION AND TRAINING

When a company makes a change to a very formal environment, such as a specific CIM technology or a new manufacturing planning and control system like MRP II, extensive retraining will be needed. Employees must learn about:
- New policies (company rules and philosophies).
- New procedures.
- New processes.
- New forms and documents.
- New planning techniques.
- New tools (systemic, rather than manual).
- New expectations.

Smoothness of the transition, and the ease with which people adapt to the new system, will relate directly to the extent and quality of the training provided. Companies installing formal systems frequently fail to provide the retraining or education needed to ensure a smooth transition. Experience has shown that, if the implementation is to be successful, the company will have to—sooner or later—invest in education and training. When education and training programs are carried out properly, the resulting benefits will more than offset the costs incurred. Even if the company decides not to implement the system for which retraining was performed, the process will still provide better results for the company.

Companies frequently fail to show sensitivity to the need for employee education and training. An example is the way a new employee (or an existing employee being transferred or promoted) is integrated into the company's workforce. The induction process typically consists of a one- or two-hour cultural orientation that introduces the employee to the company, its general practices, and the location of various facilities. The employee or transferee is then placed in the new job assignment with only past experience and assumptions about the new responsibilities as a guide.

Such a situation places record accuracy at risk, since information being generated can be contaminated easily by a new employee as a result of inexperience or lack of knowledge. One way to minimize the possible damage (in addition to training) is to establish a standard of minimum acceptable quality level on a job-specific basis. To make the standard effective, a means of certification is needed. The certification program validates the employee's competence and ability to meet the quality standard for accurate entry of information. Certification implies periodic recertification, evaluation of acceptable test scores, and the awarding of certificates (an opportunity for management to demonstrate its "commitment").

ENGINEERING DATA CONTROL

The creation, organization, and maintenance of records used by all segments of a business to identify, manufacture, and market its products are functions of *engineering data control*. Fig. 6-6 shows engineering responsibilities in a manufacturing company. The documents they generate include:
- *Bills of material*, or *parts lists*, that describe the product's configuration.
- *Manufacturing routings*, or *process sheets*, that describe the sequence of operations to be performed on a specific part.
- *Manufacturing resource records* that describe machines, work centers, and other physical resources.
- *Item master data* that describe the characteristics of each unique manufactured and purchased part.

Fig. 6-7 shows how files are typically organized in Engineering and Manufacturing databases.

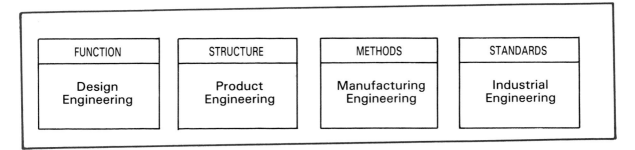

CIMLINC, INC.

Fig. 6-6. Engineering responsibilities in a manufacturing company are often divided among different engineering specialties.

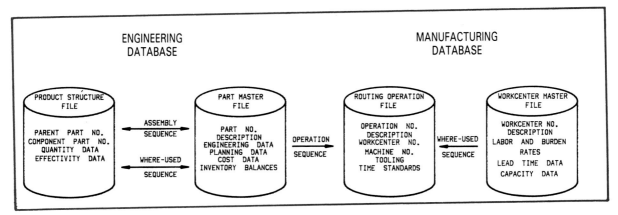

CIMLINC, INC.

Fig. 6-7. Typical file organizations for Engineering and Manufacturing databases.

BILL OF MATERIAL TYPES

The bill of material (BOM) lists all parts needed to build a product. At the lowest level are all purchased parts that must be procured. The bill of material can be viewed as a single-level BOM or as a multiple-level BOM. The *single-level bill of material* is a display listing the parent item and its immediate components only; the *multiple-level bill of material* displays a list of the parent and its components through all intermediate levels until the purchased items are reached. Fig. 6-8 shows how bills of material can be viewed in an *explosion* (top-down) form or an *implosion* or "where-used" (bottom-up) form. To be complete, a bill of material must be properly documented, including general specifications, test specifications, procurement specifications, theory of operation, schematics, logic diagrams, and other necessary information.

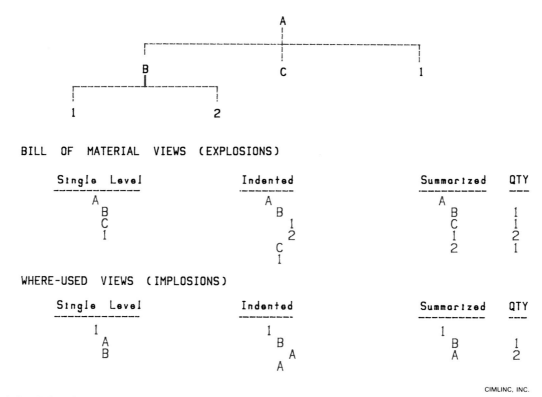

BILL OF MATERIAL VIEWS (EXPLOSIONS)

WHERE-USED VIEWS (IMPLOSIONS)

CIMLINC, INC.

Fig. 6-8. Bills of material can be viewed in either an exploded or imploded form, depending upon planning needs.

Bill of material structuring is a subset of a function called configuration management, Fig. 6-9. *Configuration management* entails documenting of the configurations used throughout an entire product life cycle. It is a means of managing engineering change. It is important for Engineering to recognize its relationship to Manufacturing through the *as-designed* ("latest and greatest") structure.

The actual release for production of an as-designed bill of material initiates planning activities. Manufacturing uses the *as-planned/as-built* structure to produce a salable item, and needs to communicate back to Engineering the materials they are capable of using and the processes they are capable of performing.

FUNCTIONS OF CONFIGURATION MANAGEMENT

Product Definition ———————————————————

- Elements
  Documents
  Drawings
  Parts

- Characteristics
  Revisions
  Interchangeability
  Relationships
  Approved Parts List

Document Control ———————————————

- Document Revisions
- ECO Incorporation
- Document Management
  General Specifications
  Procurement Specifications
  Test Specifications

  Wiring Diagrams
  Schematics
  Approved Vendor List

Specifications ———————————————————

- As Designed
  Engineering Product Structure
  Product Life Cycle
  Configuration Control

- As Planned/Built
  Manufacturing Product Structure
  Authorized Substitutions
  Authorized Deviations

Configuration Change ——————————————

- Change Control Board
- Authorization for Change

- Product Liability
- Establish Parts Effectivity

Configuration Management has a key relationship to:

Quality Assurance ————————————————

- Receiving Inspection
- In-process Inspection
- Quality Engineering
- Product Baselines

- Product Statistics
- Product Deviations
- Acceptance Test Procedures

Fig. 6-9. *Configuration management involves defining, specifying, and documenting all configurations of a product through its entire life cycle.*

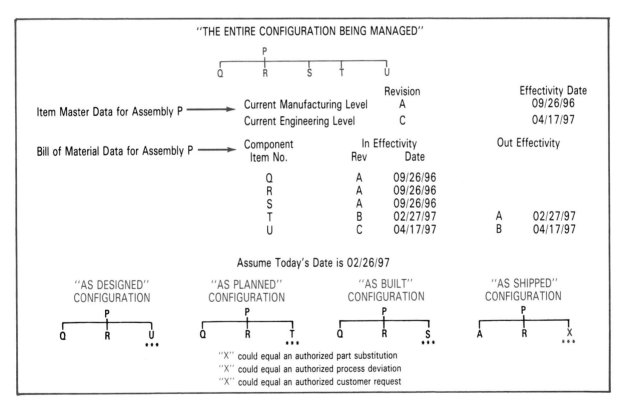

Fig. 6-10. *Various configurations of a product can exist simultaneously as a result of revisions. The "as designed" configuration is the designer's concept of the product to be built; "as planned" is the configuration being used for near-term requirements planning. The "as built" configuration is the product as it actually is being built today; "as shipped" is the configuration actually being sent to fill orders.*

Finally, to support product integrity issues, the *as-shipped* structure identifies the actual configuration built. This configuration may include authorized bill of material deviations. A review of the various product structures in bill of material format is shown in Fig. 6-10.

The concepts of engineering change and effectivity are also illustrated in Fig. 6-10. The example shows *effectivity* in terms of a *date*, but it also can be used in a *serial* manner. Determining the effectivity date for an engineering change requires the cooperation of a number of functions within the company. The managers of these functions, or their representatives, normally meet as a **Configuration Change Board**. At meetings of this board, differences of opinion are common. For example, Material may want an effectivity date far enough in the future to use up the inventory of an old part number; Engineering may desire immediate effectivity for cost-effectiveness. Value Management would have to enter the discussion to determine if savings derived from an immediate effectivity will be equal to, or greater than, the inventory of parts being made obsolete. Similar tradeoffs will have to be made on the basis of such considerations as tooling availability or supplier capability. Depending upon the severity of the problem, any product integrity changes will be made expeditiously.

Companies constantly change product designs for reasons such as those shown in Fig. 6-11. New products are also added to a company's line, and old products discontinued. Failure to promptly and accurately reflect these changes will defeat an integrated systems approach. Fig. 6-12 shows the evolution of a product's configuration, incorporated with the change process. The change process occurs repeatedly, and can take place at any point in the evolution of the product configuration; chances are that more than one change will be in process at any given time. Note that, as depicted in Fig. 6-12, the CIM process (using CAD/CAM/CAE) enhances the ability to process changes by shortening the time required.

By turning a bill of material sideways and superimposing lead time increments, an interesting perspective can be achieved, Fig. 6-13. The **critical path** (another name for product cycle time) for the item is easily seen in this view.

INVENTORY CONTROL

The control of the inventory asset in a manufacturing company is central to meeting the goals and objectives of that business. It is an area where specific management actions (both organizational and physical) and better system tools can make immediate improvements in efficiency.

Inventory control implies that an accurate "on-hand" balance of a part number is available in a timely manner. While quantity balance is the essential element of the inventory control function, such additional information as location of the inventory and levels of work-in-process is often included. Physical security of the inventory and protection from damage are also included under inventory control.

ENGINEERING CHANGES		
CHANGE	REASON	RESPONSIBILITY
FORM	PRODUCT ACCEPTANCE	MARKETING
FIT	PRODUCIBILITY	MANUFACTURING
FUNCTION	PRODUCT LIABILITY	ENGINEERING
COST REDUCTION	PROFITABILITY	ALL

Fig. 6-11. Engineering changes in a product design can take place one or more times during that product's life cycle. The list above shows typical kinds of changes and the reasons they are made.

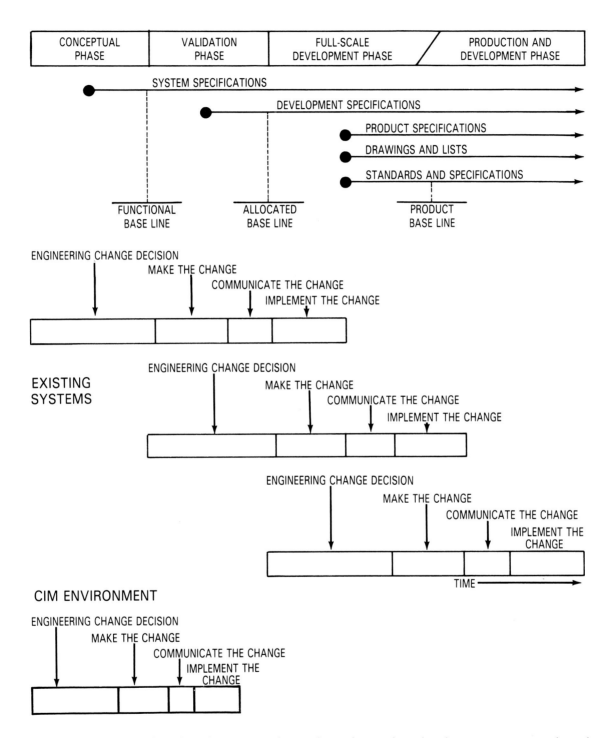

Fig. 6-12. Several engineering changes can be made as the product development process takes place. In a CIM environment, such changes can be implemented more rapidly as a result of the electronic linking of all phases of the design and manufacturing processes.

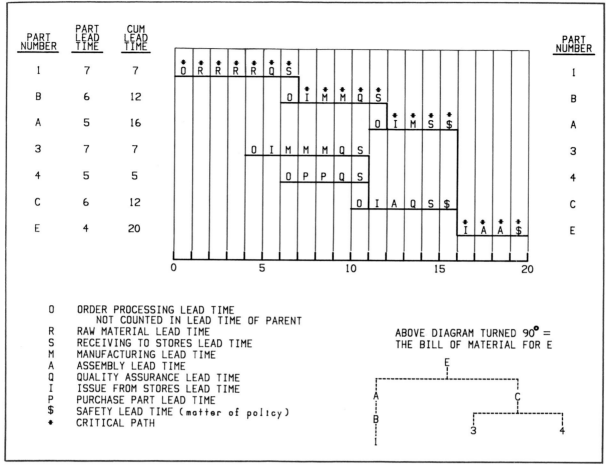

Fig. 6-13. A critical path diagram results when a bill of material is rotated 90 degrees and lead time increments applied.

Inventory control is the *execution* portion of ***inventory management***. Typical inventory management objectives are to:
- Provide a level of inventory consistent with the Master Production Schedule (MPS).
- Optimize inventory investment through lead time management and planned order policies.
- Pursue cost reductions while improving product quality.
- Minimize the frequency and magnitude of parts shortages.

Inventory frequently represents 40% to 60% of a manufacturing company's assets. The conflicting objectives of inventory management are best understood by reviewing the inventory levels desired by different company departments. Marketing traditionally wants product on the shelf, so that orders can be filled immediately. Accounting prefers low inventory levels, since it views inventory as an accounts payable (payment to vendor) drain on working capital. Manufacturing agrees with Marketing, in part: it favors high inventories at the subassembly level to avoid material shortages. Planning attempts to integrate the output of MRP II or similar tools with practical knowledge to establish optimum inventory levels. In effect, Planning tries to strike a happy medium between the optimism of sales-related functions and the conservatism of finance-related functions.

```
┌─────────────────────────────────────────────────────────────────────────┐
│                          INVENTORY FUNCTIONS                              │
│                                                                           │
│  • Lot Size: balances the cost of acquiring inventory with the cost of    │
│              carrying that inventory.                                     │
│              Acquisition costs include ordering, receiving, transportation,│
│              inspecting, etc.                                             │
│              Carrying costs include material handling, warehouse space,   │
│              insurance, taxes, etc.                                       │
│                                                                           │
│  • Demand Fluctuation: ensures against demand variation.                  │
│              Variability in demand may be caused by:                      │
│                  Forecast errors                                          │
│                  Sales promotions                                         │
│                  Seasonality                                              │
│                  Urgent customer requirements                             │
│                  Action from competitors                                  │
│                  New product introduction                                 │
│                  Recall                                                    │
│                                                                           │
│  • Supply Fluctuation: ensures against supply variation.                  │
│              Variability in supply may be caused by:                      │
│                  Late deliveries                                          │
│                  Variable lead times                                      │
│                  Scrap                                                     │
│                  Rejection                                                 │
│                  Rework                                                    │
│                  Engineering changes                                      │
│                                                                           │
│  • Anticipation: levels production and prepares for expected events, such │
│                  as seasonal changed in demand or a potential strike.     │
│                                                                           │
│  • Transportation: fills the distribution pipeline.                       │
│                                                                           │
│  • Hedge: covers changes in product mix (options) and plans for internal  │
│           price changes.                                                   │
│                                                                           │
│  • Speculation: covers potential cost increases for raw materials,        │
│                 special alloys, etc.                                       │
│                                                                           │
└─────────────────────────────────────────────────────────────────────────┘
```

Fig. 6-14. Some inventory management methods require viewing inventory by function. There are a number of recognized inventory functions.

INVENTORY PLANNING METHODS

There are two popular methods of viewing inventory: by *function* and by *stage of process*. Fig. 6-14 shows various inventory functions; inventory transitions between process stages are depicted in Fig. 6-15. Basic inventory planning methods are Material Requirements Planning (MRP I) and *statistical order point*. Other inventory planning methods in general use include the *two-bin system*, the *min/max system*, *periodic review*, and *visual review*.

Two-bin system

Bin one contains inventory issued for use; bin two is reserved for future use. When the reserve supply (bin two) begins to be issued for use, a replenishment order is placed. An example of this might be a typical household situation: coffee is added to the shopping list when the container in the pantry is opened.

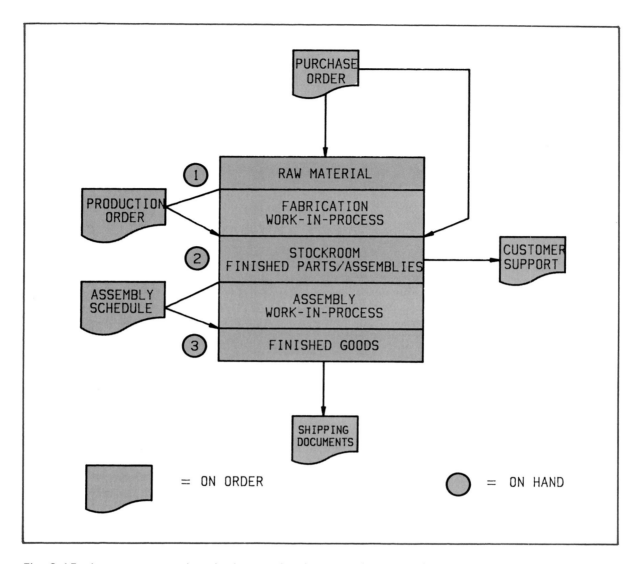

Fig. 6-15. Inventory control methods may view inventory by stage of process. The transitions between process stages are shown.

Min/max system

The estimated demand during replenishment lead time, plus a reasonable buffer (which serves as a trigger point for reorder), forms the *minimum* level. The *maximum* level is the on-hand quantity that management does not want to exceed. An example: the minimum inventory of office copier paper is one case, plus a buffer of two 500-sheet packages; the maximum inventory (for storage space reasons) is three cases.

Periodic review

Inventory records are reviewed at fixed intervals (often weekly or monthly), and sufficient material is ordered to restore the on-hand plus on-order quantities to specified levels. An example would be monthly balancing of your checkbook to make sure there is sufficient money in the account to prevent a "bounced check" (inventory shortage).

Visual review

This method is similar in most respects to the periodic review, but involves checking the actual physical inventory, not just inventory *records*. This is the type of review you would do at home prior to a shopping trip to the supermarket: visually reviewing the items in kitchen cabinets or pantry and the refrigerator to determine replenishment needs.

ABC ANALYSIS

In a manufacturing environment, Inventory Control may have to monitor thousands of part numbers. Since it is economically unfeasible (and may be physically impossible) to analyze every part number in detail, some system of determining **criticality**, or importance, is necessary. The method used is called **ABC analysis** or **inventory stratification**. It is based on the work of the nineteenth century Italian economist Vilfredo Pareto, who arrived at the general conclusion that a small percentage of a population accounts for the largest fraction of an effort or value. Placed in Inventory Control terms, Pareto's conclusion can be applied as the necessity for separating the vital few from the trivial many. This is shown in graphic form in Fig. 6-16, where it is obvious that the 5 parts in Class A account for 80% of annual usage value, while the 85 parts in Class C account for only 10% of annual usage value.

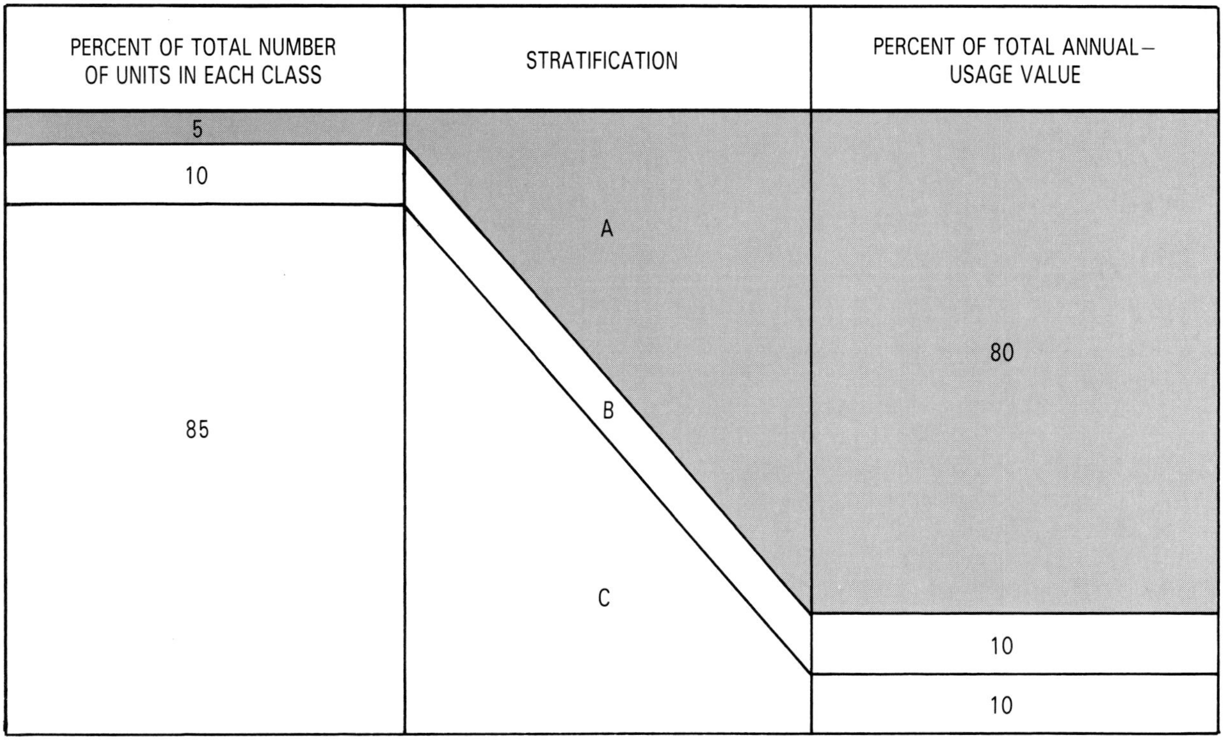

PERCENT OF TOTAL NUMBER OF UNITS IN EACH CLASS	STRATIFICATION	PERCENT OF TOTAL ANNUAL— USAGE VALUE
5	A	
10		80
85	B	
	C	10
		10

Fig. 6-16. In this typical distribution resulting from ABC analysis, it is easy to see that Class A, which represents 5% of total parts in quantity, is responsible for 80% of annual usage value.

Classifying inventory with ABC analysis is done as described below:

1. Determine the annual usage for each item in inventory. (Inventory also can be classified by assigning a dollar value to other criteria, such as current on-hand balances, past usage, slow-moving parts, a mixture of past usage with future demand, and so on.)
2. Multiply the annual usage of the item by its cost. This will yield its total *annual usage value*.
3. Add together the annual usage values of all items in inventory to determine the aggregate annual inventory expenditure.
4. Divide the total annual usage value of each item by the aggregate annual expenditure. This will identify the percent of total annual usage value for each item.
5. Rank all inventory items in order of percent of total annual usage value.
6. Group items on the basis of their ranking. Items that are at or above the 80th percentile are classified as "A" items. They will normally represent less than 20% of the total part numbers.
7. Divide the remainder of the items into "B" and "C" classes, or even smaller subdivisions, depending upon your situation.
8. For various reasons (long lead time, size, or shelf life are examples), specific items may be placed in classifications different from those they would be assigned to on the basis of their mathematically arrived-at value.

Other applications for ABC analysis

ABC analysis is a fundamental management tool that can be applied in many forms. For example, a simple review can be made of **discrepant lots** (parts lots that do not meet standards) submitted to a company's **Material Review Board** (MRB) for disposition. This review will reveal the small number of part numbers (and by association, their suppliers) that account for the preponderance of MRB activity. In the same way, a list of specifications that have created producibility problems can be matched with the names of the MEs (manufacturing engineers) who wrote those specifications. Analysis will reveal which Engineering personnel are providing specifications that are unnecessarily difficult to meet.

The inventory *stratification* resulting from ABC analysis can be used by Planning or Production Control to prioritize their workload. More time can be spent analyzing such considerations as lead time, order policy, lot size, or safety stock levels for an "A" item than for "B" and "C" items.

PHYSICAL CONTROL OF INVENTORY

Typical reasons for an inaccurate inventory balance are bad physical counts or transactions that are unprocessed or erroneously processed. Inventory Control can implement programs to monitor and correct such discrepancies. Good physical control of inventory will also enhance the accuracy of the inventory *record*. Prerequisites for effective physical control include:

1. A simple part-numbering system.
2. Easy part identification.
3. Secured storerooms, receiving areas, and shipping areas.
4. Move authorizations and audit trails.
5. Adequate storage space and facilities.
6. Adequate material handling equipment and systems.
7. Good housekeeping.
8. Properly trained personnel.
9. A commitment to quality.
10. A conviction that errors in the inventory balance are unacceptable and reflect on individual performance.

Periodic Count	Cycle Count
All stores locations counted on a regular basis (i.e., annually).	Continuous count of a few items throughout the year.

RESULTS

Plant/warehouse shutdown.	Timely detection and correction of errors.
Introduce record errors.	Minimal loss of production time.
No correction of existing error causes.	Systematic improvement.

Fig. 6-17. An inventory is taken to balance the actual physical quantities on hand with the quantities shown in accounting records. The two most common inventory methods are the periodic count and the cycle count.

Inventory methods

There are two primary methods of taking inventory: the ***periodic count***, or annual physical inventory, and ***cycle counting***. They are compared in Fig. 6-17.

The real advantage of cycle counting is that it identifies the causes of an inaccurate balance and allows appropriate action to be taken. Cycle counting recognizes that the inventory balance can be affected by actions such as:

- A buyer who fails to process a debit memo.
- An enterprising shop foreman who "scrounges" a needed part.
- A material handler who forgets to move the last pallet or box of a shipment.
- An engineer who "borrows" a part to use on a development project.
- A salesperson who takes a part as a visual aid for a customer presentation.

By identifying these causes throughout the year, action can be taken systematically. The annual physical inventory will simply quantify the damage to the inventory balance.

Safety stock

Another inventory control principle is ***safety stock***, which inflates inventory and is protected from allocation by both formal and informal systems. Safety stock can be expressed as either a specific quantity or as a specified length of time. Like many principles, safety stock has both benefits and shortcomings.

The primary benefit is that it anticipates (provides coverage for) purchasing delays, engineering changes, capacity constraints, product shelf life, variable yield factors, seasonality, and rejection of raw materials or purchased parts from suppliers. For an item subject to independent demand, the only way to maintain desired customer service levels is to maintain sufficient safety stock.

The primary shortcoming of using the safety stock principle is that management is not forced to determine a specific quantity based upon discrete events and transactions. It is convenient to say, "add a month to the lead time" (safety stock expressed as lead time) or "increase the order by 10%" (safety stock expressed as a quantity). However, this can lead to increased inventory investment and a greater probability of obsolescence. Management should actively review safety stock factors to avoid abuse of this principle.

Fig. 6-18 correlates some inventory control concepts with information presented earlier. It should be apparent, from looking at Fig. 6-18, that efforts to properly structure a bill of material, to better

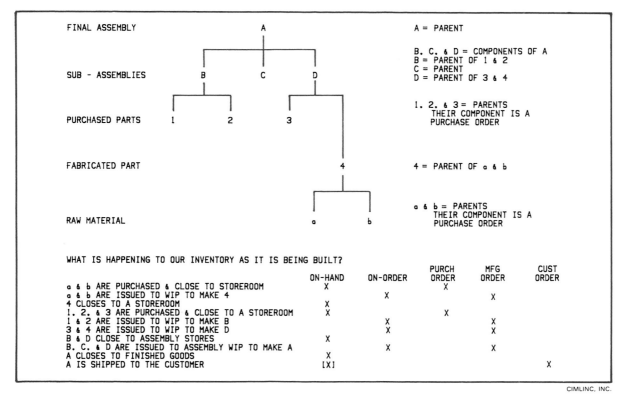

Fig. 6-18. The relationship of inventory control to concepts taught earlier in this chapter.

forecast demand for an end item, or to more efficiently flow work through the shop are interrelated and mutually supportive.

ORDER RELEASE AND CONTROL

MRP generates planned orders as far into the future as the period covered by the master production schedule (MPS). These orders prepare the business for the procurement of raw materials and the acquisition of purchased parts and other goods and services. They also provide Manufacturing with needed information about what is to be built, processed, fabricated, or assembled.

Purchase orders and subcontract orders are usually processed outside the company, while *manufacturing orders* and rework orders (MRP does not plan for rework orders, however) are completed inside the company. In large corporations, sister companies may provide a component used in a product. Orders used to control this type of activity are referred to as *intercompany orders* (or sometimes as *interdivisional* or *interplant* orders). Whether these related businesses are treated as inside (Manufacturing) suppliers or outside (Purchasing) suppliers is a management decision. Once the decision is made, the correct order type can be issued.

The order release function is essentially the same for either a purchased (bought from outside) item or a manufactured (made inside) item. The primary difference is in the department to which the order is released. A planned purchase order is released to Purchasing; a planned manufacturing order to Production Control within either the Material or Manufacturing department (depending upon company organization). Authority for the release of any supply order is the MPS. Companies operating without a master schedule must delegate release authority to Planning. Typically, a planner will be able to release orders up to some dollar limit; above that figure, authorization from a supervisor or higher-level manager is needed. These grants of authority, or approval levels, introduce delays into the order release process. They can be avoided by implementing and rigorously adhering to a master scheduling philosophy.

MRP planned orders specify the part to be ordered, the quantity required, and the date the material should be on hand (available in inventory). Since company personnel have previously loaded appropriate values into related data elements, the system knows how long it will take to firm (establish), cut (prepare for release), and release a manufacturing order, pick (issue) material, and actually manufacture an item. In the same way, the system can plan for a purchased part. Lead time values indicate the length of time it will take to firm and release the requisition, issue the purchase order, fill and ship (by the supplier), and receive and inspect the purchased item. The relationship of lead time to typical MRP reports and the order release process is depicted in Fig. 6-19.

PURCHASE ORDER	LEAD-TIME INCREMENTS				
	Planner Time	Buyer Time	Vendor Time	Receiving	Closing to Stores
	FIRM	CUT	PLACE	DOCK	DUE DATE
DATES					

MANUFACTURING ORDER	LEAD-TIME INCREMENTS				
	Planner Time	Release Time	Build Time		
	FIRM	CUT	START		DUE DATE
DATES		PICK			

	ORDER STATUS				
Planned	Firm	Cut	Released (Mfg) Placed (Purch)	Dock	Closing Process

TYPICAL MRP REPORTS				
PURCHASE ORDER	ORDER ACTION	PUR. REQUISITION PUR. DECISION	ORDER ACTION FOLLOW-UP	ORDER STATUS EXCEPTIONS DAILY RECEIVING
MANUFACTURING ORDER	ORDER ACTION	MATERIAL LIST SHORTAGE REPORT	ORDER ACTION DISPATCH LIST	ORDER STATUS EXCEPTIONS

Fig. 6-19. Lead-time increments are the basis of the MRP reporting and order release process. Both manufacturing orders and purchase orders can be affected by lead time changes.

MANUFACTURING ORDERS

Production Control must compile related documentation before releasing a manufacturing order. This documentation may include drawings, routings, bills of material, inspection instructions, tooling (such as jigs, fixtures, or NC tapes), tickets to indicate completion of a particular event or operation, and material handling pick lists. Before release of a manufacturing order, a check of material availability must be made, since the right mix of *matched sets of parts* is necessary to build a product efficiently. See Fig. 6-20. If matched sets are not available, steps must be taken to overcome the imbalance or shortage condition. Possible strategies are to substitute another part for the one in short supply, reschedule the order, release another order early, or allow a machining center to stand idle.

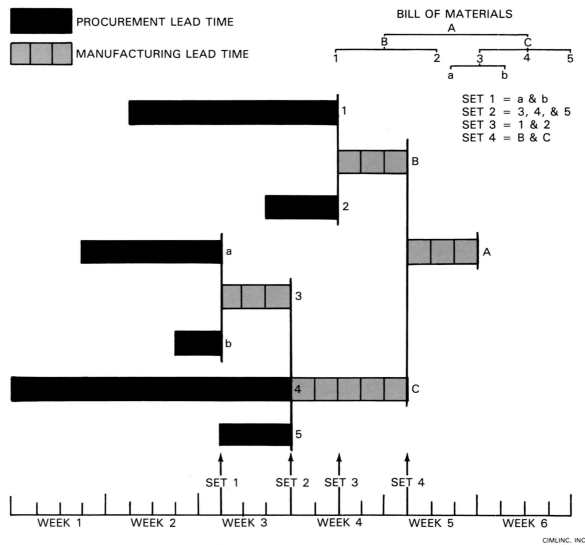

Fig. 6-20. Before an order is released, it is important to be sure that a matched set of parts is available to build the product. Varying lead times needed to build the sets of parents must be taken into account.

PURCHASE ORDERS

Before a purchase order is released, documentation similar to that required for a manufacturing order must be obtained. In addition, Purchasing also must obtain such items as a purchase history for the part, listings of qualified and approved suppliers, quotations from suppliers, letters of credit, and (where necessary) import or export licenses. To validate that the purchased product is being made according to standards or specifications, documentation at the supplier may have to be audited.

SPECIAL SUPPLY ORDERS

Rework and subcontract orders, known as *special supply orders*, have some unique characteristics. Generally, the entire *pick list* (bill of material) does *not* have to be reissued in the case of a rework order, since some or most of the items it contains are suitable for further processing. The specific components needed to make the product usable will not be known until the product is disassembled. Also, special instructions may have to be generated, since the rework order may have been created at any point in the routing.

Subcontract orders are different because of their source. For example, a manufactured item that is normally fabricated inside may have to be procured from an outside source because the shop is overloaded. Within a shop, a particular operation or machining center may become overloaded or be taken out of service for maintenance or repairs. As a result, a subcontract order may be written to offload the work in process.

SHOP FLOOR CONTROL

Sometimes termed Production Activity Control, the *Shop Floor Control* function executes the priority plan calculated by MRP while using the capacity determined by Capacity Requirements Planning (CRP). The scope of Shop Floor Control is shown in Fig. 6-21. The "building blocks" of Shop Floor Control include:

- Attainable Master Production Schedule.
- Adequate capacity.
- Detailed workcenter definition.
- Accurate routings.
- Order control system.
- Scheduling rules.
- Priority planning and execution systems.
- Input-output control.
- Timely reporting and follow-up.
- Well-trained, dedicated workers.

The importances of the Master Production Schedule was discussed in earlier sections of this book. As MRP breaks down the MPS into planned orders, a load is created on each internal manufacturing facility. The load must be aligned with available capacity. If the MPS is not attainable, one or more of the internal shops will be out-of-phase with each other, or with the external factory. Matched parts sets will not be built and confusion will reign. In the same way, if the MPS is unstable and changing within the "firm" time fence, the internal shops will be working on items that are no longer required, and *not* working on items that surfaced recently because of a schedule change at the end-item level.

Capacity on the shop floor is not considered by MRP. Planned orders created by MRP are firmed by planners and released to the floor by production control personnel. Management has several options available to adjust capacity, but each option has its related cost. Fig. 6-22 is a representation of these relationships.

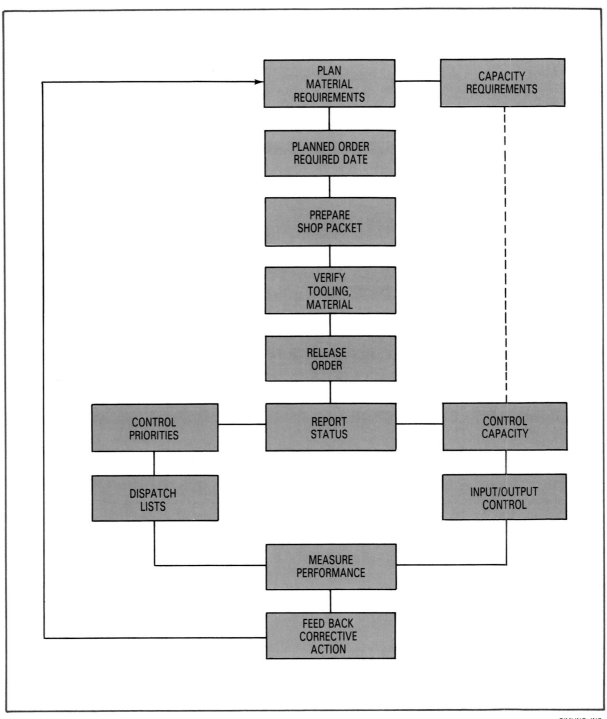

Fig. 6-21. The sequence of activities involved in shop floor control results from execution of the priority plan calculated by MRP. Shop floor control must make best use of available capacity.

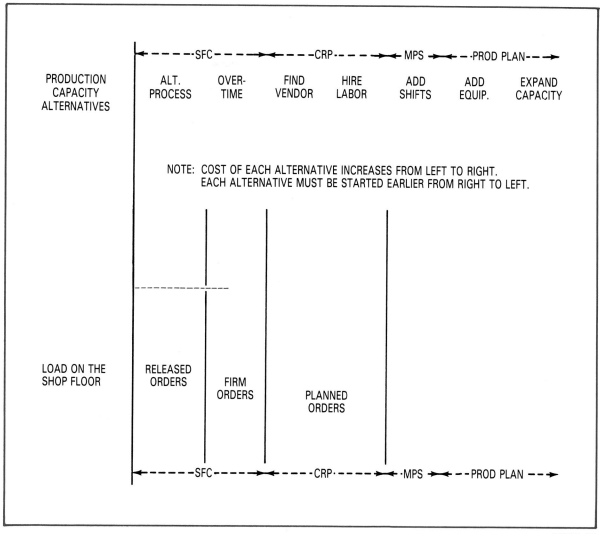

Fig. 6-22. *If demand exceeds capacity, management has a number of alternatives available. In this illustration, the alternatives increase in cost as you move from left to right.*

CAPACITY

How can the demonstrated capacity of a particular workcenter or shop be determined? Three performance measures may be calculated:

Efficiency equals standard hours earned divided by actual hours worked on product.

Utilization equals actual hours worked divided by total available hours.

Productivity equals standard hours earned divided by the total available hours. This is sometimes called the *load factor*.

Fig. 6-23 provides some sample calculations. The planned capacity used for the illustration is calculated by multiplying the total available hours by the load factor.

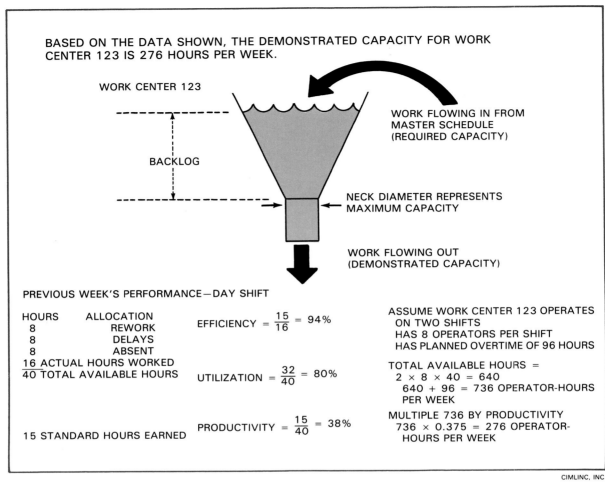

BASED ON THE DATA SHOWN, THE DEMONSTRATED CAPACITY FOR WORK CENTER 123 IS 276 HOURS PER WEEK.

WORK CENTER 123

BACKLOG

WORK FLOWING IN FROM MASTER SCHEDULE (REQUIRED CAPACITY)

NECK DIAMETER REPRESENTS MAXIMUM CAPACITY

WORK FLOWING OUT (DEMONSTRATED CAPACITY)

PREVIOUS WEEK'S PERFORMANCE—DAY SHIFT

HOURS	ALLOCATION
8	REWORK
8	DELAYS
8	ABSENT
16	ACTUAL HOURS WORKED
40	TOTAL AVAILABLE HOURS

$$\text{EFFICIENCY} = \frac{15}{16} = 94\%$$

$$\text{UTILIZATION} = \frac{32}{40} = 80\%$$

15 STANDARD HOURS EARNED

$$\text{PRODUCTIVITY} = \frac{15}{40} = 38\%$$

ASSUME WORK CENTER 123 OPERATES ON TWO SHIFTS
HAS 8 OPERATORS PER SHIFT
HAS PLANNED OVERTIME OF 96 HOURS

TOTAL AVAILABLE HOURS =
$2 \times 8 \times 40 = 640$
$640 + 96 = 736$ OPERATOR-HOURS PER WEEK

MULTIPLE 736 BY PRODUCTIVITY
$736 \times 0.375 = 276$ OPERATOR-HOURS PER WEEK

CIMLINC, INC.

Fig. 6-23. The demonstrated capacity of a given workcenter can be calculated as shown. In this example, the demonstrated capacity of Workcenter 123 is 276 operator-hours per week.

WORKCENTERS

The definition of a ***workcenter*** is changing as CIM technologies begin to modify views of how best to perform work on the job floor. The classic definition says that,

A workcenter consists of one or more machines that have interchangeable capacities.

When recent developments in such areas as robotics and group technology are taken into account, the definition alters to read,

A workcenter consists of a group of different machines that produce an item.

A ***pacing resource code*** will normally specify whether a particular workcenter is machine-paced or labor-paced. The assembly line approach to shop layout outlined in the Group Technology section of Chapter 4 balances worker and machine so that neither is pacing. Regardless of a company's degree of involvement in CIM or MRP, definition of the workcenter is critical to planning capacity, costing value-added work, and collecting other relevant information. Workcenters are defined and their relationship in a routing shown in Fig. 6-24.

WORKCENTER DEFINITION

DEPT. 10 FABRICATION	DEPT. 20 MACHINING	DEPT. 30 ASSEMBLY
101 – SHEAR	201 – LATHE	301 – SUB–ASSEMBLY
102 – FORM	202 – BORE	302 – FINAL ASSEMBLY
103 – WELD	203 – DEGREASE	303 – INSPECT & PACK

ROUTING OPERATIONS

PART NO.   DESCRIPTION
12345   PIECE PART

OPER	DESCRIPTION	DEPT W/C	MACH NO	TOOL NO	SET-UP HRS	LABOR HRS
010	LATHE	201				
020	BORE	202				
030	DEGREASE	203				

Fig. 6-24. Definition of workcenters is essential to being able to develop efficient routings.

ROUTING

Accurate routings are necessary for shop floor control. A *routing* identifies the sequence of operations that a part will follow as it is being processed on the floor. The sequence must be correct, or the part will be processed incorrectly, making the plan generated by MRP meaningless. Each operation is linked to a workcenter for data collection and consolidation purposes. Standard set-up and run times are measured and calculated for each operation, with the assumption that a single part is being processed. Current *shop floor scheduling systems* capture actual performance information that can be used to dynamically update the standard times. The same systems can monitor the move, wait, and queue times and adjust their values as improvements are made.

The shop floor is a busy place, with constant change taking place. It is quite possible for a part to follow an alternate sequence or be processed at a different workcenter from the one originally intended. If these options are also available on the routing and in the computer system, shop floor performance will be monitored correctly.

SCHEDULING RULES

Block Scheduling Rules

1. Allow one week for releasing order and drawing material from storeroom.
2. Allow six weeks for screw machine operations.
3. Allow one day for each 400 pieces in the milling department; round upward to next full week.
4. Allow one week for drilling, tapping, burring-and similar operations using minor equipment.
5. When operations are especially short, combine within the same week.
6. Allow one week for inspection and delivery of completed material.

Simple Scheduling Rules

1. Multiply hours per thousand pieces by number of thousands on order.
2. Round up to nearest 16-hour day (two shifts) and express time in days, round down to nearest day when excess hours are less than 10% of total (minimum one day for operation).
3. Allow five days to withdraw stock from stockroom.
4. Allow one day between successive operations within the same department.
5. Allow three days between successive operations in different departments.
6. Allow one day for inspection.
7. Allow one day to get material into stockroom.
8. Allow two extra weeks for screw-machine parts.

Comparison of Block and Simple Scheduling Rules

	Block Scheduling		Simple Scheduling	
	Time Allowed	Week	Time Allowed	Day
Release Date		37		402
Opn No. 01	1 week	38	5 days	407
Opn No. 02	6 weeks	44	12 days T = 3 days	419
Opn No. 03	3 weeks	47	3 days T = 3 days	425
Opn No. 04	1 week	48	1 day T = 3 days	429
Opn No. 05	1 week	49	2 days T = 3 days	434
Opn No. 06			1 day T = 1 day	438
	1 week	50	1 day T = 3 days	440
Opn No. 07				
Opn No. 08	1 week	51	1 day	444
Opn No. 09			1 day	445
Total Time	14 weeks		43 days	
Date Available		Week 51		Day 445

(T = transit time)

$$\frac{43}{5} = 8.6 \text{ weeks}$$

Fig. 6-25. Simple scheduling rules and block scheduling rules are useful in helping an engineer quickly develop ''rule of thumb'' capacity estimates and standards.

SCHEDULING RULES

If not all data needed to identify an operation or workcenter is available, how is a schedule determined? Two methods are used to help the industrial engineer develop quick "rule of thumb" standards and capacity estimates. These methods, called **block scheduling rules** and **simple scheduling rules**, are defined and compared in Fig. 6-25. By refining the information, the throughput time for a shop can be shortened—a valuable objective for both customer service and the dollar value of the inventory.

CAPACITY LOADING

MRP assumes floor capacity to be infinite, and loads work accordingly. This is called **infinite capacity loading**. The load is shown in the period required to support the plan. MRP uses a process called **backscheduling** to determine a start date for each operation described in the routing, based on the due date of the work order for the *parent* being built. Fig. 6-26 shows an example of backscheduling.

An alternate method for placing work on the floor is called **finite capacity loading**. The planned orders developed by MRP are loaded into a workcenter up to the level of that workcenter's demonstrated capacity. Planned orders in excess of that capacity are rescheduled into future time periods. This causes a problem in terms of priority, but offers feedback that a capacity check at a higher level plan was inadequate. Finite capacity loading uses *forward scheduling* to determine when a part will be completed. It does so by working forward from a known start date.

OPERATION NUMBER	DAYS OF MOVE/QUEUE	DAYS OF SET-UP/RUN	START DATE	COMPLETE DATE
005	1	2	92-115	92-117
010	2	3	92-119	92-122
015	2	1	92-124	92-125
020	1	2	92-126	92-128
025	2	2	92-130	92-132

ORDER DUE DATE IS 92-132

PRIORITY IS BY OPERATION START AND DUE DATE WITHIN ORDER DUE DATE

E.G. - OPERATION 025 MUST BE COMPLETED BY 92-132
IT TAKES 2 DAYS TO SET-UP AND RUN OPN 025 = 92-130 IS THE START DATE

IT TAKES 2 DAYS TO MOVE FROM OPN 020 AND WAIT AT OPN 025 = 92-128

OPERATION 020 MUST BE COMPLETED BY 92-128

Fig. 6-26. *The start date of each operation in the routing is determined by backscheduling from the due date of the work order. An example is shown.*

INPUT-OUTPUT CONTROL

A plan and its execution can differ considerably. There is no question that MRP can properly prioritize work orders issued to a given shop. How, then, can the execution be managed so that the plan can be adjusted and improved? The concept used on the shop floor is called *input-output control*. It is shown in simplified form in Fig. 6-27. The cumulative deviation identified in the input portion of the chart affects the planning side of the process. The cumulative deviation identified in the output portion of the chart is affected by shop floor execution. The objective is to balance the load so that both cumulative deviations are zero. Once that level is attained, a specific program to adjust backlog can be implemented.

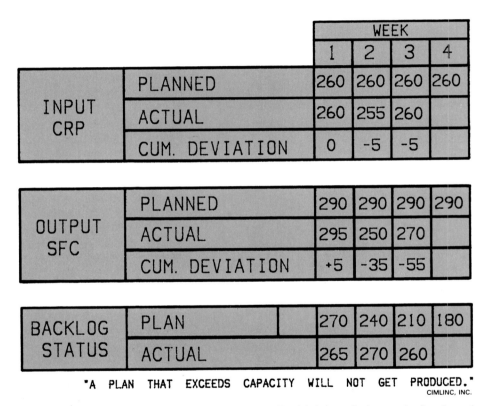

		WEEK			
		1	2	3	4
INPUT CRP	PLANNED	260	260	260	260
	ACTUAL	260	255	260	
	CUM. DEVIATION	0	-5	-5	
OUTPUT SFC	PLANNED	290	290	290	290
	ACTUAL	295	250	270	
	CUM. DEVIATION	+5	-35	-55	
BACKLOG STATUS	PLAN	270	240	210	180
	ACTUAL	265	270	260	

"A PLAN THAT EXCEEDS CAPACITY WILL NOT GET PRODUCED."
CIMLINC, INC.

Fig. 6-27. An example of input-output control, the goal of which is to balance the load on the manufacturing operation.

DISPATCH LIST

The *dispatch list* is usually produced on a daily basis from the planning system. It identifies the priority of each job at each workcenter on the shop floor, and should be treated as a guideline for the day's activities. The first-level supervisor in the shop has the ultimate responsibility for execution, but this should be done in cooperation with the dispatcher. If the dispatch list is reviewed daily, chances are the execution will serve the plan. The dispatcher and first-level supervisor should think of themselves as teammates.

Given the above, why do plants operate so often in an "expedite" mode? Fig. 6-28 lists some of the reasons why expediting is required. The question arises, "If the listed problems are difficult to solve in the company's own manufacturing operation, how much more difficult will it be to solve them externally, in the operations of the company's suppliers?" There are many similarities between Shop Floor Control and Purchasing Control.

WHY IS EXPEDITING REQUIRED?

1. Inaccurate Records
   - Routings
   - Bills of Material
   - Inventory
2. Inadequate Capacity
3. Failure to Reschedule Required Dates
4. Inadequate Priority Control
5. No MPS "Time Fence"
6. Shop Problems
   - Scrap
   - Rework
   - Tooling
   - Machine Breakdown
   - Absentees
7. Material Shortages
8. "Good Customer," "Management Expeditor"
9. High Value Item
   - Meet Shipping Target
   - Reduce Inventory
10. Engineering Changes

Fig. 6-28. Ideally, product flow through a manufacturing operation would be smooth and according to schedule. Often, however, orders must be expedited, rather than following the schedule. Some typical reasons for expediting are shown.

PURCHASING CONTROL

Activities involving contact with a company's external suppliers are the responsibility of *Purchasing Control*. In legal terms, the person who serves as the company's Purchasing Agent, or Buyer, has the authority to contract with other individuals and companies and incur a liability (purchase goods or services). In actual practice, unauthorized persons in an organization may commit the company to a *vendor* (external supplier) for goods or services. These "handshake agreements" made by persons other than the Purchasing Agent, can lead to difficulties.

To avoid problems, it is crucial that any contract with a vendor be executed only by the Purchasing Agent. This will help to assure that the purchased materials meet specifications, and that the vendor is properly paid for supplying the contracted goods or services.

As noted earlier, there are many similarities between the functions of Shop Floor Control and Purchasing in issuing production orders and purchase orders. These similarities are depicted in Fig. 6-29.

Purchasing must balance the seemingly conflicting objectives of *price* (as low as possible), *quality* (as high as possible), and *service* (as timely and responsive as possible). Most companies today would like to say to suppliers that quality is no longer negotiable, that a perfect product is expected each time (see TQC/JIT), and that poor quality performance will ensure removal from the company's list of approved sources.

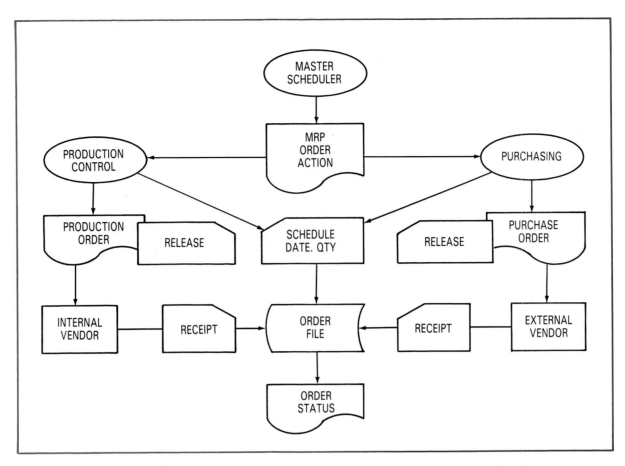

Fig. 6-29. Production Control and Purchasing perform parallel functions: Purchasing issues purchase orders to an external vendor for components or products, while Production Control issues manufacturing orders to an internal vendor. Both functions are concerned with control of material flow into the manufacturing process.

Unfortunately, for many products and many companies, there are no alternate suppliers ("second sources") available. This highlights the importance of the Purchasing Agent, or Buyer. If the Buyer's role is to expedite near-term deliveries, and sufficient time is not allotted to finding second sources or cultivating existing suppliers, current unacceptable practices will continue. When an MRP II system is implemented, the time allocations in the Buyer's day change significantly, Fig. 6-30.

BUYER RESPONSIBILITIES

As noted earlier, the Buyer (Purchasing Agent) is the *legal agent* of the company, able to commit to contracts and incur liabilities. An additional important role played by this person in a manufacturing planning and control system environment, is to secure an efficient flow of parts that enables the end item to be built as scheduled. The objective should be to have the Buyer negotiate contracts with each vendor for capacity and price and allow MRP to identify delivery expectations. Various types of contracts are identified in Fig. 6-31.

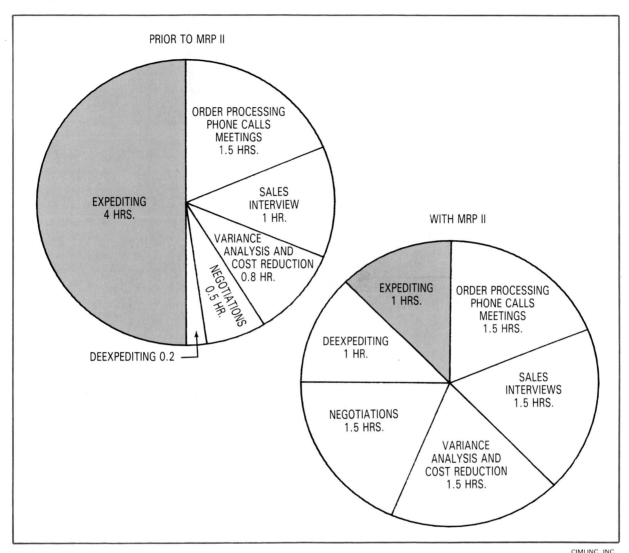

Fig. 6-30. The adoption of MRP II greatly reduces the time a Buyer must spend expediting orders. This permits more time to be devoted to negotiating with vendors and analyzing costs.

CONTRACT TYPES

SINGLE BUY	**ITEM BLANKET**	**DOLLAR BLANKET**	**OPTION AGREEMENT**
• Single item	• Single item	• Multiple item	• Single item
• Fixed price	• Fixed price	• Quoted price	• Scheduled price
• Fixed delivery	• Variable delivery	• Variable delivery	• Variable delivery

Fig. 6-31. There are various types of contracts used for purchasing raw materials, components, and products.

Developing congruence between the manufacturing company's part requirements and the vendor's sales objectives is mutually beneficial. The manufacturing company must identify the capacity level (output) desired from the vendor, and be willing to incur liability for that capacity. The vendor needs to commit to that capacity and be willing to accept mix changes (within his MPS time fences), as well as offer stable pricing. This is classic "win-win" methodology. Although it requires some effort, it ensures the willingness and ability of a vendor to perform as desired.

Single source or multiple sources?

The question of whether more than one source is necessary for a given part or process is not easy to answer. The development of a professional working relationship between companies takes time. Each side has its own expectations: the vendor wants well-defined capacity utilization and relatively stable schedules; the manufacturer seeks on-time delivery of defect-free parts and relatively stable pricing. Achieving performance levels that meet these expectations is not something that happens overnight. Respect cannot be based on reputation or history; it must be based on performance and execution. It is important to have as a goal the establishing of such a relationship. While building that relationship, however, the manufacturing company's schedule must be protected by developing multiple sources. Each vendor should be made aware of the competitive situation as a means of increasing their motivation to perform well. Often, one of the multiple sources will emerge as the best single source.

Many companies doing business in North America are multinationals with manufacturing facilities in several countries. Other companies, with facilities essentially located in one country, may have international contracts to process. In such instances, the concept of offset must be considered as a sourcing issue. *Offset* is the percentage of a product's components or parts that must be manufactured in a given location as a condition of a contract. For example, assume that your company is located in the United States and wins a contract to sell its product to a foreign government or company located in a foreign country. The contract with your customer is likely to specify that a specific percentage of the product's component parts must be sourced from companies in that country. This situation compounds the problems of single-sourcing already described. However, the process of developing a close working relationship with a vendor, based on mutual benefit, is as applicable in the international arena as it is in a domestic setting.

THE PLANNER-BUYER

Faced with monthly shipment goals, daily delivery problems, and hourly crises, how can a manufacturer ever hope to free the company's Buyer to develop the necessary long-term relationships with vendors? An important factor to recognize is that it cannot happen by decree, and it is not going to happen tomorrow. You should identify it as an objective, and work toward it using whatever tools you have available. A useful concept to employ, particularly as the company modernizes its planning and control systems, is that of the *planner-buyer*.

One of the benefits of a sound Master Production Schedule is that the requirements developed through MRP processing are *real*: the quantities and due dates relate to the company's business plan. A Planner-Buyer can be granted the authority to authorize individual releases (for specific quantities and dates) against a Buyer-negotiated contract, rather than merely issuing an order or purchase requisition to advise Purchasing that a requirement exists. The Planner follows the parts by day and quantity; the Buyer spends time negotiating for external and economical capacity.

The purchasing control process

As shown in Fig. 6-32, the purchasing control process involves a sequence of steps. Each of these steps requires specific blocks of time. See Fig. 6-19 for a graphic presentation of the lead times involved in order processing. Remember that MRP can time-phase an activity if it is advised of the *value* (in this case, the amount of time needed to complete a task).

Request for quotation. A unique activity in the procurement function is a process called the ***Request for Quotation*** (RFQ). If a part has not been built before, an RFQ must be prepared, mailed, received back, analyzed, and awarded. If the formal system is not set up to acknowledge this special case, the *informal* system will, by default, control acquisition of the part.

A special lead-time increment is commonly identified as ***dock-to-stock processing time***. This is the amount of time it takes to remove a purchased product from the delivery vehicle and process it to the storeroom or point of use. In RFQ situations, dock-to-stock processing time can be significant because of the practice of ***first article inspection***. This practice involves thorough checking of the initial shipment of purchased parts to make sure that the vendor can actually build the part to specifications. Once in production, the part should not have to be inspected as thoroughly; in some cases — where a vendor has demonstrated ability to consistently meet quality standards — little or no inspection may be done. This significantly decreases the dock-to-stock processing time.

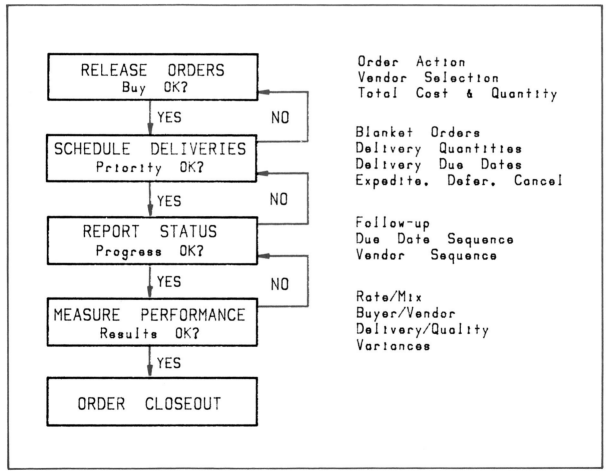

CIMLINC, INC.

Fig. 6-32. The purchasing process proceeds through a definite series of steps from order release to order closeout.

Vendor performance measurement

Measurement of vendor performance is crucial. The "external factory" (vendor manufacturing facilities) must be subject to the same measurement standards as the "internal factory" (company facilities). Fig. 6-33 lists typical measurements of vendor performance.

SERVICE	• DELIVERY • RESPONSE • PROBLEM-SOLVING • TECHNICAL INTERFACE
PRICE	• COMPETITIVE • AMOUNT OF INCREASE • FREQUENCY OF INCREASES • OPENNESS TO DETAIL BUILDUP
QUALITY	• RETURNS • TOOLING MAINTENANCE • PRODUCE RELIABILITY • COMMITMENT TO PERFECTION

Fig. 6-33. Various methods and standards of measurement can be used to assess vendor performance in the areas of service, price, and quality.

Vendor tooling maintenance

The maintenance of vendor tooling has traditionally been a responsibility of the procurement (Purchasing) function, with the assistance of Manufacturing Engineering and the formal planning system. Maintenance of vendor tooling presents another opportunity for the close relationship between manufacturer and supplier to be used effectively. A good manufacturing planning system knows the *expected life* of a tool (in terms of parts, lots, or hours) and notifies the vendor in adequate time for action to be taken. In the same way, if a vendor notes a tool problem (such as the need for frequent operator adjustment to hold tolerances), the vendor will notify the manufacturer so that the situation can be corrected.

TOOL CONTROL

Traditionally, the importance of tooling in the manufacture of a product has been a "given." Because they were more visible, such resources as personnel, material, and machines dominated the initial planning efforts that went into formal systems. With the advent of CIM, and recognition that tooling is a critical link in the integration process, *tool control* was transformed into a major discipline in manufacturing. In some operations, tool control has even achieved its own department designation. The place of tooling in the overall manufacturing picture is depicted in Fig. 6-34.

A detailed analysis of manufacturing costs shows that tooling represents a substantial portion of those costs. In a manufacturing situation, *tooling* includes such items as perishable tools, fixtures, gages, and production supplies. The cost of insuring tool availability is significant, but so is the cost

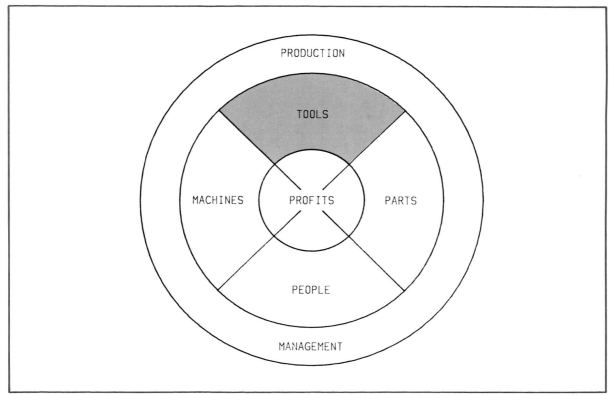

Fig. 6-34. *Tooling is one of the vital components in a manufacturing enterprise. In some companies, Tool Control has been set up as a separate department.*

of tool shortages in terms of the demoralizing effects they have on workers. Because of the limited life of a tool, its potential for obsolescence is greater than that of a specific part. Engineering changes lead to obsolescence of both tooling and inventory, but the more dramatic effects on tool life come from use and handling. Storage requirements for sensitive gages and precision tools should be *at least* as stringent as those used for detail parts.

WHAT IS TOOLING?

In simplest terms, a *tool* is anything that allows something else to be done. The wood-splitting wedge, the rope and pulley used to lower a bucket into a well, and the mortar and pestle used to grind grain are all examples of simple tools that humans have used for thousands of years. A modern example would be a machine tool—the jigs and fixtures that allow a workpiece to be positioned properly on production equipment.

Today the definition of "tool" has been extended to include anything that is required to permit a person or machine to successfully produce a part. Numerical control (NC) tapes are an example of recent additions to the list of typical manufacturing tools. With CIM, we think in terms of accessing machine control software through factory networks, so the program itself, and the *network* (systems used to transmit program instructions and other data between computers and production machines) can also be identified as tools.

TOOL PLACEMENT

Manufacturing and control systems have the embedded logic needed to handle tool control. The database structure is available to properly define, describe, and classify a tool. The decision that must be made is where to place a specific item of tooling. A tool that is unique to a specific part, for example, can better be placed in a bill of material than in a routing file. One that is needed every time a piece of equipment is used (lubricant, for example) should be structured with the operation or machine center. The use of CAD/CAM systems for tool design provides the same potential benefits as those documented for the use of CAD/CAM for product design.

The same ability to identify requirements through a combination of sales orders and market forecasts that culminates in the Master Production Schedule can be used to specify tooling needs across the whole MPS horizon. The time-phasing of the tool order can be correlated exactly to the production order (internal) or the purchase order (external).

Inventory control of a tool is not essentially different from the inventory control of a part, especially if that part has a "shelf life." Various inventory control methods may be used for tools. Packaged or bulk materials, such as lubricants, solvents, and chemicals, can be controlled with a simple periodic physical check of their storage areas. An *Automated Storage and Retrieval System* (ASRS) might be used to store and access critical tools and fixtures. Cost accounting systems are usually capable of monitoring and accumulating the costs of a specific tool.

FORECASTING TOOLING NEEDS

Although the logic is in place to control tooling, it is important to note that the disciplines required are the same as those needed to control the material plan. Are planners required to analyze a "planned tooling order" and decide if it should be firmed? *Yes!* Is an accurate inventory balance necessary? *Yes!* Does a tool have the same problems in relation to a pre-released drawing, as opposed to an in-effect blueprint? *Yes!* Is the procurement of tooling very different from the procurement of a purchased part? *No.* The benefits of integrating the planning of tooling needs with parts planning are listed in Fig. 6-35. It is now possible to establish a systemic relationship between parts forecasting as the *independent* demand item and the forecasting of tooling needs as a *dependent* demand item.

BENEFITS OF INTEGRATING TOOLING PLANNING AND PARTS PLANNING

1. Improved machine/operator utilization and minimized production delays caused by stockout occurrences.
2. Minimized excessive and obsolete tool inventories.
3. Improved schedule/shipment performance through accurate tool ordering and scheduling.
4. Reduced lead times and reduced manufacturing times through accurate tool forecasting and availability.
5. Enhanced cost accountability of tooling by tracking tool use from point of issue.
6. Enhanced cost management for tooling through accurate requirements forecasting and budgeting.
7. Improved supplier performance by having the capability in place to monitor the status of tooling at vendor.

Fig. 6-35. *Greater efficiency and improved cost accountability are among the benefits of integrating the planning of tooling needs with parts planning.*

LOT TRACEABILITY/SERIAL NUMBER CONTROL

The ability to maintain lot integrity is becoming a requirement in many industries. Pharmaceutical manufacturers have met such requirements for years. Computer manufacturers and other makers of capital equipment are finding serial-number identification of products important as the pace of system upgrades (and the need for field installation on installed equipment) becomes ever more rapid. The transportation industry, with public attention focused on product integrity and safety issues, must be able to identify specific items for replacement when a malfunction causes a product recall.

The purpose of *lot traceability* is to maintain the relationship between specific lots of parts and the products they are used on. Similar to the relationships identified in a bill of material, lot traceability allows the user to view single- and multiple-level explosions and implosions of lots.

A variant of lot traceability is **serial number tracking**. The purposes are frequently the same; however, a company at times may elect to use them in tandem. Serial number tracking is most often applied to an *end item*, eliminating the need for the customer order number or customer name. Sometimes, not all components of a product require traceability. For example, only one part may be subject to product safety recall requirements, while other components need only a limited tracking or traceability program.

INSPECTION STATUS REPORTING

The status of incoming and in-process material, based on inspection, determines whether a material is available for MRP planning. Typical material status identifications are *approved* and *released*. A status of *quarantined* indicates that a lot is being subjected to acceptance testing. Quarantined material must be segregated from released material; normally, it is not available for consumption. A status of *conditional* limits use of the material to a particular product, restricted amount of time, or other specified conditions. A lot with a *rejected* status is not available for use; it will either be returned to the vendor or scrapped. Material identified as *uninspected* has been recently received, and is unavailable pending quarantine and testing.

RECEIPT AND ISSUE CONTROL

Material that requires assignment of a lot number for tracking purposes is flagged in the receipt and issue control process. The ways that various inspection status identifications relate to receipt and issue control are shown in Fig. 6-36.

RECEIPT AND ISSUE CONTROL
LOT TRACEABILITY

Receipt control establishes the fact that a lot requires traceability.
The lot will NOT be allowed to close to a storage location until it has received an inspection status of:

Approved   Restricted   Conditional   Quarantined

A lot with a status of "Uninspected" or "Rejected" may NOT be used.

Issue control establishes the fact that a parent requires controlled lots.

The component lot will not be allowed to be issued until it has received an inspection status of:

Approved   Restricted   Conditional   Quarantined

A lot with a status of "Uninspected" or "Rejected" may not be issued to its parent.

Fig. 6-36. Lot traceability involves assigning a lot number and determining its inspection status.

EXPIRATION TRACKING

An item with a specified life expectancy ("shelf life") must have its expiration date noted. Commonly, companies will follow the stock rotation principle of "first in, first out." *Expiration tracking* is an extension of this principle. It identifies a specific date after which an item can no longer be used. After that date the item may be retested to determine whether it is still suitable for use, or should be destroyed.

DATA INTEGRITY CONTROL

Lot traceability requires further data tracking and maintenance. It may involve further audits when multiple testing is required, tolerance factors are added, actual-to-theoretical consumption is compared, or last-of-lot/location reporting is done.

Retrievals are similar to those done with bills of material. A typical bill of material structure for "Part A" is shown in Fig. 6-37. The added item is the approved lot numbers that can be used to build the configuration. Reports can now be produced to show the lot numbers (200 through 500, in this case) that were used to build "Part A." These same reports could be inverted to show in which products a specific lot (such as 200) was used.

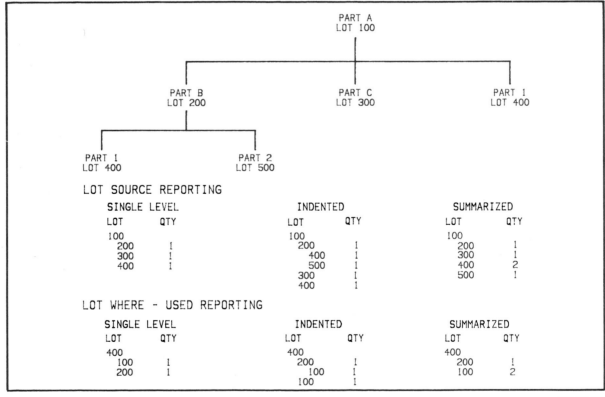

CIMLINC, INC.

Fig. 6-37. Lot traceability allows development of a report showing the lot numbers of parents used to build a product. A similar report can be constructed to show where (in which product) a specific lot was used.

COST CONTROL

Managing and accounting for cost information is an integral part of operating an organization. Companies that invest the effort necessary to maintain accurate cost information have a distinct competitive advantage when quoting jobs and accounting for variances. There are two distinct facets involved in cost control: cost accounting and cost management. *Cost accounting* tracks execution of the business plan, and is historical in nature; *cost management* is related to managerial performance and projects the future. Each has an important role in an organization.

COST ACCOUNTING

The tracking of cost information and the accounting for variances that result from previous activities are functions of cost accounting. It is important to identify the cost of each action taken in a business, and—if the cost is out of line—to determine the source of the problem so that corrective action can be taken. Clearly defined goals are necessary; so are objectives stated in terms of standards, budgets, forecasts, and schedules. If continuity and maximum performance are to be attained, these goals and objectives must be established by top management and reiterated down through each level of the organization. Accounting for performance to the goals and objectives is the essence of cost control.

COST MANAGEMENT

Providing information that will help the management team plan for the business is the primary cost management activity. MRP II includes an information feature that allows it to provide management with a projection of future variances based upon aggregation of the Master Production Schedule into the Production Plan, the Financial Plan, and ultimately, the Business Plan.

Data on current performance from Cost Accounting can be projected into the future. Specific items can then be addressed to determine whether there is a better way to conduct business. For example, it allows management to ask such questions as, "Is it more economical to make or to buy a specific component?" or "Based on current shop rates, is it time to consider offshore assembly?" The appeal of such information is obvious. Nothing can be done about what has *already* happened. However, if it can be demonstrated that something *will* happen if the present course is continued, management has a powerful tool for improving performance. In today's situation of "information overload," it is very helpful to be able to focus managers' attention and effort on activities that provide the greatest return for the time invested.

Value analysis engineering. Techniques that may help Management do a better job of cost control include projecting future variances, as noted earlier, and using value engineering and analysis. *Value engineering* involves investigating reasons for supply and demand fluctuations, identifying waste, and recommending alternative solutions to the problem being addressed. Value engineering requires that all departments, from Design and Product Engineering onward, add value and eliminate waste in their activities. By relating these activities to the current production configuration, incremental improvements can be planned and executed.

Costing methods. The two most popular costing methods in use today are actual costing and standard costing. Other costing methods, such as average, moving average, and target, are variations of those two. The *standard cost* method, Fig. 6-38, establishes (usually at annual intervals) a performance measurement baseline as the standard. The standard is then used to compute variances. Common variances that can be monitored in a standard costing system are listed in Fig. 6-39. Proponents of standard costing believe that this method provides the most consistent measure of management performance. They maintain that the use of *actual cost*, a moving target, does not provide sufficient information to measure performance, and provides a subjective excuse in the case of poor performance. The two costing methods are compared in Fig. 6-40.

OBJECTIVES OF A STANDARD COST SYSTEM

1. Integrate with common databases.
   a. Bills of Material.
   b. Routings.
   c. Work Center Data.
   d. Part Master File.
2. Integrate data collection with operating system transactions.
   a. Purchasing.
   b. Receiving.
   c. Labor/Machine Reporting.
   d. Scrap/Rework/Repair Order.
   e. Material/Operation Substitution.
3. Specifically assigned responsibility. The individual who directly affects the cost element is responsible for its validity.
4. Report costs as incurred.
   a. Daily.
   b. By function and work center.
5. Engineeered standards.
   a. Reasonable expectation.
   b. High task standard.
6. Allow cost rollup from bottom to top.
   a. Material.
   b. Labor.
   c. Burden.
7. Provide for inventory valuation.
   a. Obsolescence.
   b. Cost of engineering changes.

Fig. 6-38. *The standard cost method is one of two costing methods in wide use, and is believed by its proponents to provide the most consistent measure of management performance. It establishes a baseline against which variances can be computed.*

EXAMPLES OF COST VARIANCE ANALYSIS

1. Material Usage, calculated by part number.
   Material Usage Variance = (Actual Usage minus Standard Usage) times Standard Cost
2. Purchase Price, calculated by part number.
   Purchase Price Variance = (Actual Unit Cost minus Standard Unit Cost) times Quantity Received
3. Labor Rate, calculated by part number within work order and within work center.
   Labor Rate Variance = (Actual Pay Rate minus Standard Pay Rate) times Actual Hours
4. Labor Efficiency, calculated by part number within work order and within work center.
   Labor Rate Efficiency Variance = (Actual Hours minus Standard Hours) times Standard Work Center Rate
5. Order, calculated against all part numbers and work centers used.
   Total for each operation within each work center.
   a. Set-up Variance = Set-up hours at standard minus actual.
   b. Run Variance = Run hours at standard minus actual.
   c. Material Usage Variance.

Fig. 6-39. *Standard costing can be used to monitor a number of variances. A number of the most common are listed.*

COMPARISON OF ACTUAL VS. STANDARD COST SYSTEMS

	COST SYSTEM	
ITEM	ACTUAL	STANDARD
COMPARE PERFORMANCE	VERY LIMITED	EXCELLENT
IDENTIFY PROBLEMS	NO	YES
PROVIDE GOALS	NO	YES
HIDE INEFFICIENCY, WASTE	YES	NO
CURRENCY OF INFORMATION	HISTORIC	PREDICTIVE
BASIS FOR PRICING	POOR	GOOD
MEASURE OF PROFITABILITY	POOR	GOOD
COST—RELATIVE	INEXPENSIVE	EXPENSIVE

Fig. 6-40. *Comparison of the actual and the standard cost systems shows how the capabilities of each can be rated.*

The concept of a *cost roll-up* is important to understand. A simplified example is shown in Fig. 6-41. An actual, average, target, standard, or any other type of cost can be rolled up. Modern systems allow the roll-up of multiple cost types. For example, last year's standard cost could be used to quantify current-year improvements. This year's actual cost could be used to develop a customer's Request for Quotation. Current standards on future configurations can be compared with target costs.

What method should be used? That is management's decision to make. Two considerations should be part of the decision:

- With the currently available resources, the organization should be capable of accurately entering the desired cost information.
- With the currently available resources, the organization should be able to make use of the cost information being provided.

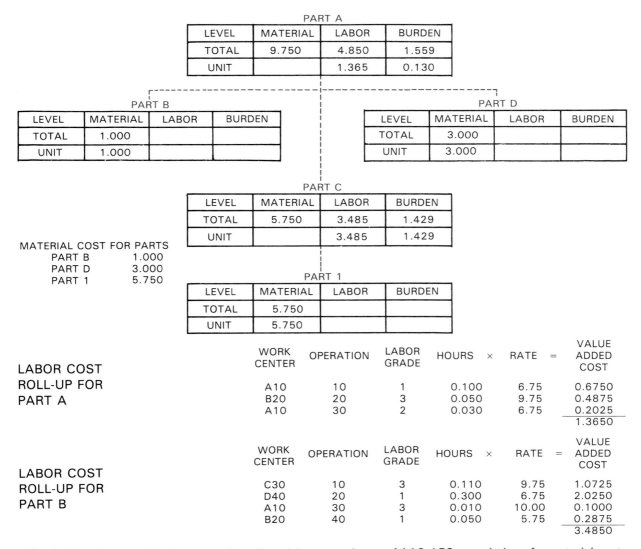

Fig. 6-41. *An example of a cost roll-up. Part A has a total cost of $16.159, consisting of a material cost of $9.750, a labor cost of $4.850, and a burden of $1.559.*

TQC/JIT CONCEPTS

This and preceding chapters have described an evolution taking place in the planning and control of manufacturing businesses—changes that support business objectives of the company and that (if properly implemented), will better position the company for later success in the marketplace. It should be apparent by now, however, that the process of evolving from the current method of doing business to a new method is *not* a simple task. It is possible, though, to order the process in such a way that it is achievable.

A manufacturing company should not attempt to implement change all at once. Key concepts of *Total Quality Control* (TQC)/*Just-In-Time* (JIT) implementation are:
- Continued improvement is better than postponed perfection.
- Action is better than "analysis paralysis."
- A less-than-optimum decision, executed thoroughly with energy and excitement, is better than the optimum decision executed lethargically and without commitment.

The management/operating philosophy of TQC/JIT is summed up in Fig. 6-42.

TQC/JIT—A MANAGEMENT/OPERATING PHILOSOPHY

A total commitment to quality is paramount:
Focus is on continuous process improvement.
- Every activity is a process.
- Uses data, scientific methods.
- Perfection is the goal.

Requires universal participation.
- Everyone, everywhere.
- Teamwork.

Results in customer satisfaction.
- Exceeds expectations.
- Both the *internal* and the *external* customer.

Fig. 6-42. Implementing a TQC/JIT program involves a strong commitment to quality by management.

TOTAL QUALITY CONTROL

Sometimes referred to as "Quality at the Source," TQC is based on the concept that it is possible to produce a *perfect part every time*. It places the responsibility for quality on the individual operator. Under TQC, Quality Assurance departments are no longer necessary. Their personnel do not *add value* to the product being produced, so they should be eliminated. Recently, TQC has been termed TQM (Total Quality Management) to expand its meaning to all levels of management and all company functions.

The "customer" for each person working in a TQC/JIT environment is the *next function* (or operation, worker, or piece of information) that is dependent upon what that person does to the product. TQC does not allow a *rework* routing; instead, the entire line is shut down until a problem is corrected.

Total Quality Control assumes that even an outsider should be able to ascertain what is happening in a shop by merely walking past and observing what is going on. Nothing is hidden: tools, graphs of performance, lists of remaining wasteful activities at a work station, are all visible. TQC assumes that a machine is *always* ready to produce a perfect part. Is TQC impossible to implement? *No.* Is it difficult? *Yes.*

Does TQC serve the goal of achieving the "factory of the future" (the factory operating in a CIM environment)? The answer is an emphatic *Yes*! MRP II defines the factory of today. The plan it produces is cohesive, logical, and prioritized. It can be aggregated and disaggregated as necessary. TQC/JIT forces a focusing of attention on the shop floor (including both internal and external shops). Fig. 6-43 compares MRP II with TQC/JIT.

COMPARISON OF MRP II WITH TQC/JIT

FUNCTION	MRP II	TQC/JIT
MPS	Must be realistic	Must be realistic and balanced
Plan Capacity	Rough cut capacity planning and capacity requirements planning	Line balancing algorithms
Control Capacity	Input-Output control adherence	100% schedule to meet specific demand
Plan Priority	Material Requirements Planning (MRP)	If required, depends on MRP output
Control Priority	Shop Floor Control (SFC)	Pull from downstream resource

Fig. 6-43. A comparison of MRP II with TQC/JIT, showing similarities and differences of the two systems.

JUST-IN-TIME

What are the key strategies of Just-In-Time? They are:
• Always produce to the exact demand of your customer.
• Eliminate waste in every process.
• Add value to the product being worked on, or don't continue.
• Produce one at a time.

One of the most useful characteristics of computers is their ability to consider one as a value — they store information in the form of *binary digits*: 1 or 0, on or off, customer demand or no customer demand. Complex algorithms to calculate economical, optimum, or classical lot sizes are no longer needed. JIT assumes a production run of *one* is standard, so a set-up to produce a single part is rational.

JIT promotes the idea of continuous improvement, Fig. 6-44. An efficiently produced lot size of any number is better than a wastefully produced lot size of one, but efforts to improve must be maintained continuously. JIT also emphasizes respect for people and their abilities. In a traditional manufacturing atmosphere, it would be considered heretical to think that a machine operator could handle material, organize the workstation, set up the operation, run the equipment, monitor the quality level by maintaining Statistical Process Control (SPC) charts, and produce a perfect part every time. JIT *assumes* that each employee is capable of doing all these things.

Implementing TQC/JIT requires a long-term commitment to excellence, as depicted in Fig. 6-45. Although the desired outcomes from implementing the Total Quality Control/Just-In-Time concept might seem difficult to achieve, the underlying principle is *the simpler, the better*. It cannot be purchased "off the shelf" like a computer software package. It is, instead, a *state of mind*, a "cultural mind-set" that encourages entrepreneurial activities and bottom-round consensus rather than top-down dictates and resulting bottom-up frustration. Objectives of TQC/JIT are listed in Fig. 6-46.

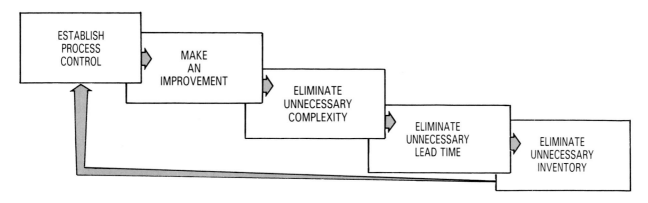

Fig. 6-44. Just-In-Time is a reiterative process, stressing the concept of continuous improvement.

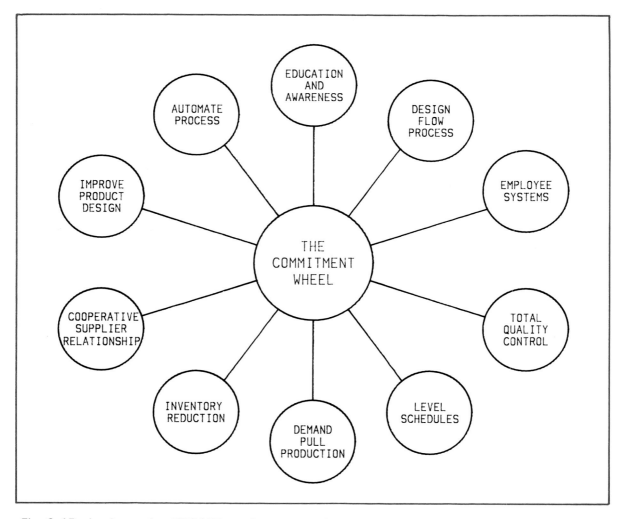

Fig. 6-45. Implementing TQC/JIT requires a commitment to excellence at all levels of an operation.

174 *Computer-Integrated Manufacturing*

TQC/JIT OBJECTIVES
◆ Produce products the customer wants
◆ Produce products only at the rate the customer wants
◆ Produce products with perfect quality
◆ Produce products instantly (Zero unnecessary lead time in the product purchased or manufactured.)
◆ Produce products with no waste of labor, material, or equipment (Every move with the purpose of never having idle inventory.)
◆ Produce products by methods that allow for the development of people.

Fig. 6-46. TQC/JIT has a number of objectives that describe the "mind-set" involved in implementing

SUMMARY

All elements of *manufacturing control* depend on the integrity of performance measurement data and on cycle time analysis. All elements of control must be addressed for a successful system within the total CIM plan. Education and training are also essential elements of the control function.

The *engineering data control* function is concerned with the creation of the organization and maintenance of engineering documents used by all segments of the company. These documents include a complete *bill of materials*, *process plans*, and *manufacturing resource plans*. Usually, this entire control function is known as *configuration management*. The bill of materials (BOM) is usually a subset of the configuration management system.

Another control element is *inventory control*. This function is vitally important to the success of a company, since 40%–60% of company assets are in the form of inventory. The inventory control function is the execution portion of *inventory management*. Inventory management provides a level of inventory consistent with the master production schedule (MPS). It also optimizes inventory levels through control of lead time and planned ordering and receiving policies for purchased materials. *Order release and control* is another control function that is essential to a successful operation. The order release system prepares the business for the procurement of raw materials, purchased parts, or manufactured product within the factory. The order release system releases orders as planned by the full MRP II system.

Shop floor control executes the priority plan that was calculated by MRP while using the capacity planning model for the factory. Shop floor control relates to the "inside" factory while the purchasing control relates to the "outside" factory. Accurate process routings are a necessity for effective shop floor control. A routing identifies the sequence of operations and the work center for each job. It is used in scheduling work in the shop floor system.

Other elements within the manufacturing control system are *tool management control*, *cost control*, *serial number/traceability control*, and *total quality control* (TQC). This control function assumes that a perfect part is produced each time at every operation. TQC puts the responsibility for quality on the individual operator. Poor quality will tend to destroy good manufacturing control systems.

IMPORTANT TERMS

ABC analysis
actual cost
Automated Storage and Retrieval System
backscheduling
bills of material
block scheduling rules
Configuration Change Board
configuration management
cost accounting
cost management

cost roll-up
critical path
criticality
cycle counting
cycle time
discrepant lots
dispatch list
dock-to-stock processing time
effectivity
efficiency

IMPORTANT TERMS (continued)

engineering data control
expiration tracking
explosion
finite capacity loading
first article inspection
forward scheduling
implosion
infinite capacity loading
input-output control
intercompany orders
inventory management
inventory stratification
item master data
Just-In-Time
lead time elements
load factor
lot traceability
manufacturing orders
manufacturing resource records
manufacturing routings
matched sets of parts
Material Review Board
measurement baseline
min/max system
multiple-level bill of material
network
offset
pacing resource code

periodic count
periodic review
pick list
planner-buyer
productivity
purchase orders
Purchasing Control
record accuracy
Request for Quotation
routing
safety stock
serial number tracking
Shop Floor Control
shop floor scheduling systems
simple scheduling rules
special supply orders
standard cost
statistical order point
tool
tool control
tooling
Total Quality Control
two-bin system
utilization
value engineering
vendor
visual review
workcenter

QUESTIONS FOR REVIEW AND DISCUSSION

1. Describe the steps involved in establishing a baseline for performance measurement, and state the benefits of establishing such a baseline.

2. Explain what is meant by the statement, "Inventory control requires a constant balancing of the optimism of sales-related functions against the conservatism of finance-related functions."

3. List the five increments that make up manufacturing lend time. Relate each of these to the corresponding actions or operations that make up lead time in nonmanufacturing areas (such as administrative or manufacturing support departments).

4. What is the basis of the inventory control tool described as *ABC analysis* or *inventory stratification*? How is it used? Cite an example.

5. Describe the three performance measures that may be calculated to determine the capacity of a shop or workcenter. How is planned capacity calculated, using the information given in Fig. 6-23?

6. What is the concept of *offset*, and why is it important to a company that has international contracts?

7. Explain what is meant by the statement, "JIT is a reiterative process that promotes the idea of continuous improvement."

7

Automation Technology

by Joe Richardson

Key Concepts

- [ ] Automation in the factory has evolved in response to market forces and profit opportunities. Its primary benefits are increased productivity, improved quality, and reduced unit costs.
- [ ] Organizing a batch production shop through the use of Group Technology disciplines can allow the shop to approach the efficiencies of continuous-flow production.
- [ ] Flexible Manufacturing Systems permit rapid changeovers to produce different products or parts in response to market demand.
- [ ] A computer network is vital to automated production: it distributes production and scheduling information, receives production data feedback, and adjusts process controls as necessary.
- [ ] Sensor feedback permits automated control of production equipment to adjust for tool wear or other variables.

Overview

Automation is the ultimate expression of the human desire to lessen the need to use their own energy to do work. It involves the use of machines to do work previously done by human hands, in ways that are faster, cheaper, and often better.

Automation is an *evolutionary* process, not a *revolutionary* one—when a tree limb was first used as a hammer or club, humanity took its first step toward automation. The roots of automation can be traced to the development of machines that used animal, water, or wind power to grind grain and pump water. See Fig. 7-1.

Modern automation attempts to replace not only the skill and dexterity of the human worker, but also that worker's decision-making power. In most factories today, *skilled workers* monitor and adjust the manufacturing process as necessary to make a product that meets standards. Modern automated systems use *sensors* and *computer-based intelligence* to perform the same functions. This permits the unattended operation of production lines within a factory and possibly, the unattended operation of the entire plant.

The preceding chapters have been devoted to the various planning, design, and scheduling steps necessary to achieve a manufacturing system that is truly integrated through the use of computers. From this point, the focus will be on the *hardware* aspects of CIM—the automated machine tools and equipment used to produce goods.

Joe Richardson is a Project Leader for Garrett Auxiliary Power Division, Allied Signal Aerospace Company in Phoenix, Arizona.

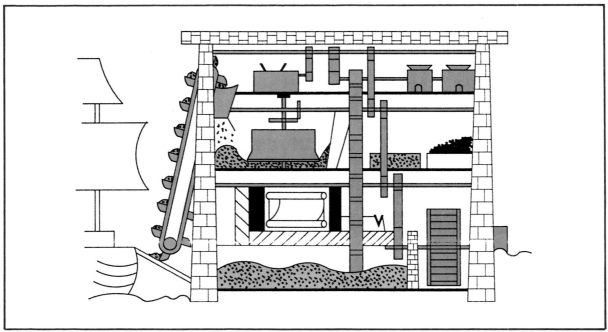

Fig. 7-1. The earliest example of an automated factory in the United States is Oliver Evans' flour mill in Philadelphia, built in 1783. The water-powered mill featured continuous processing. Grain was unloaded from ships by means of a conveyor, then cleaned and ground without need for human intervention. Evans' mill incorporated many of the material handling processes that are common in industry today.

AUTOMATION IN THE FACTORY

While **service industries**, such as banking, insurance, and merchandising have grown significantly in recent decades, that same period has seen the marked decline of wealth-producing industries in the United States. Service industries merely transfer wealth among members of the economy. Wealth-producing enterprises, such as mining, manufacturing, and agriculture, on the other hand, *add* to the wealth of the country by increasing the supply of goods available for consumption by its citizens and for trade with other nations.

The evolution in manufacturing has paralleled the changing of agriculture in the United States during the last century. At the birth of the United States, almost every person in the nation was engaged primarily in agriculture. Now, less than 3% of the total American workforce produces 120% of domestic needs (the excess accounts for a major part of U. S. exports). Manufacturing accounted for 67% of the U.S. workforce in 1950; today, that portion is less than 12% and decreasing rapidly. As automation technology is implemented in many factory operations, output per worker (**productivity**) is rising significantly, while costs and quality exceed the levels needed to compete effectively with foreign companies.

Changing market demand has given impetus to the adoption of computer-based automation that can reconfigure manufacturing processes quickly and cost-effectively. See Fig. 7-2. Reconfiguration is necessary to quickly respond to market changes that result from shifts of personal taste, as well as changes dictated by safety, environmental, or technological factors. Factories that have computer-based systems and trained personnel in place to react quickly to customer demands will grow and prosper in this new market-driven environment; those that do not, may not survive.

Fig. 7-2. Flexible machining systems, such as this one, can be rapidly reconfigured to adapt production to changing market demands. Computer-based automation makes such quick changes possible.

REASONS FOR AUTOMATION

Automation technology has evolved over the years in response to market forces and profit opportunities. During the 1980s, modern manufacturing operations in the United States changed markedly. With the rise of manufacturing prowess in Japan, Korea, Taiwan, and other Pacific rim countries, as well as continuing competitive pressure from Europe, American manufacturers find they must operate in a global market.

To maintain a meaningful presence in this world economy, U.S. companies must seek ways to maximize productivity and quality while lowering unit costs. Automation and computer technology are the primary viable means for companies to improve factory productivity and remain competitive. These tools also will make it possible to achieve the levels of return on investment required to maintain and advance the capability and capacity of United States manufacturing.

Automation provides a number of benefits to manufacturers who are seeking to reduce unit costs and thereby increase profits. The primary benefits of automation are reduced labor costs, sales growth, better quality, reduced inventory, and increased worker productivity.

Reduced labor costs

Replacing people with machines reduces the cost of direct labor attributed to each product, since automated systems are generally more productive than the individual machines and people that they replace. There are also savings in floor space, utilities, supplies, and the other costs involved in production.

Sales growth

With the reductions in labor costs and the associated savings arising from improved efficiency and productivity, product prices can be reduced to gain an increased market share.

Better quality

By reducing or eliminating defects caused by human error, the resulting product is more uniform and consistent, and thus is perceived to be of better quality. A better quality product at lower cost will increase product demand.

Reduced inventory

Funds tied up in materials and labor for work-in-process (unfinished products) represent a significant investment for many manufacturing operations. Automated manufacturing systems move material through the production process in a smooth, efficient flow that virtually eliminates the conventional storage and waiting between processes. Automated systems convert raw material into finished products in a shorter period of time than conventional methods, reducing the need to tie up capital in high levels of work-in-process inventory. This permits available capital to be more advantageously employed in acquiring new equipment, increasing advertising, supporting additional research and development, or taking advantage of the many other options that would enhance operations and improve long-term profitability.

Increased worker productivity

Automated systems enable one worker to produce many times what that person could produce using conventional methods. Automation can be applied to hazardous, monotonous jobs. This can result in better working conditions and a job that is more challenging and often, higher-paying. Automation can provide lower-cost, better-quality products for everyone and is largely responsible for the continually rising standard of living in developed nations.

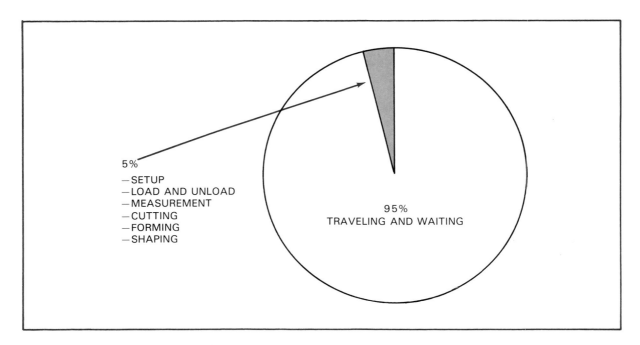

Fig. 7-3. The results of a study of batch-type metalworking shops showed that 95% of the total time parts were in the manufacturing cycle was spent traveling between operations or waiting to be worked on. The 5% of time spent in a manufacturing process was about evenly divided between set-up or related functions, and actual cutting, shaping or forming activities.

ORGANIZING FOR AUTOMATION

Automation is usually associated with high-volume *continuous-flow production* of standard products. However, aside from automobiles, home appliances, and a small number of similar consumer products, most manufacturing is done in small lots. This is usually referred to as *batch production*. Airplanes, locomotives, construction equipment, and much of the hardware we see and use in daily life are manufactured in small batches, not in high-volume production.

To achieve the *economies of scale* provided by continuous-flow production, the manufacture of products in batch lots must be carefully organized. If the manufacturing plan merely schedules the shop work in random sequence as orders are received, a great deal of time will be spent unproductively on tooling, machine setup, and waiting during in-process travel.

A study of batch-type metalworking shops found that, for a given part, only about 5% of total production time is actually spent on the machine tools. The other 95% of time in the shop is spent traveling between departments and waiting to be loaded on a machine. Further, the study found that only about *half* of the time on a machine is actual *productive* time spent cutting or forming metal. This relationship is shown in Fig. 7-3. These findings indicate that reorganizing the production process to reduce nonproductive work-in-process time would be a more effective means of achieving manufacturing cost reduction than would developing more powerful machine tools and better cutters.

GROUP TECHNOLOGY

One technique used to improve shop productivity is called *Group Technology* (GT). By using GT disciplines, the chaotic batch production shop can be organized to approach the efficiencies of continuous-flow production.

GT provides a structured analysis of *what* a shop makes and *how* it makes it. By classification and coding of parts, GT can identify those having similar shapes, features, and manufacturing processes. This greater visibility of shop operations provides an opportunity to improve processing throughput through the use of two strategies. One of these is to combine like parts and processes; the other is to schedule similar parts for sequential processing to take advantage of their similarity of tooling and operations. The GT approach can also have a significant impact on product design. It encourages design standardization and reduces the proliferation of design variations that raise tooling and machining costs (and thus adds unnecessarily to product cost).

Parts coding

Group Technology uses an alphanumeric *coding system* that captures the significant characteristics of a machined part. Part codes can be compared to form groups of similar parts or part families for planning and manufacturing operations. The underlying concept of this approach is that parts with similar codes should be manufactured using similar methods and have manufacturing costs that are equivalent. This analysis can also form the basis for a *multi-use tooling* design. These are tools that, with only slight modification, can be used for machining or fabrication of a number of similar parts. This eliminates the cost of designing and building a tool for each different part. This multi-use approach can significantly reduce the cost of fixtures, which are expensive to design, build, and maintain.

GT can also assist in shop scheduling by processing similar parts sequentially through a workstation, Fig. 7-4. When only a few tools or only a few machine settings need to be changed to process the next job, this can considerably reduce the cost and time for changeover or set-up for the new part. In batch production, set-up time can be as high as 50% of available machine time; reducing such nonproductive time can generate significant savings in manufacturing cost.

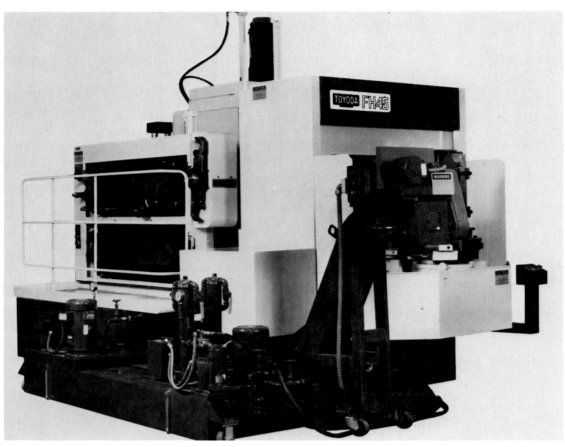

Fig. 7-4. Horizontal machining centers like this one are ideal for use in a GT manufacturing environment, since tooling and set-up can be changed rapidly to process members of a parts family. The machine has a 32-tool random access toolchanger, and a direct path twin-pallet shuttle that can interface with AGVs or other material handling equipment.

Specialization

Once part families are identified by GT analysis, the shop can be organized into workcenters, with each dedicated to processing a certain type of part. This can add the efficiencies of specialization to batch production. Tools and machines can be optimized to process a narrower range of part types; the flow from workstation to workstation can rival the efficiency of continuous-flow production.

As a result of the continuing development of computer technology, artificial intelligence and expert systems, sensor technology, and information network technology, the automated factory is becoming more common. Workcenters, or *cells*, are integrated into manufacturing subsystems. Through the factory computer network, these subsystems can be served by automated material handling systems. The fully automated factory of the future will be composed of *intelligent subsystems* that will process work as directed by the central factory computer, Fig. 7-5. The factory communication network will coordinate factory operations so that raw materials will move through required processing operations to emerge as finished goods in batches that will approach the unit-cost-efficiencies of continuous-flow production.

1. Four Milacron T-30 CNC Machining Centers

2. Four tool interchange stations, one per machine, for tool storage chain delivery via computer-controlled cart

3. Three computer-controlled carts, with wire-guided path

4. Cart maintenance station

5. Parts wash station, automatic handling

6. Automatic Workchanger (10 pallets) for online pallet queue

7. One inspection module — horizontal type coordinate measuring machine

8. Three queue stations for tool delivery chains

9. Tool delivery chain load/unload station

10. Four part load/unload stations

11. Pallet/fixture build station

12. Control center, computer room (elevated)

13. Centralized chip/coolant collection/recovery system (----- flume path)

⌒ Cart turnaround station (up to 360° around its own axis)

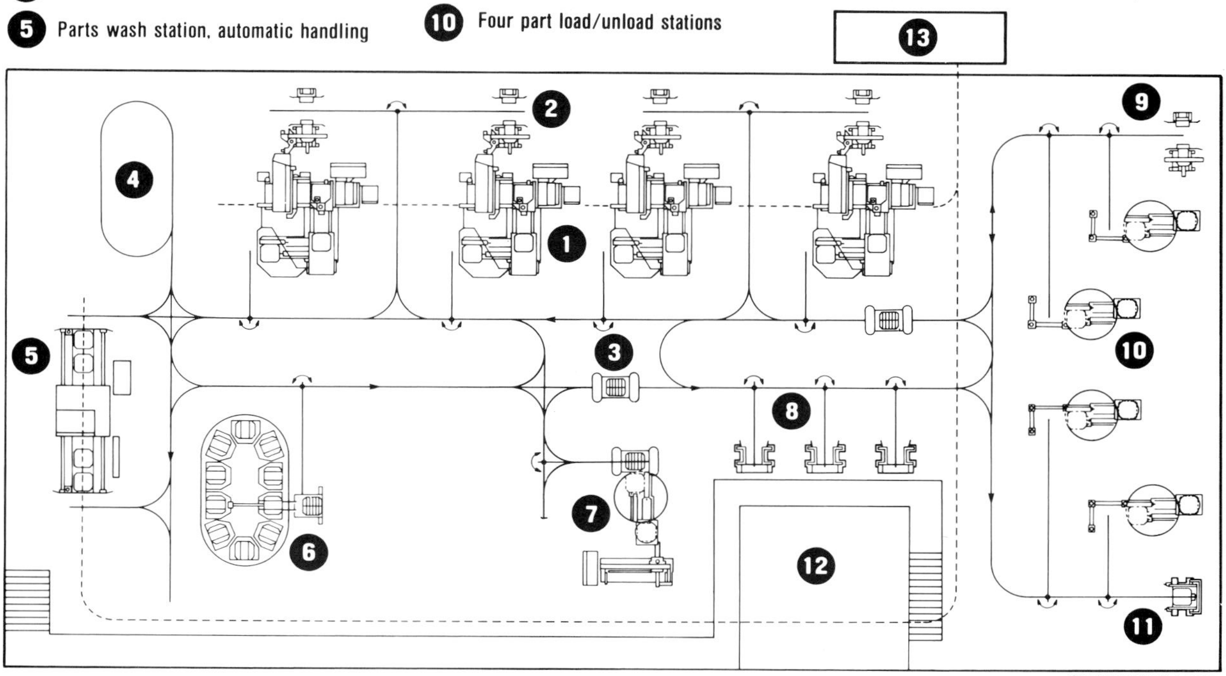

CINCINNATI MILACRON

Fig. 7-5. A Flexible Machining System consisting of four CNC machining centers and four load/unload stations, connected by means of wire-guided, computer-controlled carts. The carts deliver both workpieces and tooling.

SELF-REGULATION

A characteristic that most people associate with automation is self-regulation. *Self-regulation* is the ability to recognize a deviation in the production process and automatically take appropriate action, without relying on human intervention or decisionmaking, Fig. 7-6. The concept is not new: James Watt invented the flyball governor for use on his steam engine more than 200 years ago. That device, shown in Fig. 7-7, automatically regulated the speed of the steam engine to keep it constant as the load varied.

A modern example of a self-regulating system is the thermostat used on a household furnace, Fig. 7-8. This device maintains room temperature by sensing when it falls below the desired level and providing a signal to activate the furnace. When the room temperature again reaches the desired level, the thermostat senses this condition and signals the furnace to shut down. Thus, this simple device exhibits the capabilities of a self-regulating automatic system that operates without human intervention.

Every automated system contains some kind of subsystem that senses when the process under control begins to deviate from desired performance, Fig. 7-9. In a machining system, these subsystems

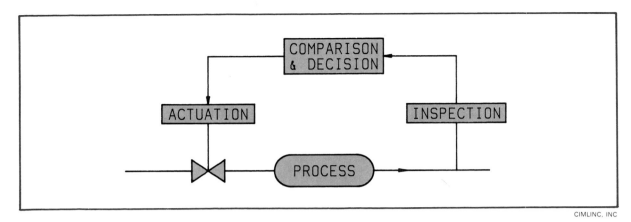

Fig. 7-6. A closed-loop automation system is self-regulating. In a self-regulating system, the process or product is continually monitored (inspected). If information gathered through inspection is outside specifications, a signal is sent to an actuating device to make necessary changes. The inspect/compare/actuate cycle is continuous.

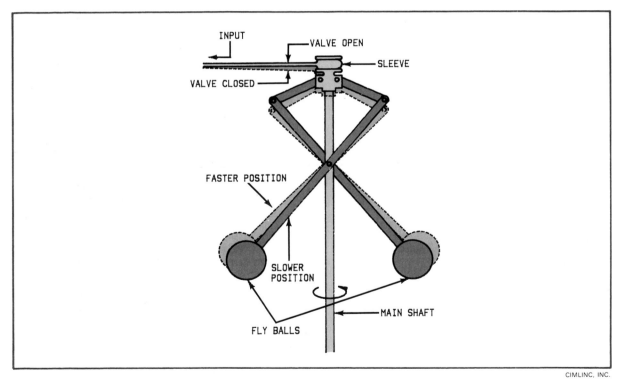

Fig. 7-7. The flyball governor, developed by James Watt in 1788, was one of the earliest automatic process control devices. The governor was connected to the output shaft of Watt's steam engine, and controlled engine speed by opening or closing the throttle valve. When a decreased load caused the engine to start speeding up, centrifugal force moved the flyballs outward, closing the throttle valve and thus slowing the engine. The opposite effect resulted when the engine slowed under load: the balls moved inward, opening the throttle valve to speed up the engine.

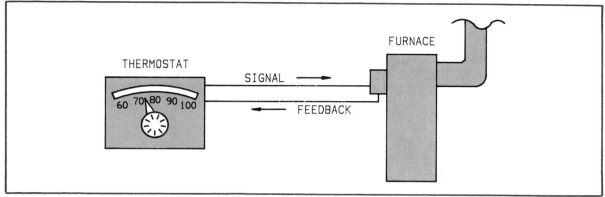

Fig. 7-8. A familiar automated control system is the household thermostat, which switches the furnace on or off to maintain room temperature within the desired range.

Fig. 7-9. Sensors of different types are used to monitor the production process and feedback information. This diameter measuring sensor is designed for in-process measurement in lathes and grinders. This sensor uses contacts on the workpiece, like a traditional in-process grinder gage, but uses a large range digital optical sensor to monitor the location of the contacts. Use of the optical sensor provides a larger range and improved stability.

may include in-process gages that sense when a part dimension is drifting out of tolerance, or temperature sensors to detect overheated spindle bearings. When sensors detect a problem, they send a signal to the machine logic unit, which decides on the appropriate remedial action. In the case of wear on a cutting tool, the appropriate action may be to automatically adjust that tool or to select a new tool. For such problems as a broken tool or overheated bearing, the system may stop the process and signal for operator assistance. The development of microprocessors and artificial intelligence, has allowed modern systems to make complex decisions, anticipate potential problems, and virtually eliminate unplanned system shutdown. See Fig. 7-10.

All modern automated systems are self-regulating and include sensing and decisionmaking elements to make possible continuous unattended operation.

COMPUTER IDENTICS

Fig. 7-10. In-process gages, sensors, and other devices are used to automate and regulate many processes. Barcode readers, for example, have many applications. They may read and verify lot number tags, product codes of finished goods, or (as shown here) component identification labels.

CHANGING TYPES OF AUTOMATION

Automation is usually associated with high-volume manufacturing of a standardized product. This approach was appropriate in earlier days when market demand was for simpler, essentially low-cost products. With changing consumer demands, market segmentation, and increased competition, however, manufacturing requirements have changed. Now, the emphasis is on a greater variety of products, smaller lot quantities, and shorter production runs. This is the challenge facing manufacturing management today.

To meet this challenge, modern manufacturing systems have become more flexible and able to change quickly to meet market demands. Rather than the cams, mechanical timers, and large relay panels of previous automated systems, control of current automated production facilities is done by using software and expert systems. Changeover to produce a different product can be accomplished in a matter of minutes by loading new software from a central factory computer, Fig. 7-11. With the previous mechanically controlled systems, changeover to produce a new part design could take weeks or months. The increasing power of computers and the decline in the cost of computing power changed the rules of manufacturing; these factors will continue to change the concepts and approaches involved in implementing automated production systems.

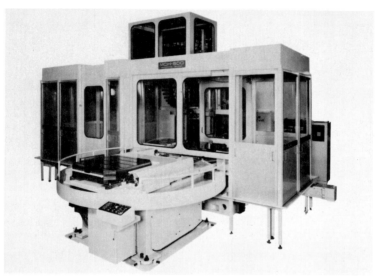

Fig. 7-11. Modern machine tools like this CNC horizontal machining center are highly flexible and able to quickly change over from making one part to manufacturing another.

FIXED VS. FLEXIBLE SYSTEMS

Fixed automation can be defined as a collection of machines and tooling designed to achieve high-volume production of a single part at low unit cost. *Transfer lines*, Fig. 7-12, represent the highest form of fixed, or hard, automation. To take advantage of all the specialized tooling and techniques available to minimize production time for a part, the machines and tooling are designed around the configuration and characteristics of that part. If the part design is changed or a new part is to be produced, the whole line must be redesigned and rebuilt. Usually, very little of the old tooling is directly usable for the new part. With fixed automation, changeover to produce a new part involves significant capital expenditure and lead time.

Flexible manufacturing systems

A *flexible manufacturing system* (FMS), Fig. 7-13, is designed to produce a family of similar parts. Machinery and tooling provide for adjustments and modifications to accommodate size and configuration differences of the parts in the family. Changeovers to produce different parts may be partially or fully automated. This quick-change capability adds to the initial cost of the flexible system,

Fig. 7-12. Automatic welding equipment is used on this transfer line producing catalytic converters for automobiles.

but can be readily transformed into profits when changes are needed in the product model or volume to meet market demand. When such changes must be made, the flexible manufacturing system is able to respond quickly. This allows a manufacturer to take advantage of customer preferences, market swings, and the inability of competitors to meet the changing market conditions.

Smart tooling

A flexible manufacturing system relies on software and smart tooling for part processing, rather than the hard-wired controls and hard tooling used on the high-production transfer line. *Smart tooling* involves cutting and part-holding tools that readily reconfigure themselves to accommodate a variety of shapes and sizes within a given part family. This eliminates the costly changeover of fixed tooling systems and makes possible the quick response and small economic lot sizes that are the basis of the newer just-in-time, low-cost manufacturing methods. With the trend toward more product variety and shorter product life cycles, production runs of parts are much smaller. Also, the need to meet changing market demands has become a critical factor in the design of new manufacturing facilities and the overhaul and renovation of older ones.

JIT manufacturing

The just-in-time (JIT) philosophy being implemented in automotive and other industries requires small quantities of a variety of parts on a continuing basis. This approach imposes a new operating environment on manufacturers who must supply parts to meet this demand. Part lot sizes are smaller; production runs are shorter; changeover is more frequent. This new operating environment is accelerating the shift from fixed automation to more flexible systems.

For example, one farm equipment manufacturer uses an FMS to sequentially produce all the parts needed for a gearbox. Instead of producing a large lot of parts for each of the pieces required in the assembly and sending them to storage for later withdrawal and assembly, a flexible machining

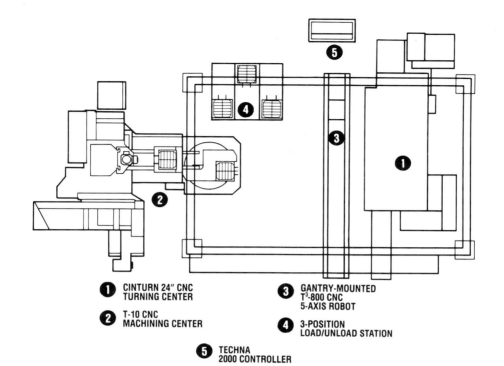

1 CINTURN 24″ CNC
TURNING CENTER

2 T-10 CNC
MACHINING CENTER

3 GANTRY-MOUNTED
T³-800 CNC
5-AXIS ROBOT

4 3-POSITION
LOAD/UNLOAD STATION

5 TECHNA
2000 CONTROLLER

CINCINNATI MILACRON

Fig. 7-13. Flexible automation, represented by this manufacturing cell, uses computers and program-mable equipment to readily change the production process when the part configuration changes. This cell includes a CNC machining center and a CNC turning center, both served by a 5-axis gantry-mounted robot.

Automation Technology   189

system produces *one* of each part required. The matched set of parts is then sent to assembly. This eliminates the need to tie up company funds in the material and labor costs contained in the completed parts awaiting use, as well as inventory storage costs. It also eliminates the potential costs of scrapping or reworking obsolete parts in inventory if the design changes or market projections do not materialize.

CURRENT FACTORY TECHNOLOGY

The most advanced manufacturing firms today apply computer technology to every aspect of operations, from market projection and product planning, through production and testing, to service and aftermarket support. A *computer network* links terminals and mini/micro computers in each function and location in the factory, integrating the inflowing data and providing detailed and timely information to speed the flow of products through each production operation. This *information control* makes it possible to produce and deliver the appropriate product at the right time to satisfy a rapidly changing market demand.

Some of these computer-directed and computer-monitored technologies include:
- factory-wide communication networks.
- automated material handling.
- computer-controlled cells and workstations.
- in-process gaging and statistical process control.
- automated test and inspection.
- automated marking and identification.
- automated storage and retrieval systems.

In the flexible factory, each of these systems is connected to the factory information network so that they function synergistically to produce both a quality product and a financial profit.

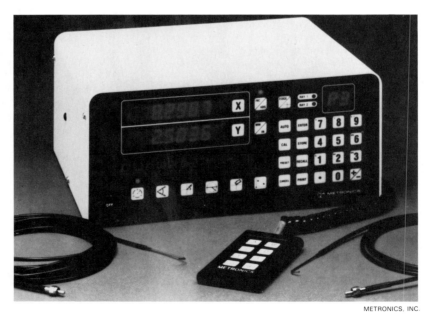

METRONICS, INC.

Fig. 7-14. Data collection equipment, such as this two-axis geometric readout unit, can be connected to a central computer network to provide real-time input. This unit can be used with two-axis coordinate measuring systems, and has edge-sensing capabilities that allow it to be used with optical comparators.

COMPUTER NETWORKS

By distributing production and scheduling information, and receiving status data from each department and function in the factory, the computer network provides control and feedback to monitor production operations. This monitoring recognizes potential or actual problems in near-real time, and issues warnings so that appropriate corrective actions can be initiated to prevent delays or unplanned interruptions of manufacturing operations. *Real time* refers to events happening now, at this instant, in contrast to *historical time* (events that happened hours or days ago). Real time can be compared to a "live" TV news broadcast, in comparison to coverage of an event on the evening or next morning's newscast. The central computer is kept aware of the status of each manufacturing operation by inputs from each workstation and data collection point, Fig. 7-14. These inputs come primarily from *direct input* devices, such as sensors or code readers, that provide real-time instantaneous status data. Sometimes, manually assisted input devices may be used as well.

The real-time data-gathering capability of computer systems provides the ability to minimize or eliminate defective or damaged products by detecting and correcting deviations in the production process. This real-time control contrasts markedly with previous reporting systems that required days (or even longer) to recognize and correct manufacturing deviations. This delay was due primarily to the use of manual data collection methods and batch processing computer procedures. An important result of the development of computer networks and real-time data gathering and processing technology has been to enhance the knowledge and decisionmaking ability of the persons responsible for the factory's productivity and quality goals.

DATA COLLECTION AND REPORTING

Automated data collection and monitoring begin at the start of the production process, with the receipt of raw materials, parts, and supplies. *Machine-readable coding* and automated scales and sensors allow the type, quantity, and quality of incoming materials to be verified and logged into the computer-based inventory control system. The materials are transported by conveyors or guided vehicles as part of an automated storage and retrieval system. See Fig. 7-15. In the automated factory, the ASRS is the primary means of controlling the flow and status of products. With such a system, parts and materials can be delivered just-in-time (JIT) to the fabrication and assembly operations as required to maintain a smooth, efficient flow of products through the factory. This continuous flow from material receiving to shipping of finished products maximizes productivity by reducing levels of work-in-process (WIP) and increasing the rate of return on the capital invested in the enterprise.

AUTOMATED MATERIAL HANDLING

Automated material handling technology has many forms and applications in the factory. It may take the form of simple conveyors to transport parts and materials between workstations, *automated guided vehicles* (AGVs) that can roam plant-wide, Fig. 7-16, or intelligent robotic devices that can load and unload workstations as well as verifying part quality during transfer between operations.

Automated material handling reduces touch labor and all the costs associated with this element of factory operations. *Touch labor* is the worker time and effort required to transport, position, adjust, reorient, and relocate material and parts as they pass through operations in the factory. This labor adds no value to the product. It merely prepares and positions it for processes that *do* add value, such as cutting, bending, welding, machining, or painting. Reductions in touch labor can lower manufacturing costs significantly, produce more uniform and consistent product, and free workers to apply their time and talents to more productive operations.

GMC

Fig. 7-15. Automated storage and material handling systems are used in manufacturing operations to move materials from storage areas to production equipment. This specially designed conveyor system provides a continuous supply of bore liners to an automated engine assembly operation.

AGV

CINCINNATI MILACRON

Fig. 7-16. Automated guided vehicles are used to move pallets containing fixtures, parts, and tooling from one operation to another.

CELLS AND WORKCENTERS

The batch production factory is organized into product-oriented departments. Each department contains a mixture of machining cells and individual workcenters connected by appropriate types of material handling systems and integrated into the overall factory operations by a computer network.

The product-oriented department is organized to produce a specific type of product or family of parts. This orientation maximizes the efficiency and quality of the production process by permitting the use of specialized machinery, tooling, and skills. The *machining cell* consists of two or more machines or workcenters that are served by a local material handling system and controlled by a *dedicated computer* (a computer assigned strictly to this task). This arrangement provides for the use of a number of devices that extend the span of control of the individual worker and enable that worker to increase output many times over conventional methods. See Fig. 7-17.

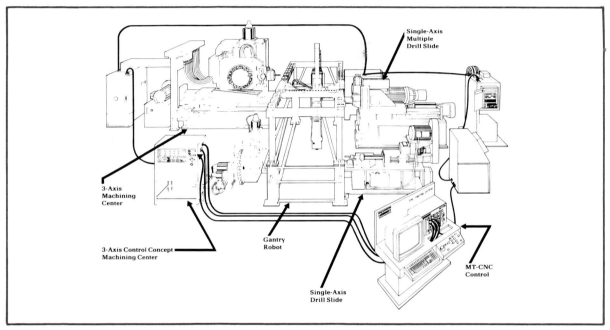

INDRAMAT

Fig. 7-17. A machining cell consists of a number of individual machines, linked by some form of material handling device, and controlled by a computer. The cell computer is linked with the factory computer network. It receives information on part and processing priorities from the factory computer, and feeds back to it performance and machine status data.

SENSORS AND CONTROLS

Monitoring and control devices consist of sensors and similar mechanisms that extend the eyes, ears, and muscle power of the skilled worker. As an example, the combination of a bar code reader, a digital scale, and a machine vision system can be used to verify the type, quantity, and condition of parts or materials entering a workstation or cell from an automated material handling system. A pallet shuttle device or robot, the part-handling system used within the cell, loads the material into the machining or fabrication zone for processing. During processing, sensors monitor progress to ensure that process variables are maintained within pre-determined limits.

These sensors include thermal, acoustical, optical, and other physical state monitors that measure the status of equipment and components and of the workpiece while it is being shaped and formed. Equipment monitoring includes machine spindle overheating, lubrication failure, machine overloading, and similar malfunctions. Such problems could be detected and prevented by a skilled operator, but now are automatically sensed and corrected by the computer-based monitoring system.

Sensing devices also are used to prevent damage to the workpiece or to detect (and often correct) deviations from allowable process tolerances. Sensing devices are used for location and clamping verification, broken tool detection, cutting force sensing, and in-process dimensional verification. These sensing systems may either take corrective action automatically, as defined by the program control logic, or send a signal to a central monitoring area to request assistance of the system operator.

The system adjusts machine speeds and feeds as it encounters oversize material, changes tools as dull tools are detected, or stops operations and signals for assistance if the problem is one that can not be compensated for automatically. Resistance and fusion welding operations have traditionally required skilled manual operators. With the emergence of machine vision systems and sophisticated electronic controls, however, many production welding operations have been partially or fully automated to improve the quality of this common joining technology.

Sensors in assembly automation

Automation has improved the quality and lowered the cost of assembly operations. Automatic character and bar code readers are used to verify that the correct part numbers and models are present. Laser, photodiode, and vision systems verify that material thickness, orientation, and placement are proper before fastening or joining components. Screw and nut torque sensors verify that fasteners are tightened to specified tolerances; weld current controllers verify that welded joints are secure and sound. Ink jet and laser markers code-mark finished assemblies so that documentation is provided for field service diagnosis and traceability. Package coding of food and beverages is universal. For example, a 12-character packaging code on the bottom of beverage cans can be applied by an automated ink-jet marker at 200 cans per minute.

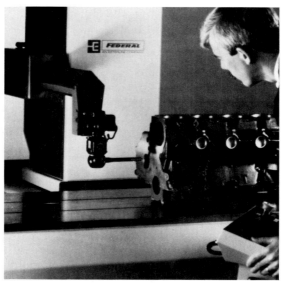

FEDERAL PRODUCTS

Fig. 7-18. This cantilever-style coordinate measuring machine can be operated manually, as shown here, or under direct computer control. It measures dimensions in the X, Y, or Z axes.

AUTOMATED INSPECTION AND TESTING

Testing and inspection is a major area of automation in the modern factory. Although emphasis is being placed on **in-process inspection** (the process of verifying that parts are made to specifications at each operation), computer-directed testing of the final product is usually necessary to assure that the manufacturing process is producing a product that meets advertised specifications. Computer-controlled dedicated measurement and test stations are interspersed among the manufacturing operations to inspect and test major parts and sub-assemblies. Computer-controlled **coordinate measuring machines** (CMM) are a major development in the technology of precision measurement, Fig. 7-18. These machines permit the fast and accurate measurement of machined parts, with data recorded automatically for analysis and reporting. Measuring is done in a fraction of the time previously required by conventional surface plate and height gage methods.

THE PAPERLESS FACTORY

For rapid response and close control, the flexible factory requires quick and accurate provision of production data. Paper or punched-card data collection methods and batch processing data analysis are not appropriate to the decisionmaking and control needs of the modern factory. Machine readability and direct data input are required for the computer to analyze, process, and rapidly respond to changes in production conditions. Newer technologies, such as bar coding, machine vision, and voice input provide direct input of data for recording and analysis, Fig. 7-19.

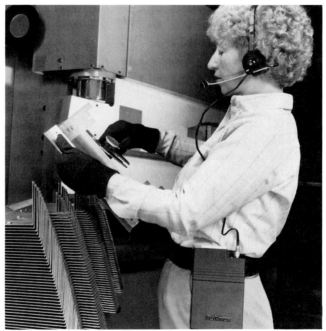

VOCOLLECT, INC.

Fig. 7-19. Voice data collection leaves an operator's hands free to perform inspection or production tasks. Some systems are connected directly to a computer for real-time data acquisition. Other systems, such as the one shown here, store the collected data in memory. At the end of a shift (or other period) the information is uploaded to the computer through an optical communications link. This system can be adapted for real-time data collection with an optional radio transmitter.

Each of these technologies illustrates a different level of analytical sophistication required to make data meaningful for factory control. The bar code input merely records the presence of an identified object, its location, and probably the date and time. This is stored for later retrieval and reporting, as required. The **machine vision** input records the shape, size, and position of some object, compares it to a stored image, and provides an accept/reject decision based on established tolerances. For decisions requiring judgment and experience beyond simple control logic, the **voice input** system allows a human operator to observe and report product data in real time for recording and response by the factory control system. Voice input technology allows a worker to communicate with the computer, using a limited and specific vocabulary, to report status of observations of a production process. This technology is a major step toward the **paperless factory**, in which production information is available from computer-generated video display and hardcopy terminals, rather than on reams of historical computer printout.

ROBOTS

No discussion of automation would be complete without mention of the industrial **robot**, the principal component in many concepts of the factory of the future. **Robotics** is becoming more common in factories as engineers become more knowledgeable about its capabilities and its applications.

The automotive industry has been the leader in the use of robots for welding, painting, and assembly work, Fig. 7-20. These "steel collar" workers have many advantages over their "blue collar" counterparts when performing monotonous, tedious, or dangerous tasks. They perform at top efficiency,

DIAMOND-STAR MOTORS

Fig. 7-20. Painting robots at an automotive assembly plant can quickly and evenly apply protective finishes to car bodies.

whether it is the beginning of the shift or late in the day. They don't get tired, take breaks, call in sick, or go on strike. Robots can function effectively in hot, noisy, dusty, and hazardous environments that would be difficult or impossible for human workers, Fig. 7-21. The use of robots frees humans to perform the supervisory and monitoring functions that require their intellectual and decisionmaking capabilities.

In a true sense, robots cannot "think" or use "judgment." However, with advances in sensor technology (especially in the tactile, acoustic, and machine vision areas) and the parallel development in *artificial intelligence* and *expert systems*, robots are acquiring a rudimentary decisionmaking capability. These advances provide major new fields of technical specialization, and provide job opportunities for new knowledge workers who are needed in increasing numbers to program, monitor, and maintain these machines of modern manufacturing.

GMC

Fig. 7-21. Industrial robots are ideal for performing repetitive, difficult, or dangerous tasks. They can be fitted with special end-effectors (grippers) to handle odd-shaped parts or assemblies. This robot is used to load parts on a machine at an automobile engine plant.

CAREERS IN AUTOMATION

The automation engineer or technician of today has dramatically different skills and responsibilities from his or her counterpart of twenty years ago. The current engineer has skills in the areas of microprocessor programming, sensor technology, artificial intelligence, and computer networking. The operating and control functions performed by cams, relays, and switches of older fixed automation systems have been replaced by software and computers.

With the refinement of electric motors and solid state controls, the role of hydraulics in machine movement has been taken over, to a great extent, by electrical drives. These drives eliminate the problems of heat, noise, and leaks encountered with hydraulic systems. Modern automation systems are more flexible, more responsive, and more intelligent than previous systems. The introduction of just-in-time production, coupled with demands for defect-free products of greater variety and complexity, have dramatically advanced the sophistication of automated production systems in recent years.

The implications for the new automation engineer are clear. He or she must have a broad range of skills and an in-depth knowledge in the newer areas of computer technology. This person must have some facility in computer programming, both in machine-level coding, and in some high-level language, such as BASIC, FORTRAN, or PASCAL. The automation engineer needs a good fundamental knowledge of mathematics, physics, and chemistry to understand the physical phenomena associated with the materials that will be fabricated and processed by an automated system. These include the various plastics and composites that are rapidly replacing the traditional metals in many new products.

The role of the maintenance engineer also will become more critical as automated systems become more capable and complex. Even with artificial intelligence techniques to aid in system diagnostis and repair, the requirement for resourceful and creative engineers to develop and implement computer-based maintenance and repair systems will be essential.

Increasing automation in manufacturing will eliminate many routine, low-skill jobs. However, it will also create more challenging, highly skilled, better paying jobs for those who have acquired the skills and knowledge needed to implement, maintain, and advance automation. See Fig. 7-22.

SI HANDLING SYSTEMS, INC.

Fig. 7-22. Maintenance and repair of automated systems requires skilled personnel. These repairers are performing maintenance on equipment used in a track-type material handling system.

SUMMARY

Automation can simply be defined as a form of work being performed by a machine or device versus that same work being performed by the manual labor of a human being. Some of the advantages of automation are reduced part costs, improved product quality, reduced inventory in the factory, and better process consistency. Better consistency in the process used to manufacture the product will generally increase customer satisfaction of the product.

Transfer lines represent the highest level of fixed automation. They are designed for high-volume part production with very low unit labor cost. Capital investment is usually rather high and is very inflexible as a result of set-up changes. A transfer line is usually set up for one product design, such as a specific automobile model, and is totally disassembled and restructured for another product.

A *flexible manufacturing system* is designed for lower-volume production and for rapid change from one product to another. The unit labor cost on a flexible automation line is usually higher than on a transfer line, but the cost to change from one product to another is lower. This type of line can readily adapt to produce a variety of similar parts. To do this, it relies on software and smart modular tooling concepts to eliminate set-up times as the line changes from one part to another. Flexible lines also fit into the Just-In-Time (JIT) philosophy, since sets for an assembly can be produced rather than a complete lot of each individual part. This philosophy also can have a major impact on inventory reduction.

Group technology (GT) is usually used in a CIM factory in conjunction with flexible manufacturing systems or cells. GT uses classification and coding to identify parts that, for manufacturing processes, have like shapes or features. With this knowledge, similar tooling and part processing routings can be used to create a rapid setup on the line. Also, group technology software can be used to create standard design features on similar parts.

The use of *robotics* with flexible automation has become a very popular strategy in modern factories. The automotive industry has become the leader in the application of robots to production and assembly. Robots function well in hot, noisy, dusty, and hazardous environments. This frees human operators from undesirable working conditions and releases them for supervisory or monitoring functions. Robots can be coupled with artificial intelligence or expert systems, evolving into very complex decision-making automated systems.

IMPORTANT TERMS

artificial intelligence
automated guided vehicles
automation
batch production
cells
coding system
computer network
continuous-flow production
coordinate measuring machines
dedicated computer
direct input
economies of scale
expert systems
fixed automation
flexible manufacturing system
Group Technology
information control

in-process inspection
intelligent subsystems
machine vision
machine-readable coding
machining cell
multi-use tooling
paperless factory
productivity
real time
robot
robotics
self-regulation
service industries
smart tooling
touch labor
transfer lines
voice input

QUESTIONS FOR REVIEW AND DISCUSSION

1. In recent decades, the United States has seen great growth in the service sector, and a severe decline in the production (sometimes called wealth-producing) sector. In terms of impact upon the nation's economy, what is the major distinction between these two sectors?

2. The primary favorable outcomes of automation are considered to be reduced labor costs, sales growth, better quality, inventory reduction, and increased worker productivity. Briefly discuss how each of these items benefits the company that automates its operations.

3. What is Group Technology? What effect can it have on the design of products? How can its implementation affect the physical organization of the shop?

4. Contrast fixed and flexible automation. What are the advantages and disadvantages of each?

5. Computers play an ever-increasing role in factory automation. List some of the computer-directed or computer-monitored technologies in use today.

6. Define *touch labor* and explain how the use of automated material handling to reduce touch labor will benefit a company.

7. Describe the skills and knowledge an automation engineer must have today.

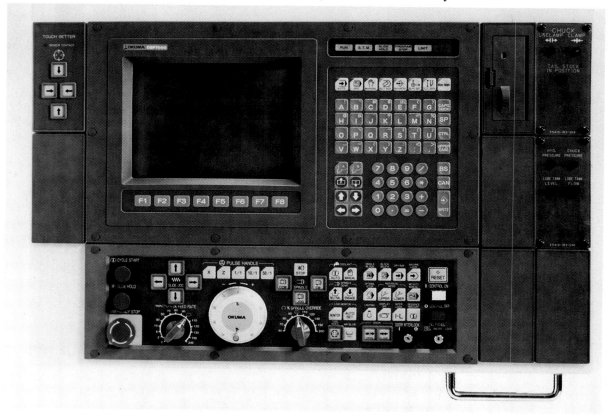

OKUMA MACHINERY, INC.

CNC machine controllers such as this one are capable of storing and running part programs in a stand-alone configuration, or functioning as part of a DNC computer network. Programs can be written or modified using the alphanumeric keyboard.

Computer-Assisted Manufacturing Technology

by Russell Biekert

Key Concepts

- ❏ Numerical control is an automated manufacturing process in which a machine tool follows an ordered sequence of operations under the control of a set of coded alphanumeric instructions.
- ❏ Standardized axis designations for the numerical control of machine tools are based on the Cartesian coordinate system.
- ❏ Absolute positioning systems employ a fixed reference point; incremental positioning systems use successive set points for reference.
- ❏ Machine tool NC instructions may be encoded on a punched paper tape, recalled from memory of a stand-alone computer, or transmitted to a machine controller through a computer network.
- ❏ Tool changers allow a machining center to perform a sequence of operations upon a single part.

Overview

Computer-assisted manufacturing (also called computer-aided manufacturing) is usually referred to by its acronym, *CAM.* In general, CAM refers to the use of computers to assist in any or all phases of manufacturing. The most widely used form of CAM is *numerical control (NC).* Numerical control concepts used in manufacturing will be described in detail in this chapter.

Russell Biekert has held a number of key positions in the manufacturing and engineering areas for Garrett Engine Division, Allied Signal Aerospace Company in Phoenix, Arizona.

NUMERICAL CONTROL

There are various definitions of numerical control, but the most basic way of describing NC is to state that it is a *method of control* that involves the use of *coded, numerical instructions* to *automatically direct the operation and movements* of one or more *machine tools*. Numerical control can also be thought of as an automated manufacturing process in which a machine tool follows an ordered sequence of operations under the control of a set of coded alphanumeric instructions. The sequence of operations would involve movement of the tool along a predetermined path, at specific feed and speed rates, to achieve required dimensions.

THE EVOLUTION OF NC

Machine tools were first developed during the Industrial Revolution, and have been evolving ever since. During that time, humans have also continued to seek ways to control these tools automatically. Eli Whitney's introduction of jigs and fixtures as a means of automating some aspects of gun manufacture was the beginning of mass production. The development of hydraulic, pneumatic, and eventually, electric control devices led to increasingly higher levels of machine tool control through the 19th and early 20th centuries.

In the late 1940s, the aircraft industry realized that precision machining of structural parts would be vital to achieving the greatest possible strength-to-weight ratio for materials used in supersonic aircraft. At the same time, technological advances in the aircraft industry were creating severe problems with the then-used manufacturing methods. As a result, the industry became a leader in research aimed at providing improved methods of automatic control for manufacturing.

The beginning of the concept of numerical control can be traced to John C. Parsons, owner of a company that manufactured templates used to inspect helicopter blades. Parsons conceived the idea of coupling a jig borer with a method of control involving numerical data to achieve the degree of precision machining needed. He took his idea to the United States Air Force, which provided a study grant.

The Massachusetts Institute of Technology received a subcontract, in 1951, to develop and demonstrate a numerically controlled machine tool on the basis of Parsons' concept. The first MIT-developed NC machine tool was demonstrated in 1952, arousing strong interest from aircraft manufacturers. Three years later, the Air Force issued a purchase order for several NC machine tools. Most of these machines were in use by 1957, and numerical control was here to stay. Only five years later, in 1962, NC equipment accounted for 10% of machine tool expenditures; today, virtually every machine tool on the market features numerical control.

ADAPTIVE CONTROL

Although it is a very broad term, *adaptive control* can be thought of as a system for altering (adapting) a process, as a result of changing conditions, to maintain optimum results. One of the best examples of adaptive control is the human being. Through sensory input and reasoning power, humans are able to adapt readily to changes and generally function at optimum levels.

As applied to numerical control systems, adaptive control has three distinct functions: identification, decision, and modification.

Identification typically relies on data from the *sensors,* which is then processed and sent to the decision function.

The *decision* function of an adaptive control system is the intelligence through which adaptation is achieved. In many such systems, the decision function is based on programmed inputs. When provided with values for cutting speed, tool life, and cutter/work deflection, a preprocessor can determine maximum/minimum speeds and feed rates.

The *modification* function is the means of carrying out adaptations resulting from the decision function. Modifications usually consist of corrections output to the NC control system, Fig. 8-1. A typical modification might be one made to increase cutter speed and material feed rates when soft sections of the material are encountered. Most systems include limits to the amount of modification that can be made to the basic NC program. Exceeding the limits will usually trigger some form of operator alert.

Whether simple or complex, an adaptive control system has only one reason for existence: to improve productivity. Statistically, *stock removal* accounts for only 25% of the total machining cycle. Adaptive control has the greatest effect upon this stock removal portion of the machining cycle: it causes the machine to operate at maximum efficiency during stock removal. It also increases cutter life, thus reducing the time spent changing or replacing cutters.

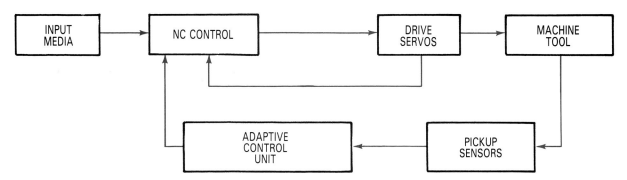

Fig. 8-1. This block diagram for an adaptive control NC system shows the feedback loop between the machine tool and the control.

CNC/DNC SYSTEMS

In recent years, numerical control has evolved into CNC (*Computer Numerical Control*) and DNC (*Distributive Numerical Control*). While both involve the use of a computer to control machine tools, there is one basic distinction: the number of tools managed by each computer.

In CNC, one computer is linked with one machine tool to perform NC functions. The computer, usually with some type of terminal hookup such as a *CRT* (cathode ray tube), monitors the machine tool from stored information, terminal input, or tape input. Some of the objectives of using CNC include:
- increasing machine tool efficiency.
- decreasing tape and reader problems.
- decreasing machine tool downtime.
- reducing per-part cost.
- decreasing the number of reject parts.
- reducing set-up time between part changes.
- reducing operator error.
- optimizing programming efficiency.
- decreasing delivery time to the customer.

In contrast to CNC, DNC involves one or more computers linked to a *number* of NC machine tools and capable of operating them simultaneously or separately. The distributive numerical control system may use a single large computer alone, a large computer and a smaller computer dedicated to handling processing procedures, or several smaller computers linked together. The trend has been toward linking several smaller computers to form a total DNC system.

A DNC system carries out many functions normally performed by individual numerical control units. Some of these functions are:
- eliminating tape and reader units.
- providing CRT terminals to let operators input and output data.
- monitoring machine tool operation.
- providing a library of NC programs and languages.
- providing a source file of NC post-processors
- keeping data processing record files.
- providing a means of verifying tape data.

Some of the real advantages of a DNC system include:
- built-in quality control inspection procedures.
- adaptive control of machine tool operations, so that optimum tolerances, machining rates, and tool life can be maintained.
- the ability to use new computer languages designed for maximum communication between NC programmers, set-up personnel, NC operators, and the computer.
- a decrease in machine downtime that results from tape, reader, or control unit problems.

The topics of Computer Numerical Control and Distributive Numerical Control will be explored in greater detail in Chapter 9.

MACHINE TOOL MOVEMENT

The basic principle of machine tool operation is movement of a tool (usually some form of cutter) across the workpiece to perform stock removal or other functions. The two tool movement systems are *point-to-point* and *continuous path*. In many cases, there is also movement of the workpiece itself during the machining operation. Tool and workpiece movements are accomplished through the use of servomechanisms of various types. *Servomechanisms* are devices that use pneumatic, hydraulic, or electromechanical means to achieve controlled motion. To ensure precision in tool and workpiece movement, machine tools make use of one of several *positioning systems*. These systems identify locations numerically in terms of a specific *reference point*, a vital consideration when writing a program for automatic operation of the machine tool.

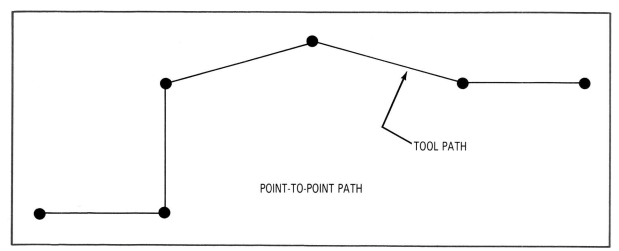

TOOL PATH

POINT-TO-POINT PATH

CIMLINC, INC.

Fig. 8-2. In a point-to-point NC system, all the moves consist of straight-line elements.

NC POINT-TO-POINT SYSTEM

Systems that allow tool movement only in straight-line segments, Fig. 8-2, are called *point-to-point systems*. These systems are designed to generate straight-line movement parallel to an axis, or at an angle of 45 degrees to an axis; no other angles are possible. Point-to-point systems *can* generate a dogleg pattern, which is useful in planning cutter paths to avoid fixtures or raised sections of the part. The dogleg movement is achieved by programming simultaneous moves of unequal distances in two or more axes. Curved contours can be achieved, in a point-to-point system, by programming tool movements as a series of short straight-line segments, as shown in Fig. 8-3.

A primary advantage of the point-to-point system is the lower cost of equipment when using this system (compared to continuous path equipment costs). Point-to-point systems are usually applied to one-, two-, or three-axis movements. The addition of a rotary table may add capabilities to the system through increased flexibility. The rotary table would be programmed as a linear axis converted into degrees, or as a "miscellaneous" function.

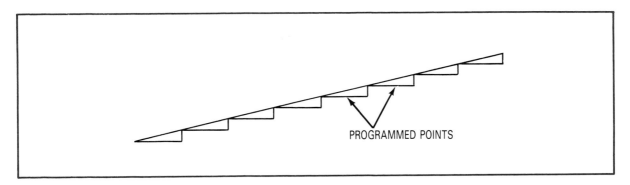

PROGRAMMED POINTS

Fig. 8-3. *Short stair-step moves between programmed points generate an angle on a point-to-point NC system.*

NC CONTINUOUS-PATH SYSTEM

If a system can generate angular distances or arcs, as well as straight-line segments, it is referred to as a continuous-path or *contouring* system. Depending upon the type of servomechanism being used, a continuous-path system may generate a true arc or an arc constructed with many short straight-line segments. A stepping motor used as a servomechanism will produce an arc using straight-line segments; a hydraulic servo or a DC variable-speed motor will construct a true contour or arc. Fig. 8-4 compares arcs generated by the two methods. Either method is capable of a high degree of accuracy, and will produce comparable finished contours.

Continuous path systems are usually applied to two-, three-, four-, or five-axis systems. See Fig. 8-5. When doing sophisticated programming for a multiple-axis system, the use of a computer is usually necessary, as well as economical.

SERVOMECHANISMS

Controlled motion along or around an axis is achieved through the use of powered devices known as *servomechanisms* or *servos*. Basically, there are three types of servomechanisms applied to NC systems: stepping motor drives, DC variable speed drives, and hydraulic servo drives.

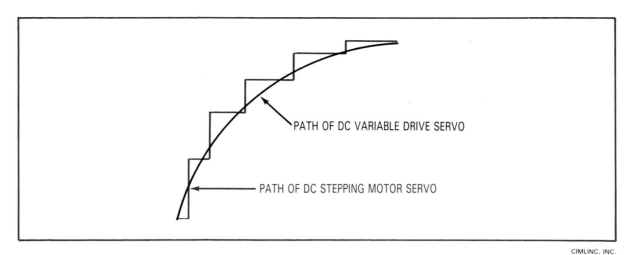

Fig. 8-4. A continuous path NC system can generate an arc with either a true contour (using a DC variable drive servo) or a stair-step series of straight-line segments (using a DC stepping motor servo).

Fig. 8-5. This horizontal boring mill has a two-axis CNC universal milling head that allows five-axis simultaneous machining capability. It can handle a workpiece of up to 55,000 pounds on its 78'' by 98'' rotary table.

Turning machines

Modern NC systems have overcome early problems with control of turning machines. Controls are now built to maintain a constant surface feet per minute, with rotational (rpm) rate changing to correspond to the diameter of the workpiece. NC controls permit programming of either point-to-point or contouring operations. Turning machines used for industrial production, Fig. 8-13, include vertical and horizontal chuckers, vertical and horizontal turret lathes, double-end and double-spindle lathes, and two- and four-axis conventional lathes.

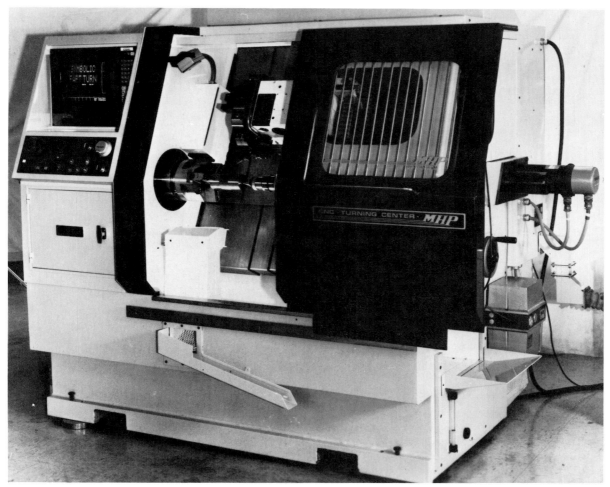

MHP MACHINES, INC.

Fig. 8-13. This precision slant bed turning center has a rotary spindle capable of speeds ranging from 50 to 4500 rpm.

Boring machines

NC boring machines offer a high degree of accuracy, combined with high positioning speed. These machine tools are available in various forms: single- and multiple-head boring machines, horizontal or vertical boring mills, and jig boring machines with digital readout devices.

Grinding machines

Cylindrical, surface, and special (such as contour) types of grinding machines are often numerically controlled, Fig. 8-14. The most prominent application of NC grinding has been in the steel industry, where the large rolls used in hot- and cold- rolling mills are ground to specified smoothness and diameter.

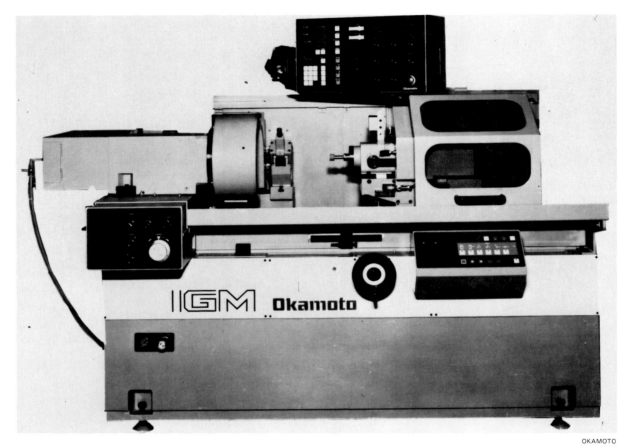

OKAMOTO

Fig. 8-14. The operator of this grinding machine can easily input grinding parameters of the controller keyboard. The machine can grind internal diameters ranging from less than 1/4'' to 8''.

MACHINING CENTERS

To improve manufacturing efficiency, ***multiple-purpose machine tools,*** called machining centers, have been developed. A machining center, using various tools mounted in an automatic tool changer, can be programmed to perform a number of operations (drilling, milling, boring, reaming, tapping, or counterboring) on a single part. Under NC control, tools can be selected from the changer in sequential or random order as necessary to complete operations on the part. Most machining centers are capable of switching from absolute to incremental positioning (or vice-versa) as necessary for a particular application. Fig. 8-15 shows a vertical machining center with a ***traveling column*** design (tool positioning is done by moving the column over a stationary workpiece).

Fig. 8-15. The traveling column design of the vertical machining center permits a low table height for improved operator access and ease of automated load/unload devices. This three-axis machine has a 32-tool random access tool magazine and a full-featured CNC controller.

OTHER NC EQUIPMENT

Numerical control has been applied to many other types of equipment in modern production operations. In addition to those discussed below, NC applications will be more extensively treated in succeeding chapters.

Presses

Steel, aluminum, and other materials in sheet form are shaped under pressure in heavy machines called *presses*. Numerical control has been applied to forming and perforating machines, high speed notching machines, turret punching machines, multiple-operation punching machines, and one-, two-, and three-axis punches.

Tube-bending machines

NC has been applied to tube-bending equipment in several areas, including the manufacture of yard furniture and replacement of automotive exhaust systems. See Fig. 8-16. Numerical control allows custom bending of standard straight lengths of exhaust tubing to the specific configuration of a customer's vehicle. This greatly decreases storage space needed for tubing.

Fig. 8-16. Programmable controls on this rotary tubing bender permit rapid changing of bend radius or other parameters with 1/1000'' accuracy.

Welding and cutting machines

Numerical control has been applied to many welding and cutting processes. Electron beam (EB) and metal inert gas (MIG) welding equipment on production lines is numerically controlled; so are gas-flame cutting systems. The U.S. Navy uses NC electron beam welding for jet aircraft maintenance.

A recent application of numerical control has been in the field of diemaking, where it has been combined with ***electrical discharge machining*** (EDM).

Inspection and test equipment

Some typical examples of inspection and testing equipment using numerical control are a four-axis ***coordinate measuring machine*** (CMM) designed for quality assurance inspection, and a programmed automatic circuit evaluator and recorder. The circuit evaluator is designed to compare the electrical components in a circuit with programmed limit values for those components.

Drafting

Drum plotters and flatbed plotting machines with NC controls, along with optical digitizers, are common in industry today. See Fig. 8-17. The sophisticated systems used for computer-aided drafting and design (CADD) offer three-dimensional, full-color displays of parts under development. With available software, such workstations can be used to develop component prototypes and write cutter-path programs for NC machine tools. See Fig. 8-18.

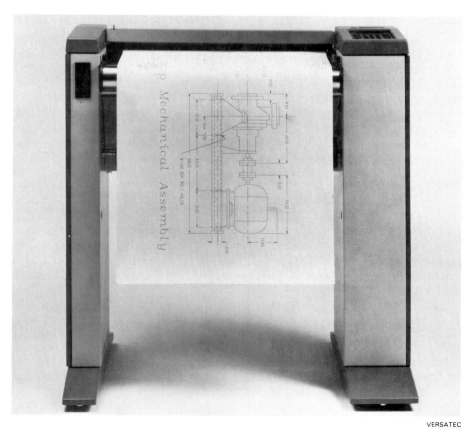

VERSATEC

Fig. 8-17. An electrostatic plotter is much faster than traditional pen plotters, especially when producing large, complex drawings.

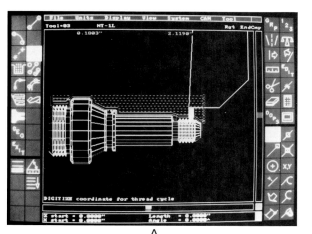

A

B

TEKSOFT

Fig. 8-18. CAD workstations with appropriate software can be used to generate tool path programs for CNC machine tools. A—A tool path for threadcutting being displayed. B—The part resulting from the program.

FUNDAMENTALS OF NC PROGRAMMING

The process of developing, verifying, and testing an NC program to operate a machine tool and perform an operation on a part can be broken down into eight major steps. Briefly stated, these steps are:

1. **Prepare working sketches or engineering drawings.** Drawings must be accurate, and must be dimensioned in a manner compatible with the type of NC system being used. Drawings for NC can be less elaborate than for conventional machining processes, especially with respect to tolerancing.

2. **Plan the sequence of operations.** A typical method used for this step is to develop a cutter path overlay. This overlay would show all NC moves: from the setpoint to the part, to various points *on* the part, and back to the setpoint after operations are completed. Also shown would be the location of each tool change. Once the route is set, sequence numbers representing programming blocks are assigned to all moves.

3. **Specify tool changes, feeds, and speeds on the program manuscript.** For each tool change sequence, tool length differences must be calculated and specified. Speeds and feeds also must be calculated and specified.

4. **Make a complete program manuscript.** The program manuscript must include all specifications and programming necessary to operate the NC machine tool and produce the part. Also included on the manuscript are any directions to the set-up person or machine operator.

5. **Perforate and proof the tape.** Working from a hard copy of the program manuscript, a CAD operator or other designated person can perforate the tape that will be used to store and run the program. *Proofing* the tape by comparing a printout with the original manuscript can save a great deal on time in verification.

6. **Verify the tape.** This is a critical step. There are various methods of tape verification, including graphic display on a CRT (cathode ray tube or "screen"), proofing on a simulated material, plotting on a drafting machine, or making a "dry run" on the NC machine tool itself.

7. **Produce a "final part."** Following verification, a single production part is produced. This part is subjected to critical inspection to determine whether it is within tolerance. If it is, production can proceed. If errors are detected, the tape must be revised to correct them. A new "first part" is then produced and inspected, with production allowed to proceed if there are no errors.

8. **Make the production run.** Despite verification of the program and careful inspection of the first part, production must be monitored to assure that tolerances are maintained. Periodic inspection will keep to a minimum reject parts resulting from mechanical problems.

COMPUTER-AIDED PROGRAMMING

One of the major contributions to the growing acceptance of numerical control has been the development of *computer-aided programming*. NC programming software has made it possible to develop programs for complex shapes and parts that would otherwise require tremendous amounts of time to accomplish, Fig. 8-19. Computer use has made machining operations on even the most difficult parts relatively easy to program.

Programming languages

Many languages are available for NC programming, but the most common are *APT* (Automatically Programmed Tools), a smaller version of APT known as *AD-APT*, and *COMPACT*, a language that provides greater flexibility than APT in some areas.

APT is, by far, the most powerful of the computer programming languages, but requires the use of a large *mainframe computer*. Programming with APT is typically a two-step process:

1. Features are defined as *geometric figures*, using common (but sometimes abbreviated) words

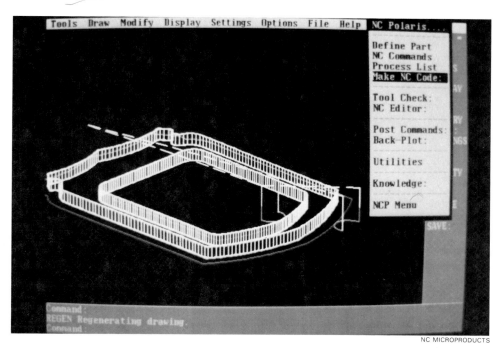

Fig. 8-19. Emulation of a router toolpath. Programming for complex shapes has been made easier with computer software such as this.

such as *line*, *circle*, *plane*, or *point*. The definition must also describe the position of the figure *in relation to* a known figure. A *name* is also assigned to each definition of geometry. An example of such a geometric definition is shown in Fig. 8-20. The figure named **C3** is defined as a *circle* with its center at **PT5** and *tangent to* (tanto) **L1**. Point **PT5** and line **L1** must have been defined earlier in the program.

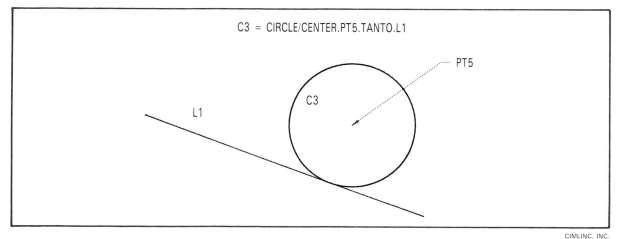

Fig. 8-20. In this sample of a geometric definition, a circle (C3) is defined tangent to a line (L1) and a center point (PT5).

2. Part features and pseudo-geometry are defined as ***machining statements*** used to direct a cutter around the previously defined geometry, following a predetermined sequence. Instructions consist of ***English-like words***, such as:

APT "Word" Meaning
GORGT Cutter go right
GOLFT Cutter go left
GO/TO Cutter go to point or piece of geometry
INDIRV Cutter go in the direction of vector
GOFWD Cutter go foward

Directions such as GORGT or GOLFT are relative to the cutter's direction of travel at the time they are received.

In *milling* operations, ***check surfaces*** are used as turning points. A check surface is another piece of geometry. A typical group of instructions would look like this:

FROM/SETPT
INDIRV/V1
GO/TO,L2
GORGT/L2,Tanto,C3
GOFWD/C3,to,L3

This programming language known as *compact* uses similar statements, but is considered more flexible, since definition statements can be combined with the machining statements. Some typical programming statements in this language are:

ATCHG,TOOL3,OFFSET3,5.813GLX,1.078GLZ,.1875TLR,
100FPM,.005IPR,.05STK,NOX,NOZ
MOVEC,OFFLN(8.4325D)/XL;PARZ,,OFFLN13/.01ZS;CON
⟨3⟩CUT,PARZ,OFFLN13/ZS
ICON,CIR(LN9/.75XL,LN13/.75ZS),CW,S(TANLN13),F(TANLN9)
⟨4⟩CUT,ONLN/.01ZS;.01X
MOVEC,OFFLN11/.O4XL,.01STK;PARZ,OFFLN13/.01ZS

The latest trend in NC programming is the use of ***graphics***. This technique eliminates the geometry phase of programming, because the part definition has already been specified to the computer as a ***graphic element***. Thus, all the programmer must do is direct the cutter around the graphic element on the screen to create an NC tape program of the cutter path. Most graphic NC software on large computers (and increasingly, on smaller computers) uses APT or one of its variations. Fig. 8-21 shows an NC program (using the APT language) that was generated with the use of graphics.

```
%
N10G70
N20G90
N30G92X0Y0Z0
N40G0X0Y0Z0D0
N50S594M40
N60M3
N70X-2628Y3161
N80G1Z-45000F1200
```

```
N90Z-20000
N100X-1399Y1840
N110G2X3750Y3850I5149J-5590
N120G1X60625
N130G2X68225Y-3750J-7600
N140G1Y-25000
N150G0Z0
N160X0Y0
N170M30
```

Fig. 8-21. This NC/APT graphic program shows coordinate moves for producing a geometric shape.

POST-PROCESSORS

Software programs, called *post-processors*, must be used to interface NC programs with the controllers used in specific machine tools. For each different computer-to-machine tool interface, a different post-processor transforms program data into machine motion commands in the code and format accepted by the specific combination of machine and control system. Also included may be calculations for feed rate and spindle speed, along with commands to execute various auxiliary functions. Larger NC users often write their own post-processors, but software can also be purchased commercially.

SUMMARY

The majority of work done in modern factories is done with machines under some level of computer control.

Numerical Control (NC) was the first evolution of machine tools, occurring in the late 1960s. NC machines are operated by punched paper or Mylar tape, and were stand-alone machines not linked to larger computers. There was one machine control per machine tool, with the punched tape loaded into the control unit for each job to be run.

Computer Numerical Control (CNC) is a more advanced version of NC. In CNC systems, a fairly powerful small computer is used to program and control the machine tool. CNC can eliminate the use of punched tape, since the programming can be done at the machine tool and stored in computer memory in the machine control. Punched tape can be loaded into the control, if desired.

Distributive Numerical Control (DNC) is very common in large companies, especially those that are using a CIM philosophy. DNC is a network system in which several CNC control units are linked to a larger computer. Programming is usually done by the larger computer and numerical control programs are down-loaded to the CNC control when needed. Programming can still be done at the CNC controller, if desired.

Other basic elements of numerical control include adaptive control, point-to-point and continuous path systems, positioning systems, machine controls, and servomechanisms.

Adaptive control systems have the ability to identify machine variables, make decisions based on these variables, then take corrective action or modify machine moves or functions. This system can approach a situation in which no operator is needed, except to possibly load and unload parts.

Point-to-point and *continuous path systems* refer to machines that make use of, respectively, straight-line moves and full contouring moves.

A *machine positioning system* is the control logic capable of interpreting dimensional data in a set format. Systems are either *absolute* or *incremental*.

Servomechanisms are the power devices used to produce motion along a machine axis. There are three basic types: stepping motor drives, DC variable drives, and hydraulic drives.

IMPORTANT TERMS

absolute positioning system
AD-APT
adaptive control
APT
binary coded decimal
CAM
Cartesian coordinate system
check surfaces
COMPACT
Computer Numerical Control

computer-aided programming
coordinate measuring machine
CRT
Distributive Numerical Control
electrical discharge machining
English-like words
even parity
feedrate (F) command
geometric figures
graphic element

IMPORTANT TERMS (continued)

graphics
incremental positioning system
machine commands
machine positioning system
machining centers
machining statements
mainframe computer
miscellaneous (M) commands
modal
multiple-purpose machine tools
numerical control

odd parity
point-to-point systems
positioning systems
post-processors
preparatory (G) commands
sensors
servomechanisms
standard codes
stepping action
tool (T) commands
traveling column

QUESTIONS FOR REVIEW AND DISCUSSION

1. What are the three functions that are involved in an adaptive control system? How do they relate to each other?

2. Distinguish between the two NC tool movement systems, point-to-point and continuous path.

3. How are the primary axes in the Cartesian coordinate system defined?

4. Commands conveyed by NC codes tell the machine where to move the tool and what to do. Groups of "what to do" commands are designated by the letters *G, M, F,* and *T.* Describe these groups of commands.

5. Describe a machining center and some of the functions it can perform.

6. What are the eight major steps involved in developing, verifying, and testing an NC program that will be stored on tape?

7. What are *post-processors*? How are they used?

9

Numerical Control Technology

by Russell Biekert

Key Concepts

❑ Computers used in CNC systems today permit operator programming of machine tools through use of a graphic display.

❑ The most common types of interpolation used by NC and CNC controllers are linear, circular, helical, and parabolic.

❑ Distributive Numerical Control, consisting of a mainframe computer linked to several CNC computers that control machine tools, permits quick updating of programs in response to changing shop floor conditions.

❑ Communication protocols, such as MAP, are agreed-upon data transfer rules that permit otherwise incompatible systems to share information.

❑ A total factory network communications system is becoming a necessity for manufacturing companies that expect to remain competitive.

Overview

Numerical control was introduced in the preceding chapter. In this chapter, NC will be discussed in greater detail, as well as being placed in the context of advanced systems that represent current technology. Machine systems do not necessarily stand alone; they are often combined with computer systems as part of a Computer-Integrated Manufacturing (CIM) concept. More advanced total factory systems, such as flexible machining cells, are discussed in later chapters.

Russell Biekert has held a number of key positions in the manufacturing and engineering areas for Garrett Engine Division, Allied Signal Aerospace Company in Phoenix, Arizona.

COMPUTER NUMERICAL CONTROL (CNC)

As shown in Fig. 9-1, Computer Numerical Control (CNC) is a system that involves use of a small computer to control one or more machine tools. This computer/machine tool combination is not linked to a large mainframe computer, but *can* be so linked to form a DNC (Distributive Numerical Control) system. DNC is described later in this chapter.

The small computer used in today's CNC systems, with its very powerful software routines, has drastically changed the role of the machine tool operator. The operator now has a tool, often with a graphic display, that permits programming right on the shop floor. The software employs a very *conversational* mode of communication, guiding the operator through the programming process with a series of questions that elicit the necessary information on process variables. The software includes mathematical formulas for determining such factors as cutting speeds. All the operator must do is fill in the cutting parameters—the program will then calculate cutting speed and feedrate.

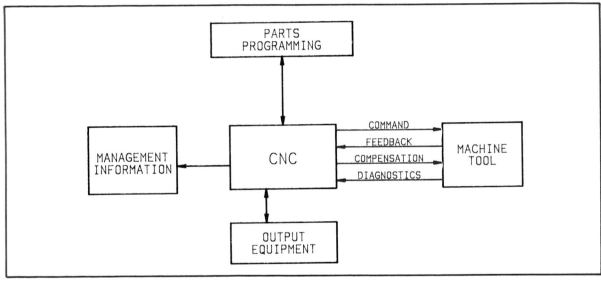

Fig. 9-1. This block diagram shows a typical CNC system used to control a machine tool.

The ability of CNC to provide an integrated approach to the production of machined parts lies in the extensive array of features available from various *controller* manufacturers and its capability of meeting new control requirements. Typical CNC controllers are shown in Fig. 9-2. The basic executive software used with such controllers is designed so that new features can be easily added. Software is usually developed as a series of modular units that can be linked together as a package that meets the needs of a specific customer.

CONTROL FEATURES

When production of a part involves machining a complicated contour surface, the controller must be capable of calculating the relative movement between the tool and the workpiece. This is called *interpolation*, and involves the addition of the *time* factor to the usual calculation and identification of Cartesian location coordinates.

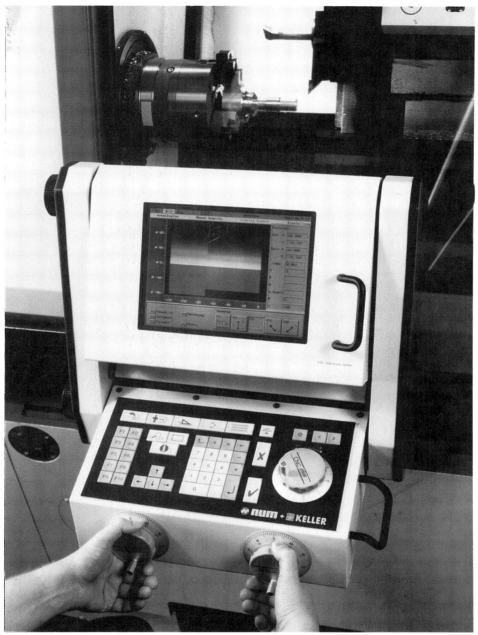

Fig. 9-2. The CNC control unit illustrated above allows the combination of manual and automatic CNC machining on a single lathe. The operator selects one of three modes according to his skills and the difficulties of the part to be produced. The production sequence is reproduced on the computer screen where production related faults are eliminated by the operator during the machining process using the handwheels. In a conventional machining mode, the machine is operated with the handwheels. In a conventional plus machining mode, software support is added for work that is more complicated.

Types of interpolation

There are a number of types of interpolation used by NC and CNC controllers, but the four most common are linear, circular, helical, and parabolic.

Linear interpolation is a mode used in contouring a two- or three-dimensional curved shape using a tool moved in straight-line segments. It involves simply designating the end coordinates of the shape.

Circular interpolation is a mode of contouring a two-dimensional curved shape by moving the tool through arcs of a circle. Some CNC controllers are limited to moving a tool through an arc segment of no more than 180 degrees in one block of programmed information, but most newer equipment permits *full circle* (360°) tool movement.

The elements necessary for circular interpolation are:

X, Y, Z	The linear distance from the arc starting point to its ending point in the X, Y, Z direction.
G 2	Circular interpolation in a clockwise direction.
G 3	Circular interpolation in a counterclockwise direction.
I	The linear distance from the arc starting point to its center location, parallel to the X axis.
J	The linear distance from the arc starting point to its center location, parallel to the Y axis.
K	The linear distance from the arc starting point to its center location, parallel to the Z axis.

Circular interpolation is more difficult when it involves arc segments of less than 90 degrees. Determining the linear values of X,Y, Z, I, J, and K requires the use of trigonometry; calculations must match the accuracy of the control system. Fig. 9-3 shows a program for an arc segment that is less than 90 degrees.

Helical interpolation literally adds another dimension (linear movement along a third axis) to circular interpolation. The tool moves in all three axes simultaneously to follow a helical (spiral) cutting path.

Parabolic interpolation (also called quadratic interpolation) is a mode of contouring a curved shape by moving the tool though a parabolic (bowl-shaped) arc.

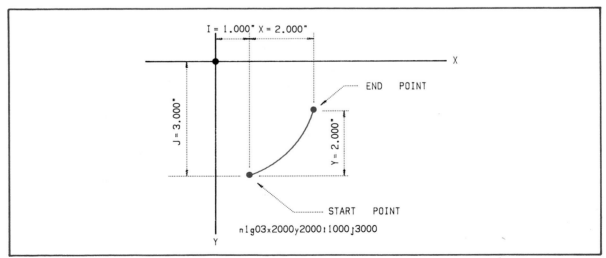

Fig. 9-3. A program for a partial arc segment, with calculations for X, Y, I, and J functions, using circular interpolation.

CSFM programming

Constant surface feet per minute (CSFM) programming is a control feature used to achieve constant cutting speed. The desired rpm of the machine spindle is specified in a block of information at the beginning of the control program, before any machining moves. The computer monitors the diameter of the cut, using this variable to continually recalculate the cutting rate in surface feet per minute. Based on these calculations, the computer adjusts spindle motor speed (increasing or decreasing rpm) to maintain a constant cutting rate.

Parametric programming

Parametric programs are used to produce parts in which only certain elements (parameters) change from one production run to the next. They are programs that can be used to produce an entire small part or that can serve as a component of a larger program. An example of such a program would be one written to produce a bolt hole circle with holes that are center-drilled, drilled, and tapped. Parameters such as the number, size, spacing, or depth of the holes can be specified, then the controller will automatically generate a program to produce the part. The parameters can be easily changed to produce a differently configured bolt hole circle in a later production run.

This routine eliminates a considerable amount of programming time, and is excellent for use with a family of parts concept. The programmer initially programs the part, assigning parameters to features that will change (length, type of material, and so on). When new parameters are assigned, the computer will alter the program to incorporate them. Fig. 9-4 shows a turning center controlled by an artificial intelligence-based NC CAD/CAM system that includes a parts-programming feature which allows the use of parametric design.

MHP MACHINES, INC.

Fig. 9-4. This turning center has a parts programming feature that allows the use of parametric design. The control system is artificial-intelligence-based.

Digitizing programming

Today's control systems have the combined capability of *digitizing* part measurements, then using that stored digitized information to reproduce the part. Using its great computational capacity, the controller translates the digitized information into a program with the necessary X, Y, Z coordinates. This capability obviously minimizes programming and calculation time.

Centerline programming

A method that greatly reduces the need for expensive post-processors for different machine tools is *centerline programming*, a feature of many current generation controls. The CNC control receives centerline outputs in programming language such as APT, and translates them directly into machine moves. A CNC control that does not employ centerline programming requires APT output to be post-processed in the mainframe computer and the resulting output converted to punched tape for use by the machine controller.

Adaptive control

A CNC feature that optimizes cutting operations by changing feeds and speeds in response to *feedback* is called *adaptive control*. Sensors constantly monitor the machining process and send signals (feedback) to the computer about favorable or unfavorable conditions. The computer decides, within set limits, whether and how much to change feed rates and tool speed for most efficient cutting. See Fig. 9-5.

FADAL ENGINEERING

Fig. 9-5. Feedback from sensors monitoring such factors as the spindle speed on this vertical spindle machining center will permit the controller to alter the process for greatest efficiency. This is known as adaptive control.

Overtravel monitoring

By constantly comparing programmed moves with stored movement limits for the X, Y, and Z axes, the control feature called *overtravel monitoring* can prevent damage to the machine tool and the part being processed. If the control senses that a programmed move will make the tool move beyond the limit set in a given axis, it will not permit the move to be made. Instead, it will alert the operator to the *error limit* condition, so that it can be corrected.

Mathematical capability

A CNC machine tool is able to store mathematical functions within a machining statement. This *mathematical capability* can provide up to 14 functions in memory, such as $(+)$, $(-)$, $>$, $<$, $=$, or $< >$. An example would be the statement **N10GOX(3/5)**. This statement first calculates the expression in parentheses (3 divided by 5), with the result .6, then makes a rapid traverse "X" move to the coordinate location X.6.

MANAGEMENT FEATURES

CNC controllers in use today are able to collect a wealth of information that can be used to aid management in both daily operations and longer-term decisionmaking. Much of this information was never available before— it is a byproduct of applying computer control to the manufacturing process. For example, modern CNC controls can gather precise information on *idle time* (time when the machine tool is not actually removing material from a workpiece). Since idle time adds to production cost, manufacturing management finds a measurement of idle time a useful starting point for attempting to reduce it. Examples of other "times" that can be measured by the CNC system, with only minimal operator input, include:

- Inspection time.
- Maintenance time.
- Waiting for tooling time.
- Waiting for parts time.
- Operator training time.
- Programming time.
- Set-up time.
- Piece part run time.

DISTRIBUTIVE NUMERICAL CONTROL

A *Distributive Numerical Control* (DNC) usually takes the form of a large mainframe computer connected to several CNC systems to form a powerful two-way communication network for NC operation without the need for tape readers. There are a number of variations upon the basic DNC format, as described below.

CONVENTIONAL SYSTEM

A fairly typical DNC system, with a mainframe computer linked to nine CNC systems, is shown in Fig. 9-6. The size of a mainframe determines how many CNC systems can be handled efficiently, but 10 to 40 is the common range.

The two-way communications capability of DNC allows relatively simple and quick updating and alteration of programs, based on actual shop floor conditions. Programs are written in APT and stored in the mainframe, then sent to CNC units as needed. On the shop floor, programming changes can be made, if necessary, and sent back to the mainframe (via the CNC) to update the source program

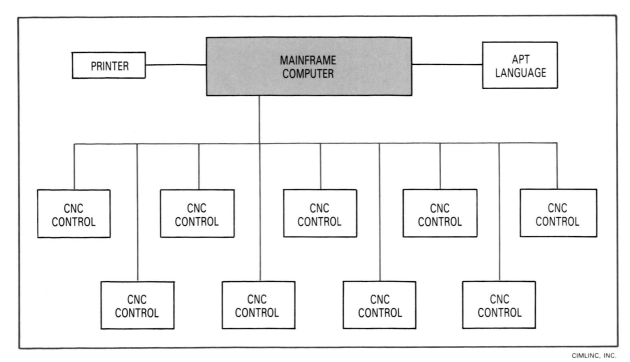

Fig. 9-6. This block diagram shows a typical Distributive Numerical Control system with a mainframe computer and 9 attached CNC units.

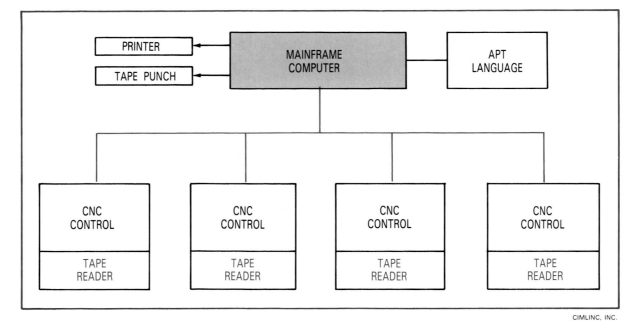

Fig. 9-7. A DNC system with "behind the reader" capability includes a tape reader as backup along with each CNC controller. It permits continued machine operation even when the mainframe is "down."

230 *Computer-Integrated Manufacturing*

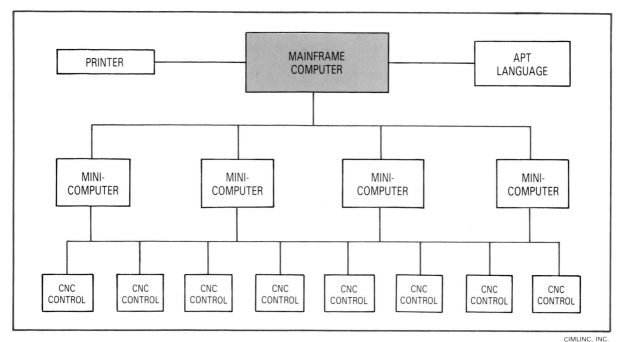

Fig. 9-8. In this system, minicomputers are used to channel traffic to and from CNC controllers and the mainframe computer.

file. Individual CNC systems can store one or more programs, depending upon length.

The major disadvantage of the conventional system is the dependence upon the mainframe computer. If the mainframe must be taken out of service for any length of time, the individual CNC systems will also "go down" when program information is depleted.

CNC "BEHIND THE READER" SYSTEM

This version of DNC incorporates an important aspect of "stand-alone" CNC systems: each CNC unit is equipped with a conventional tape reader, as shown in Fig. 9-7. If the mainframe computer is out of service, CNC systems can continue to function by using programming input from tape. Program tapes can be quickly produced from the computer-stored source programs by using a tape punch output device. Although redundant storage of programs both on tape and in computer memory adds some cost to operation, the amount involved is minor when compared to the potential for lost production time without the "backup" tape capability.

DNC MINICOMPUTER SYSTEM

This version of DNC is frequently used in large manufacturing operations involving 100 or more CNC systems. In the largest installations, two or more mainframe DNC systems could be linked together to control several hundred CNC systems.

The system shown in Fig. 9-8 shows how minicomputers are used in one large system, functioning as "schedule clerks" for traffic in either direction between the mainframe and individual CNC systems. In this system, each minicomputer handles scheduling for 10 CNC systems.

APT, the main computer language used in constructing programs, resides in the mainframe computer. Programs are stored in the memory of the mainframe, but are also kept in the memory of the minicomputer for as much as a week for downloading to the CNC units. Many CNC units are also capable of storing one or more programs in memory. Program storage at each control level virtually eliminates machine tool downtime resulting from computer failure.

With a series of minicomputer-connected systems, an entire manufacturing plant can be linked under DNC. By linking DNC to a CADD (Computer-aided drafting and design) system, the operation will approach total integration. Designs created with the aid of a computer can be automatically converted, through the use of computer software, to programs that operate tools to turn out finished parts.

INTEGRATED MACHINE TOOL CONTROL SYSTEMS

During the decade of the 1980s, computing activities became widespread as the electronics and computer industries grew and expanded their markets. In the business sector, computer use spread from accounting and management functions to manufacturing and process control operations. At the same time, the use of home computers boomed, creating a new level of computer literacy and helping users develop skills that ultimately were transferred to industry.

With a large pool of talent to draw from, a host of small computer software and hardware companies sprang up almost overnight to develop a wide range of applications. The products of these companies differed from traditional data-processing-oriented hardware and software in one significant respect: they were "user-friendly," designed for use by persons who did not have strong computer skills. These early applications were designed to handle personal computing needs on a *local* basis, with no need for computer-to-computer passing of data. Packaging of hardware and software led to the concept of the *workstation*, an individual computer (usually a microcomputer) that provides the user with access to a set of programs designed to improve productivity. See Fig. 9-9.

HONEYWELL, INC

Fig. 9-9. Designers and programmers use workstations, such as this one, with appropriate software for such functions as CAD, production scheduling, machine programming, and materials management.

Workstation users soon realized the limitations of such a "stand-alone" or autonomous installation; they began to seek access to the far greater capabilities and data storage of the mainframe. Engineers responded by designing plug-in circuit boards for microcomputers that would allow them to carry out two-way communications with mainframes. The computers could be connected directly, with coaxial cable, or use a *modem* (modulator/demodulator, a device for converting signals for transmission) and telephone lines. Once computers could "talk to each other," the development of microcomputer *networks* was inevitable.

COMMUNICATION PROTOCOLS AND MAP

For years, systems engineers have been linking large mainframes and minicomputers, using a variety of communication protocols developed to ensure proper transfer of data. File transfer problems are inherent when linking computers, for reasons ranging from incompatible cable connector designs to incompatibility of the data formats required by different *operating systems*. The problem resembles that of the United Nations, where interpreters help people speaking different languages to communicate with each other. A *communication protocol* is a set of agreed-upon rules for data transfer and standards for physical linkages of computers and associated devices. Like the United Nations translator, the communication protocol makes it possible for otherwise incompatible systems to "understand each other."

Typically, protocols were developed on an *ad hoc* ("for this situation") basis to connect *office* computer systems. When the concept of networking was extended to the factory floor, however, it had to deal with additional considerations (such as electromagnetic interference, or EMI, that can garble data transmissions). It also had to accommodate a broad range of new devices, such as CNC machine tools, programmable logic controllers, robotic equipment, and various types of data collection devices (gauges, temperature probes, barcode readers, optical comparators, and many others).

Systems engineers at General Motors recognized the need for new communication protocols that were specifically designed to interface computers with a variety of factory floor devices. The effort resulted in development of the *Manufacturing Automation Protocol* (MAP), which appears likely to become the de facto standard for factory data communications.

MAP is a collection of existing standards adopted from those established by such organizations as the Institute of Electrical and Electronics Engineers (IEEE) and the International Standards Organization (ISO). Network architectures are characterized mainly by their *topology* (configuration) and the communication protocols they use. MAP is based on the Open Systems Interconnect (OSI) model and the IEEE 802.4 standard. This standard includes a "token-passing" communication protocol and a *broadband* (wide frequency range) coaxial cabling system that allows transmission of data, video graphics, and voice signals.

FACTORY FLOOR NETWORKS

The combination of low-cost computer hardware and the networking concept led manufacturing visionaries to think in terms of factory computer networks that would serve as communication links between office and factory floor. They recognized the need to pass to the factory floor such information as NC part program code, process plan information, operating procedures, and engineering drawings in computer graphics form. They saw additional opportunities for informational flow in the opposite direction: collecting in-process data at workstations to monitor part dimensions (Statistical Process Control), monitoring machine tool performance indicators (coolant temperature, vibration, power consumption, and so on), and editing of NC part programs at the machine tool site. See Fig. 9-10.

In general, networks have come to be viewed as analogous to *plumbing systems*, with the difference that the system carries data, rather than water. In such an analogy, computers and modems are the "pumps," while plotters, printers, and video display terminals (VDTs) are "faucets."

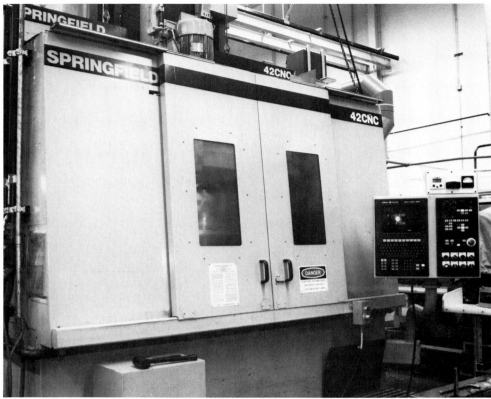

Fig. 9-10. Computer networks connecting CNC equipment (such as this vertical spindle grinder) to a mainframe computer allow information flow both to and from the machine. In addition to the transmission of process data, the information flow from machine tool to computer permits on-site editing of NC parts programs.

CELL CONTROLLERS

Some vendors include in their systems a **cell controller** and a **local area network** (LAN) for cabling. The LAN considerably simplifies the cabling job and minimizes the footage of cable that must be strung through the factory. The cell controller serves as the final distribution point for instructions to a group of 4 to 6 machine tools; in effect, it functions as a low-level DNC computer. The number of machine tools served will vary with the number of communication ports available on the cell controller.

A broadband LAN, such as Ethernet, is often used to connect the cell controller to machine tools. Whether such a broadband system is necessary, or whether a baseband system (with a narrower range of frequencies) is appropriate depends upon the data to be transmitted and whether or not electromagnetic interference is a problem. A typical functional organization for a cell controller system is shown in Fig. 9-11.

Once the cabling, computers, and related networking devices were in place, engineers came to realize that DNC was just *one* use for the system. Since data could travel in both directions through the system, they could now think seriously about additional network uses, particularly the use of total factory data communications.

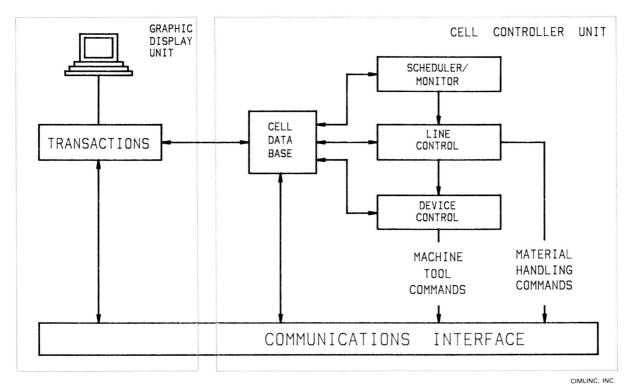

Fig. 9-11. A typical machining cell controller, which serves as the final distribution point for instructions to the machine tools in the cell. The communications interface ties it into the factory network.

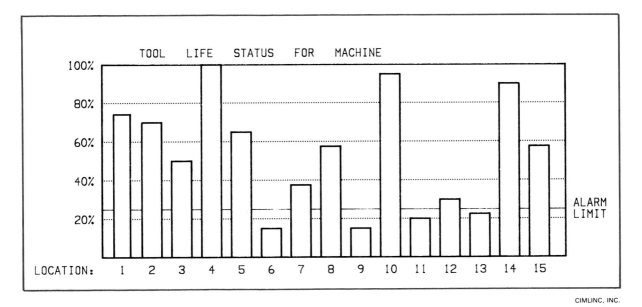

Fig. 9-12. An example of a graph of the type that might be output by a cell controller. This graph shows the tool life status for 15 machines.

FACTORY NETWORK APPLICATIONS

Large companies that have had extensive experience with computer-aided design (CAD) may download graphics information to the factory floor (and other locations) for viewing and even updating, when appropriate. Typically, the downloaded information consists of engineering drawings associated with operating instructions. Also of interest for downloading are company operating procedures, technical guidelines, tool inventories, and factory schedules.

Informational flow in the other direction (*uploading* from the factory floor) consists of statistical process control data, part counts (of both acceptable and rejected components), machine sensor data for maintenance scheduling and for scrap/rework prevention, and work-in-process (WIP) reporting to permit the easy identification of lots for expediting, labor reporting, and engineering change purposes. A graph of typical data (tool life status) that is output from a cell controller is shown in Fig. 9-12.

Computers will continue to play a key role in meeting the demands for quality improvement, increased factory throughput, and decreased cost that have been created by increased world competition. The key to success, in this so-called information age, is the ability of a business to acess information and turn it to profitable use. Some experts view manufacturing as the last great frontier for computer applications; as such, it will attract a growing number of skilled systems engineering people who are interested in factory floor applications.

Total factory network communications seems destined to play a central role in the future of any manufacturing company that desires to remain competitive. The *factory of the future* is likely to include groupings of machine tools that can communicate with each other and with external departments as easily as we now communicate with other individuals by telephone. A typical CNC machine that would be connected in a total factory network communication system is shown in Fig. 9-13.

Fig. 9-13. The machining center above can operate independently or be connected in a total factory network communication system. The workpiece loading system allows unattended operation of the machine.

SUMMARY

Computer Numerical Control (CNC) systems are very powerful with many software routines. CNC controls are full graphic systems with software that interacts in a very conversational mode with the operator. Mathematical formulas and cutting parameters are available for most materials and cutting tools. The basic executive software is designed so that new features or routines can be added easily. For sophisticated parts, contouring features for linear, circular, and parabolic movement are available.

Parametric programming is available on most CNC controls today. This feature allows the operator to store repetitive programming routines or, in some cases, entire part feature programs. When a given sequence of operations appears again on another part, the entire routine can easily be called into the new program. An example might be a bolt-hole circle of drilled and tapped holes. For a circle of a different diameter, only a new centerline dimension would need to be added. The entire bolt-hole sequence would be repeated in the new program.

Distributive Numerical Control (DNC) systems can be found in various forms in a fully integrated factory network today. In a conventional DNC system, several CNC controls are linked directly to a large mainframe computer. In this system, there is a direct download of programs from the mainframe to individual CNC controls. One disadvantage is that there is no intermediate program storage capability. If the mainframe computer goes down, CNC machining operations come to a halt.

A system that is very similar to the conventional DNC system is called *DNC behind the reader*. This system features a tape reader for each CNC control as backup to the DNC system. If the mainframe goes down, each CNC controller can stand alone, functioning as a conventional NC system. Tapes are read into the CNC control units until the mainframe comes back on line.

A third version of Distributive Numerical Control is the DNC *minicomputer* system. In this system, minicomputers are linked to the mainframe. Each minicomputer, in turn, controls several CNC controllers. Since the minicomputers are also networked with each other, there is no downtime of machine controllers in this system.

A factory network system provides communication links between workplaces and the business office. The factory network system is used not only for the DNC system, but for operating instructions, engineering drawing data and graphics, statistical process control (SPC) data, machine tool performance tracking, labor tracking, troubleshooting, and part scheduling and tracking.

The *cell controller* is an important element of a complete factory network system. The cell controller is the final computer distribution point between the overall management system and a small grouping of machine devices. The cell controller acts as the DNC computer for the cell and also as the management tool for the cell machine devices. Management includes scheduling, performance tracking of machines, labor tracking, and all other desired performance information.

IMPORTANT TERMS

adaptive control
cell controller
centerline programming
circular interpolation
communication protocol
constant surface feet per minute
controller
digitizing
Distributive Numerical Control
feedback
helical interpolation

idle time
interpolation
linear interpolation
local area network
mathematical capability
modem
overtravel monitoring
parabolic interpolation
parametric programs
topology
workstation

QUESTIONS FOR REVIEW AND DISCUSSION

1. What is interpolation? Name and briefly describe the four most common types.

2. What is the advantage of parametric programming when used with a family of parts (GT) concept? How is this form of programming normally used?

3. CNC controllers can record information about machine operations that can be used for planning or other management activities. For example, the controller can keep track of time that a machine was out of service for maintenance. List some of the other types of "time" data a controller can record.

4. An important weakness of the conventional DNC system is its dependence on the mainframe computer: when the mainframe *goes down*, so does the rest of the system. Describe one or more of the alternatives that have been developed to overcome this problem.

5. Describe the role of a communication protocol in a computer network. What were some of the problems overcome by development of the Manufacturing Automation Protocol (MAP)?

6. Name the two basic types of LAN used to connect a cell controller to machine tools. What determines the type to be used in a given installation?

7. Factory communication networks permit a two-way flow of information between the mainframe computer and the factory floor. Describe the kinds of information that might flow in each direction.

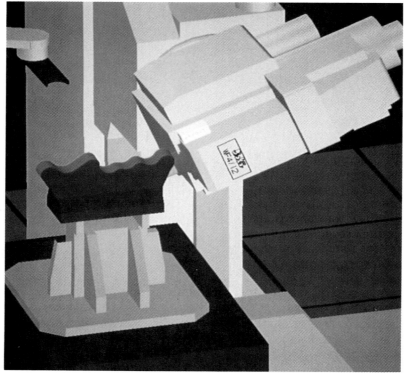

SILMA INCORPORATED

Software programs have been designed to help eliminate NC programming errors and allow manufacturers to reduce dry runs and shorten prove out cycles. The computer-generated image above was created while testing an NC program using the same tape that the physical machine would use. Through the use of these types of programs, costly errors such as this colision between the spindle and the finished part model can be avoided.

Flexible Manufacturing Systems

by David Berling

Key Concepts

- ☐ Flexible manufacturing systems incorporate many individual automation concepts and technologies into a single production system.
- ☐ The use of automated material handling devices to transfer workpieces between machines is vital to the FMS concept.
- ☐ Random sequencing of parts and alternate routings to accommodate heavy workloads or machinery breakdowns are made possible by central computer control of the FMS.
- ☐ FMS for prismatic parts are more common than those making rotational parts.
- ☐ The central control computer of an FMS is often responsible for system scheduling, as well as real-time control of the different aspects of the system.

Overview

Flexible Manufacturing Systems (FMS) have been described as the "state-of-the-art" in manufacturing and CIM technology. They are also sometimes referred to as the *factory of the future*, but in fact, many manufacturers have FMS systems already in operation. Implementation of the FMS concept can provide manufacturers with such significant advantages as more efficient use of capital equipment, increased productivity, and improved product quality and consistency. FMS also reduce work-in-process (WIP) inventory, direct labor costs, and floor space requirements.

Flexible manufacturing systems incorporate many individual automation concepts and technologies into a single production system. The three basic elements of a FMS are its **workstations** (usually CNC machine tools), a **material handling system** (to move workpieces from machine to machine), and **computer control** over the entire system.

The FMS concept can be applied to a wide variety of manufacturing processes, such as assembly, welding, and metal removal machining. Since the most common application of FMS technology to date has been the manufacture of parts by metal removal (drilling, milling, tapping, etc.), this chapter will concentrate on this application. However, the principles of FMS extend beyond metal removal parts manufacturing to many other areas.

David Berling is a Project Leader for Garrett Engine Division, Allied Signal Aerospace Company in Phoenix, Arizona.

THE FLEXIBLE MANUFACTURING SYSTEM CONCEPT

A *flexible manufacturing system* typically consists of a group of processing or workstations connected by an automated material handling system and operated as an integrated system under computer control, Fig. 10-1. It is capable of processing a variety of parts at the various workstations *concurrently* (at the same time) under computer program control.

CINCINNATI MILACRON

Fig. 10-1. This Flexible Manufacturing System is used to manufacture a family of machine tool parts. The workchanger unit in the foreground can hold up to 10 pallets with fixtured workpieces, ready for transfer (through use of wire-guided carts) to any of the machining centers in the left background.

A typical FMS has workpieces loaded and unloaded at a central location. Automated material handling devices are used to transfer workpieces between machines. Once a part is loaded onto the handling system it is automatically routed to the particular workstations required in its processing. Each type of workpiece may need to follow a different routing or process plan. The operations and tooling required at each workstation may also differ. See Fig. 10-2.

The coordination and control of the parts handling and processing activities is accomplished under command of the FMS central computer. One or more computers can be used to control a single FMS.

The sequence of parts processed in the FMS can be random. Variations in parts can be accommodated by the processing machinery, as directed by the controlling computers. The supervisory computer can be programmed to determine alternative routings for parts. When one of the machines in the FMS breaks down or is busy, the computer directs the workpiece through the alternate routing.

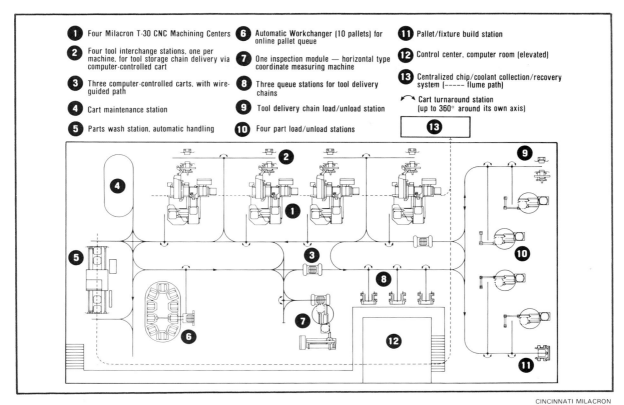

1. Four Milacron T-30 CNC Machining Centers
2. Four tool interchange stations, one per machine, for tool storage chain delivery via computer-controlled cart
3. Three computer-controlled carts, with wire-guided path
4. Cart maintenance station
5. Parts wash station, automatic handling
6. Automatic Workchanger (10 pallets) for online pallet queue
7. One inspection module — horizontal type coordinate measuring machine
8. Three queue stations for tool delivery chains
9. Tool delivery chain load/unload station
10. Four part load/unload stations
11. Pallet/fixture build station
12. Control center, computer room (elevated)
13. Centralized chip/coolant collection/recovery system (----- flume path)
⮠ Cart turnaround station (up to 360° around its own axis)

CINCINNATI MILACRON

Fig. 10-2. Floor plan of the FMS shown in Fig. 10-1. Note the wire paths followed by the computer-controlled carts as they transport workpieces and tooling around the FMS.

Although some installations approach the idea of the "workerless factory," **operators** are usually needed to support the functioning of the FMS. Operators load raw workpieces onto the system, unload finished parts from the system, change and preset tools, inspect workpieces, and provide equipment maintenance. Operators also input data, edit part programs, and perform tasks related to the computer control system. Many of these items can also be automated. The use of advanced technology, such as robotics, vision systems, sensors, and artificial intelligence programming can elevate the operator's role to that of a *system manager*.

Before a manufacturing company can consider FMS as a solution to its production needs, it must evaluate the number and types of parts that are to be produced. Flexible manufacturing systems are most beneficial to the *middle ground* of batch manufacturing. This is where the part variety is too high for dedicated processes (such as transfer lines) and too low for stand-alone machine tools, Fig. 10-3. This area of batch manufacturing represents a very large percentage of the U.S. manufacturing base today.

Implementation of an FMS requires proficiency in machine tools, in automated material handling systems, and in hierarchical computer control. Furthermore, a FMS needs to be part of a total computer-controlled manufacturing environment. All of the manufacturing control systems, such as production scheduling, inventory management, and shop-floor data collection, should be controlled and coordinated by the computer. Even the part specifications or blueprints will normally be developed on a Computer-Aided Design (CAD) system. This is Computer-Integrated Manufacturing or CIM. A flexible manufacturing system is only one piece of a total CIM System.

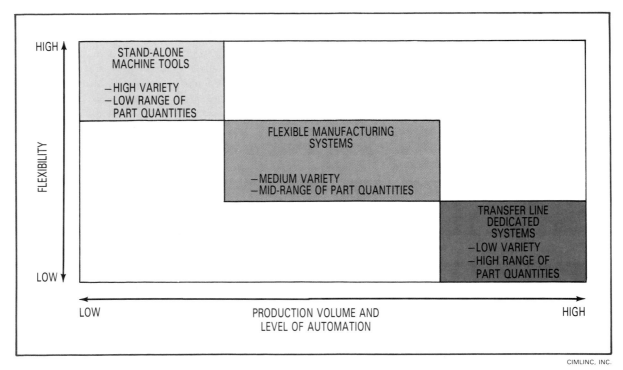

CIMLINC, INC.

Fig. 10-3. Flexible Manufacturing Systems are most useful to the manufacturer with a medium volume and medium variety of products. Stand-alone machine tools make it possible to produce a wide variety of parts in low volume; transfer lines turn out high volumes of a small variety of parts or products.

FMS BENEFITS

As noted earlier, flexible manufacturing systems are designed to fill the gap between high-production transfer lines and low-production NC machines. **Transfer lines** are very efficient when producing parts in large volumes at high output rates. The limitation on this mode of production is that the parts must be *identical*. The highly mechanized lines are seldom able to tolerate variations in parts. A changeover in part design requires that the line be shut down and retooled. If design changes are extensive, the line could be rendered obsolete.

On the other hand, **stand-alone NC machine tools** are ideally suited for variations in workpiece configuration. Numerically controlled machine tools are appropriate for "job-shop" and small batch manufacturing. They can be conveniently reprogrammed to deal with product changeovers and part design changes, but lack the high efficiency and productivity of a transfer line. The solution to this production problem for the mid-volume manufacturer is the flexible manufacturing system.

PRODUCING A FAMILY OF PARTS

A flexible manufacturing system is designed to handle a **family of parts**, as identified through the use of Group Technology (GT). Parts in a family have features in common or follow similar manufacturing processes. The configuration of a FMS is dictated by the equipment and processes needed to produce a given family of parts. In some cases, an FMS can be configured to produce several families of parts in any variety required.

242   *Computer-Integrated Manufacturing*

RANDOM LAUNCHING OF PARTS

When a part is launched onto the line, it is identified to the computer control system, which then routes it to the proper machines in the system. ***Random launching*** means that any workpiece among the parts family (or families) handled by the FMS can be introduced to the system without requiring downtime for machine set-up, Fig. 10-4. The only limitation is that the workstations must be equipped in advance with all the tooling required to process the part. A FMS is designed to process various part configurations at the same time by using its workstations concurrently. When only parts of the same type are loaded onto the system, certain workstations and tooling will tend to be fully utilized, while others may be underutilized. It is usually desirable to maintain a mix of part types in the system.

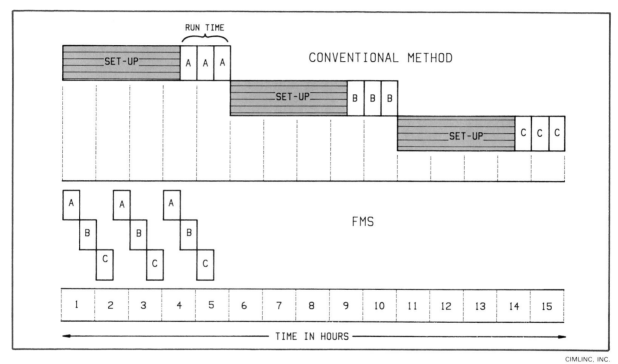

Fig. 10-4. Set-up and runtime of an FMS and conventional stand-alone machining methods are compared in this illustration. In conventional systems, set-up time is required before each batch of parts can be machined. When the runtime of one batch is completed, the set-up must be changed to machine the next batch. In an FMS, time is saved because parts arrive already fixtured and ready for machining. Setups may be saved off-line on pallets, ready for re-use whenever needed. A wider variety of parts, in smaller quantities, can be manufactured in a given period of time by the FMS.

REDUCED MANUFACTURING LEAD TIME

In traditional manufacturing, most workpieces require processing in batches through several different work centers (groups of same-type machines, such as mills or lathes). There is set-up time and waiting time at each of these work centers. With flexible manufacturing systems, the nonoperation time is drastically reduced between successive workstations on the line. Set-up time is also minimized.

The set-up for a traditional production machine consists of two main elements: tooling set-up and workpiece set-up. **Tooling set-up** means collecting the required tools from the tool crib and setting them in the machine. **Workpiece set-up** involves adjusting the workholding fixture and getting raw materials ready.

The set-up for a FMS also consists of tooling set-up and workpiece set-up. FMS tooling is *preset off-line* to minimize the set-up time on the actual machine. For a particular workstation, tool set-up consists of loading the preset tools required for the job into the **tool chain** at that station. Each tool chain may be capable of holding 60 or more cutting tools. Workpiece set-up usually consists of mounting a **fixture** (workpiece holder) on a **pallet** (carrier or platform used with automated material handling systems). The workpiece is then registered (aligned) and firmly secured on the fixture. This set-up is accomplished in the load/unload station *before* the pallet carrying the workpiece is launched into the system.

Fixtures are designed to adapt to holding various part configurations within a family. Several different fixtures may be required to handle the various part families. Good FMS design requires that the pallet/fixtures pairings outnumber the workstations (sometime by a ratio of more than 5:1). This permits some pallets to be off the system for loading or unloading, while others are carrying workpieces through the system. To avoid congestion, the *float* (number of workpieces in the system not being directly worked on) in the FMS is kept to a minimum.

REDUCED WORK-IN-PROCESS

An FMS allows work-in-process (WIP) to be dramatically reduced, when compared to WIP in a job-shop environment. Reductions of 80 percent have been reported at some installations. There are a number of factors that reduce the time a part is waiting for metal-cutting operations in an FMS. They include:

- close physical placement of equipment within the FMS, reducing travel distances.
- reduction in the number of fixturings required.
- combined operations, reducing the number of machines to which a part must travel.
- computer scheduling of parts through the FMS.

REDUCED OPERATOR REQUIREMENTS

A flexible manufacturing system can operate virtually unattended during the second and third shifts. This mode of operation is not common practice to date, but is expected to become increasingly common as better sensors and computers are developed. They will be able to detect and handle unanticipated problems such as tool breakage or part flow jams.

In cases where the system operates in the unattended mode on second and third shifts, such tasks as inspection, fixturing, and maintenance usually will be performed by people during the first shift.

EXPANDABILITY

With careful planning for the use of available floor space, the FMS can be designed for low initial production volumes. Later, as demand increases, new machine tools and material handling equipment can be added easily to provide the needed additional capacity.

INCREASED MACHINE UTILIZATION

Many CNC machines are actually operating productively far less than 50 percent of the available time, as shown in Fig. 10-5. The utilization of a flexible manufacturing system, however, may run as high as 80 percent, because of minimal set-up times, efficient workpiece handling, simultaneous processing, and other features.

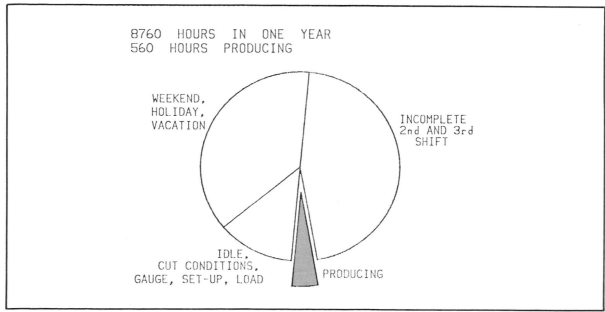

8760 HOURS IN ONE YEAR
560 HOURS PRODUCING

WEEKEND,
HOLIDAY,
VACATION

INCOMPLETE
2nd AND 3rd
SHIFT

IDLE,
CUT CONDITIONS,
GAUGE, SET-UP, LOAD

PRODUCING

CIMLINC, INC.

Fig. 10-5. Actual productive use of most stand-alone CNC machines is a small fraction of total available time, as shown in this illustration. In contrast, an FMS would have a much higher percentage of productive time, due to minimal set-up and the fact that the system can operate with little or no human input.

Downtime on a FMS is usually due to the need for scheduled and unscheduled maintenance. This may include tool changeovers, tooling problems (failures and adjustments), electrical failures, and mechanical problems, such as oil leaks. These are basically the same types of problems that are encountered when operating stand-alone machines.

The productivity or **throughput** for a FMS can be up to *three times* that of identical machines in a stand-alone (job-shop) environment. The FMS achieves such high efficiency by having the computer schedule and supervise every aspect of the system:

- The computer coordinates the automated material handling system to maximize productive time.
- The NC programs are electronically transferred to the machine just before they are needed.
- The part arrives at a machine already fixtured on a pallet, ready for immediate machining, eliminating wait time.

REDUCED CAPITAL EQUIPMENT COSTS

By keeping the FMS at a very high level of utilization, the total number of machines needed to accomplish a given amount of work is reduced. A 3:1 equipment reduction ratio is common when a flexible manufacturing system replaces machining centers in a job-shop configuration.

RESPONSIVENESS TO CHANGE

The flexibility of the FMS allows a manufacturer to respond efficiently to engineering design changes or to changes in the customer market. Also, production of needed spare parts can be mixed in without dramatically interrupting the normal FMS production schedule.

ABILITY TO MAINTAIN PRODUCTION

When a single-machine failure occurs, the FMS must be able to continue to operate. Backup plans for continued production in the event of various equipment problems are a necessary part of an effective FMS design. This is accomplished by incorporating redundant machining capability and a material handling system that allows failed machines to be bypassed. Thus, throughput is maintained at a reduced rate.

PRODUCT QUALITY IMPROVEMENT

One of the most significant, but often hard to measure, benefits of FMS production is improved *product quality*. The high level of automation results in a more consistent level of quality, when compared to the results of stand-alone machine tool production. This also results in reduced scrap, repair, and parts rework costs.

REDUCED LABOR COSTS

In a typical manufacturing situation, one machine operator is assigned to each NC machine. To operate a 6- to 10-workstation FMS, however, requires only 3 or 4 direct labor personnel. The ratio of operators to machines is significantly reduced. Other labor is also reduced, compared to a job-shop operation. This results from the use of automated material handling, rather than manual parts handling, between workstations.

Because of the high level of automation present in particular machine tools (as well as in the overall system), highly skilled machinists are not required to make good parts. Less skilled personnel can handle the more common duties of simple loading and unloading of parts. Each system will, however, require a system supervisor. Thus, FMS labor cost reductions are somewhat offset by the need for a smaller number of more skilled (and thus, more highly paid) specialists.

BETTER MANAGEMENT CONTROL

Since throughput time on the FMS is substantially reduced, parts do not have the opportunity to get misplaced in the shop. This results in better information and improved control of the parts moving through the plant.

OPERATION OF A FLEXIBLE MANUFACTURING SYSTEM

The best way to understand what a FMS is and how it operates is to track the flow of parts through the system. A typical FMS is capable of random part production within a given part mix. Using simulation, group technology, and other production analysis techniques, a family of parts is identified that utilizes the system capacity. At any given time, any or all of the parts in the family might be found somewhere in the system.

The parts start out at a load/unload station, where the raw material (rough casting, forging, or blank stock) is normally fixtured onto pallets, Fig. 10-6. The parts then move via the material handling system to queues (holding areas) at the production machines where they will be processed. In properly designed systems, the holding queues are seldom empty. There is usually a workpiece waiting to be processed when a machine becomes idle. (When pallet exchange times are short, machine idle times are quite small.)

The FMS control computer keeps track of the status of every part and machine in the system. It continually tries to achieve the production targets for each part type and, in doing so, tries to keep all the machines busy. When selecting parts to be sent into the system, it chooses part types that have the highest priority to meet the production schedule, and for which there are currently

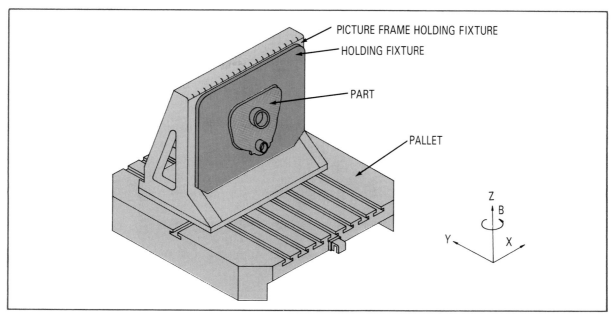

Fig. 10-6. Pallets and their accompanying fixtures are vital elements in a flexible manufacturing system. Parts are mounted on the fixture and held in the proper orientation for machining. A picture-frame-type fixture, like the one shown, permits machining of more than one side through rotation of the pallet.

Fig. 10-7. A shuttle, like the one in the foreground, is used to transfer a workpiece-bearing pallet from an AGV or other transporter to a machine tool or holding queue.

empty fixture/pallets or load stations. If an appropriate pallet/fixture combination and a desired workpiece are available at the load station, the loader will get a message at the computer terminal to load that part on its pallet. The loader will then enter the part number and pallet code into the terminal, so the computer system can update its status file. The computer will send a transporter (such as a cart) to the load station to move the pallet.

When the transporter carrying the pallet/fixture arrives at the machine, the system computer will notify the machine's transfer mechanism to activate (accept the deposit of a fresh pallet). This machine mechanism is usually called a *shuttle*, Fig. 10-7. The pallet is moved off the transporter and onto the shuttle. The transporter is then free and will leave when a new move request is assigned. The part and pallet on the shuttle must wait until work is completed on the part being machined. Next, the two parts on their pallets exchange positions. While the shuttle moves the new part and pallet onto the machine, the proper NC part program is being downloaded to the machine controller from the FMS control computer. Under direction of the newly downloaded program, machining begins on the part.

Meanwhile, the just-completed part waits on the shuttle for the system computer to schedule a transporter to pick up its pallet and carry it to the next machining station. If, for some reason, the part cannot go to that destination, the computer checks its files for alternative destinations. If an alternative exists, the computer must decide whether conditions in the FMS (parts backlog, machine availability, lateness) warrant sending the part to that destination. If the part is not sent to an alternative destination, it may circulate around the system on the transporter until the original destination is available. Another possible action would be an order from the computer for the transporter to unload it at some intermediate or storage queue. It will be automatically retrieved from storage when the destination is available.

The final destination on the material handling system is the load/unload station. This is where the part is removed from the pallet. A new part may be loaded, or the pallet may be stored until it is needed.

COMPONENTS OF THE FLEXIBLE MANUFACTURING SYSTEM

A flexible manufacturing system has three basic elements: workstations, a material handling system, and a central computer control system. The number of workstations in a system typically ranges from 2 to 20 or more. The material handling system may consist of carousels, conveyers, carts, robots, or a combination of these. The system computer is responsible for all scheduling, dispatching, and traffic coordination functions. The key to the system is that the workstation, material handling, and computer control elements *combine* to achieve enhanced productivity without sacrificing flexibility. See Fig. 10-8.

FMS WORKSTATIONS

The types of *machine tools* used in a FMS depend on the processing requirements to be accomplished by the system. This means that geometric tolerances, machined features, and other part characteristics usually determine the type of machines, or mix of machine types, to be included in the flexible manufacturing system. In turn, each FMS will be constrained by the quantities and types of equipment that are interfaced into its system.

In general, horizontal-spindle machining centers are the key metal-removing machines in an FMS. Any particular line may also employ a variety of special-purpose machines to support this basic equipment. An example would be multiple-spindle machines (such as head changers) used to most economically produce multiple hole patterns. Another example would be special, single-purpose machines used to accomplish such machining operations as broaching, planing, hobbing, turning, and grinding.

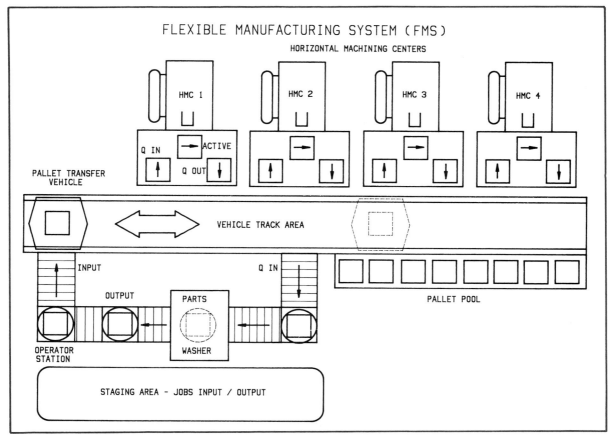

FLEXIBLE MANUFACTURING SYSTEM (FMS)

HORIZONTAL MACHINING CENTERS

HMC 1   HMC 2   HMC 3   HMC 4

Q IN   ACTIVE
Q OUT

PALLET TRANSFER
VEHICLE

VEHICLE TRACK AREA

INPUT   Q IN

OUTPUT   PALLET POOL

PARTS

OPERATOR
STATION   WASHER

STAGING AREA - JOBS INPUT / OUTPUT

Fig. 10-8. *Major elements of a Flexible Manufacturing System include workstations (labeled HMC 1 to HMC 4), an automatic material handling system, and a computer control system.*

Currently, FMS technology to manufacture ***prismatic*** (box-like) parts is more mature than the technology for manufacturing ***rotational*** parts. The choice of machines for prismatic parts essentially is between various brands and sizes of vertical and horizontal machining centers.

A CNC vertical turret lathe (VTL) may have to be added to an FMS to machine large bores or circular bearing surfaces, even in a strictly prismatic part system. Rotational parts having a length-to-diameter ratio of less than two, such as disks, hubs, or wheels, are usually candidates for inclusion in a prismatic-part-type FMS. These parts usually require additional milling, drilling, and tapping.

FMS for prismatic parts

A basic vertical CNC machining center will provide three axes (X,Y,Z) of programmed motion for machining. This basic type machine will only process one side of a part, along with some limited machining on adjacent faces. A simple three-axis machine will normally require multiple fixturings to complete the manufacture of each part. Each operation or set-up requires a separate part program. Successive operations on a part may be done on the same machining center or a different one, depending upon the work schedule and tool complement of each machine on the line.

Horizontal machining centers can be equipped with a fourth axis (usually pallet rotation) and a fifth axis (usually tilting of either spindle or part) to allow parts to be machined on more surfaces in a single operation. Vertical machining centers can also have a fourth and fifth axis (usually in the form of a tilting rotary table) to allow machining on additional surfaces.

Full five-axis machining, however, can have some problems. The NC programming can be more difficult, geometrical accuracies can be worse, and machine costs will increase. The fifth axis is normally reserved for special machining in unusual planes, contour machining, and solutions to special reach problems. There is equipment available for five-axis machining that uses a tilt head as an alternative to tilting and rotating the pallet/fixture/part. That alternative is particularly useful where the combined weight of the pallet/fixture/part is excessive.

Choosing machining centers for the FMS also requires investigating the available mechanical interfaces with the material handling system. Machining centers that are equipped with pallet exchangers as options are best suited to this interfacing task.

The characteristics of the family of parts to be produced on a FMS determines the horsepower ratings, cube-sizes, and dimensional accuracies needed from the machining centers. It is not uncommon to adopt a mix of machines for a line, covering a range of accuracies. From a scheduling standpoint, it is most efficient to provide *full redundancy* by using identical machines. However, the dictates of the parts in a family may necessarily compromise this idea. For example, if special bore tolerances are required, a single high-precision machine may be needed. However, the entire FMS operation would then depend upon the reliability ("up time") of this machine, because there would be no alternative routing available. For this reason, it might be better to produce the high-precision bores off-line (outside the FMS).

If a vertical turret lathe is to be included in the FMS, it must be equipped with a pallet shuttle feature in order to be interfaced to the FMS material handling system. Turning, facing, and boring operations can normally be performed on a VTL with little need for a tool changer. Usually a four- to eight-tool turret with dedicated tools is more than enough to complete the necessary turning work. Tool changers are available, however, if there is a need for a variety of turning and grooving tools, or if tool wear requires frequent replacements.

FMS for rotational parts

Flexible manufacturing systems for rotational parts (gears, shafts, cylindrical parts) are not quite as common or advanced in technology as those for prismatic parts, but represent a large area of activity in FMS technology. CNC lathes with both bar and chucking ability can be integrated to form a rotational FMS.

There are a number of systems in operation that could be classified as rotational flexible manufacturing systems, Fig. 10-9. These systems might be better described as turning cells operated by programmable robots and connected by a conveyor system. The system is not usually controlled by a master computer. The equipment consists of CNC lathes or other CNC turning machines. Some of these machines can change cutting tools, chucks, or both.

Most of these systems do not operate in a random part-processing mode, as do their prismatic counterparts. Instead, all of the equipment is set up to machine one distinct part through each succeeding step in the manufacturing process. Once the production quantity for the part is obtained, the system is shut down. To prepare for the next part, the chucks and tools at each machine are changed, if necessary; so are the grippers for the robots. Often, rotational parts can be grouped to take advantage of the need for similar chucks, tools, and grippers. Meanwhile, the NC program(s) for each part are electronically transferred from the mainframe to the CNC machine controllers. Finally the correct robot program(s) are activated and part production begins again.

Workpieces to be machined are placed on the conveyor to the first cell. Usually, the parts will ride on V-blocks to maintain orientation and keep them from rolling into or damaging each other. The parts line up at the first cell while the robot loads and unloads the individual machines in the cell. The queuing area provides a buffer for minor imbalances between this cell and the one before it. Balance is achieved among the cells by activating or deactivating machines, and by using the buffers. When a part has been completed at a cell, the robot places it on the conveyor to the next cell. If the part can bypass a cell, there is usually a second conveyor accessible to the robot.

Robots. Industrial robots for loading machines in a FMS can be stand-alone units that are programmed to coordinate their motions with the motions of the machine tool (door open, chuck grasp, spindle rotate, etc.). Another type of robot can be bolted onto, or otherwise made integral with, the machine tool. The major advantage of a stand-alone robot is the capability of being easily removed from one cell and placed in service in a totally different cell. The advantages of robots that are an integral part of the machine tool include reduced floor-space requirements and virtual elimination of time that the machine must wait while the robot completes its actions.

Choice of robot configuration (Cartesian coordinate, cylindrical coordinate, spherical-coordinate, or articulated) depends on the type of machine it will serve. In general, cylindrical or spherical-coordinate (sometimes called *polar*-coordinate) robots are more appropriate to machine loading. Cartesian coordinate machines are simpler to equip with tactile-feedback sensors for assembly work. Articulated (jointed-arm) manipulators have the greater dexterity needed for such processes as welding and spraying, but are also more costly.

CINCINNATI MILACRON

Fig. 10-9. This FMS for rotational parts uses a conveyor for raw stock delivery and finished parts removal. A robot loads material from the conveyor onto the CNC lathe for machining.

Fixtures and pallets

Fixtures can take many forms in a FMS. They may range from simple clamps, similar in design to those used on a stand-alone machine, to complex picture-frame and pedestal fixtures that allow machining access from several sides, to even larger fixtures accommodating two or more parts which aid in reducing the nonproductive time used by tool changing and part transportation. Fig. 10-10 shows a fixture designed to hold more than one part at a time. The use of multiple mountings can significantly improve the throughput of a FMS.

The choice of fixtures depends on many factors. A basic manufacturing process plan must be developed first to estimate the number of different fixtures required. Cycle times must be estimated. Then, scheduling through a simulation program will give some idea of how a part fits into the overall part mix. At this point, decisions on the number of parts per pallet and the access required for each machining set-up can be made more constructively. There will be several "trial runs" using the simulation before a final process plan can be developed to best take advantage of the FMS. Additionally, consideration must be given to the effects of multiple fixtures on part accuracy. An expensive picture-frame fixture that allows all the machining on a part to be accomplished in one fixturing may be significantly better than machining the part by means of sequential mounting on a number of less expensive fixtures.

CINCINNATI MILACRON

Fig. 10-10. A fixture designed to hold a number of parts for machining. The fixture is mounted on a T-slotted steel pallet for automated handling.

Pallets. In the FMS, pallets are usually those offered by the machine tool builder; the shuttle, material handling system, and load/unload stations all must be designed to be compatible with them. Almost invariably, precision-grade couplings will be used to *orient* the pallet on the machine. This mechanism provides the precise indexing capabilities necessary to perform multi-sided machining on a part or to accommodate more than one part on a fixture. An important consideration in the design of the pallet conveyances is protecting this coupler from chips and other foreign material. Even the stickiness of evaporation residue must be considered in choosing a coolant. The machining center itself normally has a protective device built in to shield the coupler when a pallet is removed from the machine. Despite these safeguards, a common manufacturing problem is *coupler contamination*, which leads to machining errors. Quick recognition of the problem through inspection is necessary to minimize scrap.

Tooling

One of the problems associated with producing a large family of parts on a flexible manufacturing system is the limitation on the tooling complement imposed by the number of tool pockets offered by the machine-tool vendor. Often, special tooling needed to prepare and finish a single complex hole can occupy several pockets. None of this tooling may be able to be used in other areas or on other parts. It is not difficult to reach the point of being severely constrained in flexibility by tooling limitations. Trying to utilize as many common tools as possible is important when developing process plans. Since tapped holes require at least two tools, it may even be worthwhile to conduct a design review to reduce the number of different bolt sizes in the end products. This will minimize the variety of drills and taps on the line.

If each part in a family requires many tools, it may not be possible to have all tools in the system at one time. The system then will be limited by its tool storage capacity. In this case, the part set will be divided into groups, called batches. The required tooling for each batch will then fit in the FMS. To process a batch, each machine will be tooled to process specific part operations. Not all parts will go to all machines. Each part will be routed to one or more specified machines to have operations performed upon it.

Tooling can be classified according to where it resides, inside or out of the FMS. Tools in a machine's tool chain are in the system, Fig. 10-11. There are several methods of maintaining readily accessed off-line tool storage to help alleviate the limited tool complement of each machine. Several vendors have developed palletized automatic tool delivery systems. The exchange is somewhat time-consuming, so it is most advantageous to reserve this option for the batch change situation rather than for each part.

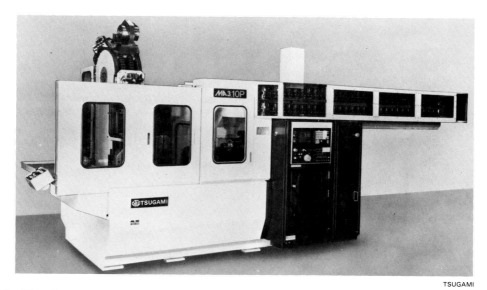

TSUGAMI

Fig. 10-11. This 4-axis machining cell has a magazine (tool chain) that will hold 126 tools for rapid changing. The tool magazine is visible extending from the right end of the machining cell.

Flexible Manufacturing Systems 253

Another technique to accommodate batch change employs an overhead automated tool storage and retrieval system. The overhead system delivers a set of tools to a machine in time to prepare for a new part batch. This dramatically extends the part mix, with a minimal manual interface at the machine.

Other techniques for effectively expanding a machine's tool handling capability to circumvent a storage limitation use a robot to transfer tools from a dense matrix located near the machine directly to the tool chain. This can be done during the machining cycle to avoid interrupting production.

Tooling for multiple-spindle heads generally cannot be shared, except in the case of identical hole patterns in a family of parts. Generally, each head is associated with one part. In designing the FMS, choosing a multiple-spindle head machine comes as the result of tradeoff studies between the productivity gains (short cycle time) of the multiple-spindle head with its part-specific configuration and the slower production rate of the more adaptable single-spindle machine. See Fig. 10-12. There is no redundancy if the line has only a multiple-spindle machine. If it fails, throughput is drastically reduced for parts produced on it. Transferring the parts to single-spindle machines in the system is usually more complicated than waiting for the machine to be repaired.

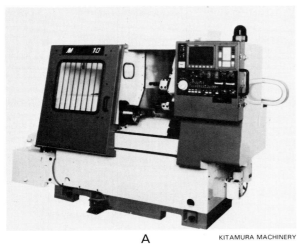
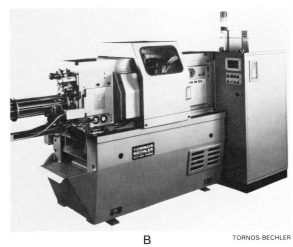

A   KITAMURA MACHINERY   B   TORNOS-BECHLER

Fig. 10-12. Single-spindle and multiple-spindle CNC lathes have differing advantages. A—A single spindle model is more adaptable, but has a slower production rate. B— A multiple-spindle lathe offers shorter cycle times (higher productivity) but is very part-specific.

Control station

Because the computer control system will normally be in a room or area separate from the FMS, local terminals must be available to allow monitoring and control of system functions. Much of this communication takes the form of status reports for each of the workstations, the material handling systems, and the load/unload stations. Each machine has its own CNC console with a display showing machining data.

The local control station for the FMS, Fig. 10-13, might have full system capability. This means that it can display individual machine information, part routing information, part status waiting, machining, cycle time to go, and other data. Also available may be information about scheduling, availability of parts, tooling, pallets, NC programs, etc. Manual intervention can be done through this control station to rectify specific problems with machines, tooling, the material handling system, or parts.

Fig. 10-13. Computer terminals used on the factory floor, such as those functioning as local control stations for an FMS, are often ''ruggedized.'' Strong, sealed enclosures for the electronics help to protect them from dirt, fumes, and other possible causes of damage.

Operators

When the FMS is operating reliably, the number of people required to maintain its operation can be reduced to a few. Typically, there would be one system supervisor, enough loaders to adequately feed the line, and one machinist to monitor up to a half-dozen machining centers.

The machinist would maintain tools and watch for tool breakage on machines. The machinist also would detect and replace worn tools by either reading outputs of a tool monitoring program, or by using personal experience and judgment. Additional duties would include maintaining machine operation during the workshift shift and initiating any special service requests made necessary by machine failure.

The loaders are responsible for keeping part mounting surfaces clean, properly mounting workpieces to fixtures with prescribed clamping forces, and communicating with warehouse workers to ensure a continuous flow of raw and finished parts to and from their workstations.

The system supervisor oversees the entire operation of the line. He or she is responsible for maintaining continuity of the line's production in the face of the various problems that may occur. The supervisor also uses the computer to implement part routing changes in situations where a bottleneck or machine failure has occurred.

It must be emphasized that the work force just described is for a fully operational FMS. The start-up crew for a *new* line would be greater, with additional machinists, production engineers, and maintenance personnel diagnosing and overcoming the problems in the shakedown phase of the installation.

There is also an *indirect labor* group associated with any FMS. This is comprised of programmers, process planners, manufacturing engineers, electrical/mechanical maintenance personnel, computer maintenance personnel, and others who support the line. These people are not usually assigned full-time to the FMS.

Quality assurance personnel will also be required if there is no on-line inspection machine in the system. They may still be required, even if there is a machine, to monitor the inspection equipment or to perform audits.

Inspection system

The inspection process can be performed on-line or off-line; each has its advantages. An *on-line inspection machine*, Fig. 10-14, can be programmed to identify machining errors and implement tool offset changes directly through the central computer. (Of course, not all machining errors can be rectified in this manner.) Perhaps the greatest benefit of an on-line inspection system is the quick identification of manufacturing problems. An *off-line inspection system*, Fig. 10-15, has inherent lags due to remote location, part fixturing or queue delays, and perhaps lack of automated inspection.

MTI CORPORATION

Fig. 10-14. This SPC (Statistical Process Control) workstation is used with on-line digital measuring instruments to provide real-time data analysis.

FEDERAL PRODUCTS CORPORATION

Fig. 10-15. Off-line inspection systems, such as this computer-controlled cantilever-style coordinate measuring machine, can provide extremely accurate measurements to determine if a part is ''in spec.''

The on-line machine is not an answer to all of these problems, however. In general, inspection is considerably slower than the production rate, so it is difficult to perform 100% inspection on an on-line system.

There are other philosophical questions regarding the utility of the on-line system. Since it is in the machine environment, can it perform adequately if this environment is not well controlled? Since the part is inspected in its clamped, as-manufactured state, is this a reasonable condition for inspection? This method is best for identifying changes in cutting tool performance, but may not be able to tell if parts are made to print.

Coolant and chip handling systems

Since the FMS represents a fairly dense cluster of machines, it is usually efficient to design a central coolant distribution and chip separation system. (An exception to this might be systems in which different materials are machined. In that case, different coolants might be used, making it inadvisable to mix them. Also, considering the potential chip reclamation returns, it may be unwise to mingle the different chip types.

An important consideration in choosing a particular coolant for the FMS is the nature of the residue after water has evaporated from it. An oily or sticky base of low volatility will create many problems due to dirt or chips sticking to tool shanks, spindle tapers, and couplings.

Cleaning stations

Many systems include an automatic or semi-automatic cleaning station as a workstation in the FMS. Some systems designate the function of cleaning parts as an external or off-line function, depending on the difficulty of cleaning, special processes required, or amount of manual attention required. It is also common to integrate the cleaning function with the load/unload stations. Cleaning is considered chip removal (washing of the part, fixture, and pallet) and not deburring. Deburring is performed off-line since it would interfere with the load/unload functions.

FMS off-line operations

Automated deburring is not currently performed on-line in a flexible manufacturing system, although certain manufacturing techniques (holes countersunk before drilling, for example) can reduce the burden of manual deburring.

Another common off-line operation would be painting or other forms of surface protection done before additional machining or shipping.

MATERIAL HANDLING SYSTEM

The material handling system of an FMS must satisfy several fundamental design requirements, as described below.

System requirements

The material handling system must provide random, independent movement of palletized workpieces between workstations in the FMS. The term *random* means that parts must be able to flow from any one station to any other station. *Palletized* means that the parts are mounted on pallet fixtures. The pallets must be able to move independently of each other to minimize interference and maximize workstation utilization. There are limitations to the degree of independence that can be achieved on any given material handling system.

The material handling system also must provide temporary storage or *queuing* of workpieces. The number of parts in the system will exceed the number of workstations. In this way, each machine in the system will have a queue of parts waiting to be processed. This helps to assure maximum use of each machine.

Another necessary feature is convenient access for loading and unloading workpieces. This includes both manual workpiece handling at the load/unload station, and automatic loading and unloading at the individual workstations. Since a given FMS may have machines on both sides of the handling system, such a system must be capable of loading and unloading from either side.

A good material handling system must have a computer or logic controller that is easily interfaced to the central control system. The system should be designed to be expandable on a modular basis.

The material handling system should provide unobstructed floor-level access to all individual workstations on the line, and should comply with all applicable health and safety codes. The system must operate reliably in the presence of metal chips, cutting fluids, oil, and dirt.

There are various ways to satisfy these design requirements. Material handling systems that easily adapt to the FMS concept include powered roller conveyors, powered and free overhead conveyors, and automated guided vehicles (AGVs).

The material handling system within the FMS serves two functions. The first is the movement of parts *between* workstations. The second function is interfacing with the individual workstations. This involves locating the workpieces in the proper orientation for processing, and may also include banking of workpieces at the workstations. To satisfy the two functions, separate material handling systems may be required.

Parts delivery

It is normally not practical to combine parts delivery outside the FMS with parts delivery *inside* the FMS. The first type of parts delivery involves moving raw parts from warehousing areas to the load/unload stations, while the second involves part, fixture, and pallet assemblies moving between workstations.

Material handling outside the FMS. Quite often, delivery of parts to the system is a manual function, performed by using carts or forklift trucks. In some cases, however, AGVs (automated guided vehicles) have been utilized both within the FMS and externally, Fig. 10-16. Large storage facilities for long-term part storage can be remote from the FMS. The transport of parts to the line by forklift truck requires only that adequate traffic ways to the load/unload area be incorporated in the FMS design.

MANNESMAN-DEMAG

Fig. 10-16. Parts handling outside the FMS may involve various types of equipment. Most common is the forklift truck, shown here loading a cart that will be towed by the AGV at left.

Various types of cranes can be employed to maneuver parts too heavy to lift manually. The technology also exists to use robots. Bins, magazines, or pallets of raw parts should be located near the load/unload stations to facilitate the loading function.

Material handling inside the FMS. The selection of a type of automatic material handling system within the FMS can take on many forms. Systems widely used today include carts, roller conveyors, automated guided vehicles, and robots. Control of carts can take a number of distinct forms. In one commonly used system, carts move along tracks, energized and controlled externally by the central computer. *Sensors* (optical, proximity, or limit switches) are located at appropriate points along the track to identify the precise location of the cart. The sensors can be used to position the cart to the required tolerance (typically 0.06 inch) for transfer of pallets to a machine or unload station. *Wheel encoders* can be used as less precise feedback for the drive system and its programmed speeds. All carts should have dead-man bumpers at each end to help prevent accidents.

Another delivery system employing carts uses a *tow chain* in an under-floor trench, Fig. 10-17. The chain moves continuously, with cart movement controlled by extending a drive pin from the cart down into the chain. At specific points along the guideway, computer-operated cam-type stop mechanisms raise the drive pins to halt cart movement. One advantage of this system is that it can provide some automatic buffering (see below) with stationary carts along the track. A variation on this method uses a floor-mounted *spinning cylinder* that imparts motion to the roller drive of individual carts. A computer controls drive engagement and disengagement, allowing the carts to be stopped at selected locations.

SI HANDLING SYSTEMS, INC.

Fig. 10-17. A frequently used delivery system for components and tooling is the tow-chain-operated cart. The tow-chain trench in the floor can be clearly seen in the foreground of this scene of a robot in an enclosure. Note the drive pin at the front of the cart. It engages the tow chain inside the trench to move the cart.

AGVs (automated guided vehicles) are normally battery-powered, and can be guided in a number of ways. One technique uses *a wire embedded in the floor* to guide the vehicle with radio-type signals for guidance. Another technique uses *rails*, much like those of a railroad, to provide a guidepath. Position sensors still must be used to control pallet transfer.

Conveyor systems can also be designed to transfer pallets from the load/unload stations to the machine tools. Most conveyor systems employ a powered roller system activated by computer control or sensors mounted along the track. Individual sections can have separate drives to control placement of pallets near machines. Limit or proximity switches can provide feedback to locate pallets along the conveyor. Side tracks and switching gear can support *buffer zones* within the system. In contrast to AGVs and wire-guided carts, a conveyor system limits access to the major elements of the FMS, because it must be raised above the floor level to align with the pallet exchangers on the machine tools.

Cutting tool delivery. Each FMS must have a means of delivering new cutting tools to the system. This can be accomplished by simply shutting down the system and manually delivering tooling to each station. Preferably some type of automatic means is used that allows the FMS to rapidly access additional tools. Some systems deliver fresh tools using the same pallet transfer system as the part. Other systems have been developed to deliver tools automatically to the rear of the machines through the use of tool racks and robots. A loading robot transports the tools to the machine, as needed, from the auxiliary rack.

As previously mentioned, robots can be an integral part of the automatic material handling system. Because most robots have a limited reach, they are most useful when machines are clustered in a circular work cell, so one robot can serve several machines. They are often used with unfixtured parts-of-rotation, in fact, robots are the crucial link in realizing an FMS for parts-of-rotation.

Load/unload stations

A major requirement for a load/unload station is to be easily accessible to both the manual operator and the automatic material handling system. The load/unload station must have good pallet access to permit the operator to mount and remove parts. Normally, compressed air and coolant wash facilities must be provided to flush residual chips and obtain clean mounting points for new workpieces. The load/unload station will normally have a computer terminal, Fig. 10-18, for operator communication with the FMS central computer.

The load/unload station provides support for the pallet, manual or electro-hydraulic rotation of the pallet on the stand, and necessary interfacing for the material handling system. If necessary, it can move the pallet from the material handling delivery spot to a safe point for operator loading and unloading.

Handling equipment. The operator may need equipment in the form of an overhead crane or hoist to handle FMS parts and fixtures. Most are too heavy to lift manually. Raw and finished part storage must be located close to the load stations to minimize complexity of the crane system. The crane system must orient the part to mate quickly with the fixture.

The station design should incorporate protection for the precision coupling mechanism on the pallet, to prevent chips or other foreign materials from lodging in it. Foreign matter in the coupling mechanism would destroy its orientational accuracy. The load/unload stations will quite often be part of a raised, open-grid platform where chips can be flushed through to a recirculating coolant trough. Any spray hoses at these stations would be supplied with coolant rather than water to prevent dilution of the coolant.

Operator control. For control, the operator will need a pushbutton box or computer terminal at the load/unload station. While loading or unloading is being done, the operator can initiate a HOLD status to prevent the material handling system from retrieving a pallet from the station. Upon completion of loading, a GO status is initiated. This tells the computer that it can send a signal to the material handling system to exchange a new pallet for the completed pallet.

Fig. 10-18. This computer terminal allows the load/unload station operator to keep the system computer up-to-date by acknowledging completion of part loading. The computer system will then dispatch a transporter to pick up the pallet and move it to a machining operation.

Multiple or redundant load/unload stations (those having the same fixturing on pallets) prevent the line from being paced by the loaders. This gives the loaders some freedom to vary their work pace.

Buffer storage

In addition to on-shuttle and off-shuttle queues at stations, several different kinds of buffer zones can be designed into an FMS. *Buffer storage* is necessary to gain flexibility in sequencing production through the system, and to allow for contingencies on the line, such as machine or tool failure. See Fig. 10-19. The most obvious form of buffer is a separate loop of track or conveyor where pallets can be shuttled to allow others to proceed past them on the direct route. There can be many versions of the side loop, including parallel tracks with multiple crossover points.

If one type of pallet/fixture combination is not required to pass another, buffer zones need not be distinct areas set aside for temporary storage. Simply circulating carts or pallets in the material handling system, while making only appropriate transfers, would serve to buffer unwanted pallets. Extra load/unload stations also can act as buffers of limited capacity.

FMS COMPUTER CONTROL SYSTEM

The FMS system computer or supervisory computer manages the total combination of devices in the system. This includes the machine tool controllers, the material handling system, the system monitoring devices, and system communications. Computer software supplies all the control management and monitoring functions that permit a high degree of system utilization.

The operation and control of a complex flexible manufacturing system requires the capabilities of a powerful computer installation, Fig. 10-20. The two primary functions of the FMS computer control system are *scheduling* and *real-time control*. In some situations, the scheduling functions and the control functions may be performed by different computers, rather than a single central unit.

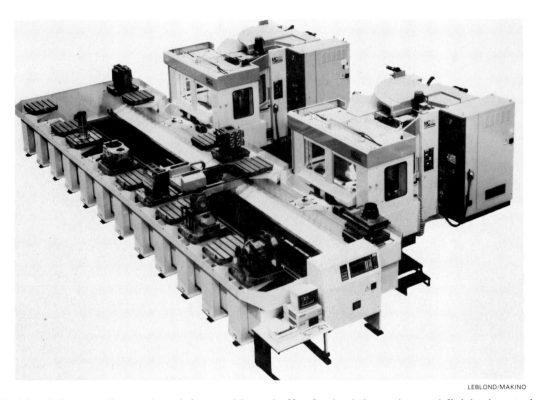

LEBLOND/MAKINO

Fig. 10-19. A linear pallet pool, at left, provides a buffer for both incoming and finished parts in this flexible manufacturing cell. A transporter moves pallets between the pool and the two horizontal machining centers.

CINCINNATI MILACRON

Fig. 10-20. A powerful system computer is needed to coordinate activities in a Flexible Manufacturing System. The computer installation is normally in a room separate from, but near to, the manufacturing area.

Database architecture

To provide the necessary monitoring and control functions, the FMS computer control system should maintain a database of files containing both control information and historical information. At minimum, the FMS computer control system should maintain the files described in the following sections.

Part master file. This file provides a complete explosion of production requirements for each unique part number to be manufactured in the FMS. The data to be maintained for each part number consists of operation numbers, revision levels, tooling requirements by operation, pallet/fixtures by operation, NC tape requirements by operation, and inspection requirements.

Machine tool status file. The current status of each machine tool in the system is provided by this file. The data to be contained in this file should include, at a minimum, the machine identification number, machine status, status of the tool in each tool pocket, and parts in the machining center. The machine status should indicate whether the machine is running, in set-up, down for maintenance, in emergency stop, or idle. The tool status data provides the number, condition, offsets, and type for each tool.

Part programs. The FMS computer control system should maintain a database of all NC part programs. These programs should be maintained and accessed by tape number, part number, operation number, and machine. The FMS computer control system may have just a local holding file for NC programs that are actually stored and distributed from a larger plant-wide DNC (Distributed Numerical Control) System. The part programs will be generated off-line on a separate NC programming system.

MHS status file. This file is used to maintain the current status of the material handling system. The data contained in this file should include transporter status, command being processed, and any alarms that might be generated. The transporter status will indicate that the transporter is running, idle, or in an emergency stop state. If the equipment is idle or in an emergency stop, then the location should be provided. The command currently being processed by the transporter will indicate the from/to movement of the unit. The alarm field will provide the alarm type and the location of the transporter when the alarm was generated.

Inventory file. The Inventory file will maintain one record for each batch of parts and the operations being performed upon them in the FMS. This record provides the part number, operation, pallet/fixture identifier, part location, and the quantity.

Pallet/fixture status file. This file is used to maintain an inventory of all available pallets and fixtures. The data for each type of pallet/fixture combination should include the pallet identifier, location, part numbers, and operations.

Standards file. The Standards file maintains the machine cycle times for each unique part number in the system. The detailed data to be maintained in this file will include the processing time for the operation.

Tool inventory file. A record of all tools that have been preset but not yet loaded onto a machine tool is maintained by the tool inventory file. The data usually includes tool number, tool type, and tool offsets.

System maintenance file. The System Maintenance file will be used to maintain historical records of system component failures, repairs, and preventive maintenance (PM). The data to be maintained should include the component name, entry type (failure, repair, or PM), date, and time of occurrence.

Inspection history file. This file will maintain historical data on probe readings from the machining centers specific to part and fixture verification. The data maintained includes part number or fixture number, operation, revision level, machine number, pass/fail quantity by part number, and the actual probe data generated for on-machine inspection.

Production sequence of current schedule file. This file is created by the FMS scheduling system. The contents of this file will be a sequenced list of parts to be produced in the FMS on a shift-by-shift basis. The data record provides the part number, operation, revision level, pallet/fixture required, start time, and stop time.

Tool requirements file. The Tooling Requirements file provides a list of all tools required for a given production schedule. The elements this file should include are tool number, tool type, machine number, tool pocket number, and time required.

Production history file. This file will maintain the audit trail of all parts produced in the FMS. The data to be tracked is the actual cutting time for a part while it is on the machining center, in addition to any machine alarms and faults that occur while a particular part is being processed. The data is stored according to part number, operation, revision level, and machine number.

Machine queue file. The purpose of this file is to maintain a machine lineup for each machine tool in the system. The data contained in this file should include a list of the part number, operation, pallet/fixture number, and location of each item in the queue. Also, each machine queue should have a desired level and a minimum level. The FMS computer control system will continue to load parts into a machine queue until it reaches the desired level. Once this desired queue level is achieved, the FMS computer control system will try to keep the queue size stable. If the queue size drops to the minimum, the FMS computer control system will generate an alarm for the system manager.

FMS scheduling

Scheduling for a flexible manufacturing system is performed using a look-ahead software package. The scheduling package may or may not run on the same computer as the FMS control software. To produce a production schedule, the FMS system needs an update of requirements from the overall factory requirements scheduling system (often MRP II). The FMS scheduling system will produce a schedule that lists the sequence of parts to be produced within the FMS.

The time window for scheduling the FMS can vary from a day to months. Due to the large number of variables that can affect the system, however, the schedule is seldom developed in detail for a period much further ahead than a few days.

Schedule update. The process of updating the current schedule on the FMS controller can be accomplished in one of two ways. First, the Real-Time Control module of the FMS computer control system can request an *automatic schedule regeneration*. This will cause the FMS scheduling system to generate a new schedule to be placed in the Current Schedule File. The primary scheduling criteria would be to achieve maximum machine utilization.

The second method of updating the schedule would be through a manual schedule update request from the system manager. This would involve the system manager manually invoking the FMS scheduling system to evaluate "what-if" scenarios and determine a new production schedule. After arriving at an acceptable schedule, the Current Schedule file would be updated with the new schedule.

Real-time control. The second, and *primary*, task of the FMS computer control system is to direct and coordinate FMS operations in real time. Real-time FMS functions can be divided into four areas: machine tool control, material handling system control, load/unload station control, and auxiliary equipment control. In addition to these four functions, there should be a supervisory function that coordinates the activities in each of these sub-systems.

Supervisory FMS control. This sub-system or module is responsible for coordinating all the activities between major functions in the FMS, such as machine tool control. The primary functions performed by this module are:

- Report alarms received from each major FMS sub-system, such as machine tools and the material handling system (MHS).
- Issue high-level instructions for the loading and unloading of parts from the FMS. These instructions will be executed by the load/unload control module.
- Generate high-level move transactions to be processed by the MHS control module.
- Generate high-level commands for initiating work on the machining centers.
- Request automatic schedule regeneration when a machine queue drops below the minimum level or a part cannot be loaded into the system.

- Update the plant-wide computer system with real-time production status as completed parts exit the FMS.
- Maintain the machine queues according to the desired levels and minimum levels. If a machine queue drops below the desired level, initiate a part load transaction. If a part load transaction cannot be completed, due to resources that are unavailable, then wait for the next part completion. If a serious limitation prevents following the schedule, then request production of a revised schedule.

Machine tool control. The primary purpose of this module is to process high-level machine control commands into detail commands that can be sent to a specific machine tool. Also, this module will receive and process all alarms and status messages from the machine tools. The primary events that this module should address are as follows:

- Tool breakage at a machine.
- Machine/controller faults (fluid levels, electrical, etc.).
- Program uploads from the controller.
- Program downloads to the controller.
- Probe data uploads from the controller.
- Manual override on a machine tool.
- Tool replacement at a machine tool.
- End of machine cycle.
- Start of machine cycle.
- Emergency shutdown, either automatic or manual.

Program downloads. The FMS computer control system should be capable of responding to requests for NC program downloads to the machine control units. The FMS computer control system should have the logic for coordinating the download of part programs corresponding to the part that is being processed on the machining center. The request for a program download could be initiated automatically by the machine control or the FMS computer control system. The FMS computer control system should also respond to a manual request for program downloads from an operator's terminal.

Program uploads. The FMS computer control system should provide the capability for uploading NC programs from the machine control units. The programs uploaded should be maintained in a separate area from the Part Program Database in order to protect data integrity. There should be a mechanism in the FMS control system to protect the data integrity of the Part Program Database and to control the revision levels of programs uploaded from machine controls. The request to upload NC programs could come from the machine control unit or manually from an operator's terminal.

Upload of probe data. The FMS computer control system should be capable of retrieving probe data (on-machine inspection data) related to part, fixture, and tool from the machine control unit. The FMS controller should maintain historical records on parts, fixtures, and tools. Tool verification data is used to generate new tool pre-sets and tool replacements.

Upload of controller faults. The FMS computer control system should be capable of receiving machine control faults, such as parity error, programmed stop, cycle start, automatic cycle, optional stop, tool change, broken tool, pallet change, and emergency stop.

Machine hour accounting. The FMS computer control system should track and maintain a record of the actual machining center utilization. The FMS computer control system should collect the appropriate data from the machine control unit to provide the actual amount of cutting time on each part manufactured. The cut time per part will then be stored by the FMS computer control system for historical purposes and traceability.

MHS control. The FMS computer control system should generate and transmit movement transactions to the material handling system. These transactions will be high-level transactions, similar to "move pallet X from point A to point B." The FMS computer control system should then provide a mechanism for receiving the results of move transactions from the MHS. There should be a mechanism in the FMS computer control system for responding to incomplete transactions and to exceptional conditions, such as system interlocks.

Load/unload station control. The FMS computer control system should send load/unload/refixture commands to the load/unload stations with the proper information for the operator to perform the operation. Operator instructions at the stations include part number, operation, pallet identifier, fixture number, clamping details, unclamping details, operator cautions, protective device/enclosure instruction, part cleaning/oiling details, and pallet destination. The pallet destination will be a machine number or a storage location, either inside or outside the FMS.

Auxiliary equipment control. The FMS computer control system should provide a mechanism for signaling any auxiliary equipment, (such as an automatic parts washing station), that a part is present in the input queue. Also, the system should receive a signal from the auxiliary equipment station when a part is complete and ready to be removed by the material handling system.

System reports

Data collected during monitoring of the FMS can be summarized for preparation of performance reports. These reports are tailored to the particular needs and desires of management. Typical categories include utilization, production, status, and tool reports.

SUMMARY

Flexible Manufacturing Systems (FMS) incorporate many individual automation concepts and technologies in a single system. These include CNC routines, cell controllers, automated material handling equipment, and statistical process controls. All are part of a total integrated factory network system.

The FMS concept provides several advantages in today's CIM environment. These include:
- Increased productivity.
- Improved quality.
- Reduced "work in process."
- Reduced direct labor.
- Reduced floor space.
- Increased efficiency in use of capital.
- Reduced lead times.
- Increased manufacturing flexibility.
- Improved management control.

The major subsystems of a typical Flexible Manufacturing System installation include *machining center* or *workstation*, the *material handling system*, and the *computer control system*.

The *machining center* or *workstation* can be designed as a dedicated FMS or a random FMS. The *dedicated FMS* is designed to meet specific machining applications on a specific family of parts. Special machines are usually used on a dedicated system.

The *random FMS* system is designed to be much more flexible and able to process a wide variety of parts. Usually parts that run on a random system require a number of different machining operations and have varied part geometries.

Most FMS systems today are built around machining centers, both horizontal and vertical. The centers include three-, four-, and five-axis machines. Coordinate measuring machines are also a very vital part of this type of system. FMS systems for rotational parts consisting of grinding and lathe turning workstations do exist but are not widely used in North America.

The second major subsystem of an FMS is the *material handling system*. The key feature of this system is the capability of random independent movement of palletized workpieces from workstation to workstation. *Palletized* is defined as the mounting of parts on fixtures located on machine pallets. These pallets are interchangeable from machine to machine, requiring no new setup of the part. The loading and unloading of pallets can be manual or automated.

The two basic functions of the material handling system are to *move parts between work centers,* and to *interface with the work centers.* The primary system often employs an underfloor towline system to pull pallet-carrying carts between workstations. The secondary system is a shuttle system at the workstation. This secondary system must move pallets from the towline cart to the machine tool table. Robots are a special consideration at the machine tool for the secondary system.

The third major subsystem of FMS is the *computer control system.* The functions of the control system include numerical control part program storage, shuttle control, production control, work handling system monitoring, workpiece traffic control, tool control, loading of part programs to workstations, and systems performance monitoring and reporting.

The control system architecture is designed with a hierarchical structure. It usually consists of four levels. The first level sets the parameters and production goals of the FMS, the second consists of general manufacturing and support tasks, the third involves general plant coordination, and the fourth controls departmental and machine operations.

IMPORTANT TERMS

automatic schedule regeneration
computer control
concurrently
conveyor systems
coupler contamination
family of parts
fixture
flexible manufacturing system
float
full redundancy
machine tools
material handling system
off-line
off-line inspection system
on-line inspection machine

operators
pallet
prismatic
queuing
random launching
rotational
sensors
shuttle
stand-alone NC machine tools
throughput
tool chain
tooling set-up
transfer lines
workpiece set-up
workstations

QUESTIONS FOR REVIEW AND DISCUSSION

1. Describe a Flexible Manufacturing System. How does it compare to high-production transfer lines and low-volume stand-alone NC machine tools?

2. A major benefit of an FMS is the great reduction it makes possible in work-in-process. List several factors that help to reduce the time parts are waiting to be processed in an FMS.

3. Discuss the ways in which an FMS usually reduces labor costs.

4. What are some limitations of a machining center with only three axes of programmed motion? What can be done to overcome these limitations?

5. Discuss the important requirements for a material handling system used with an FMS.

6. Distinguish between parts delivery from outside the FMS and parts delivery within the FMS.

7. What is termed the *primary task* of the FMS computer control system? What functions make it up?

Programmable servo-driven manipulators can be used to quickly and accurately transfer workpieces, such as these automotive body panels, between stamping presses.

Robotics Technology

by Dale E. Palmgren

Key Concepts

- ☐ An industrial robot is a reprogrammable multifunctional manipulator.
- ☐ The most desirable geometric configuration for a given robot depends on the size of the work envelope and the complexity of the tasks to be performed.
- ☐ The number of axes of motion and the number of degrees of freedom available in a robot are usually (but not always) equal.
- ☐ In closed-loop control of robot movement, sensors feed back position and velocity information to the controller, allowing error correction.
- ☐ The type of path programming used for the robot (point-to-point or continuous) is determined by the application.

Overview

The use of robots in manufacturing is becoming more prevalent as industry continues to automate various operations for improved efficiency, quality control, and cost reduction. To understand how robots fit into the industrial picture, you must comprehend what an industrial robot is and how it can be used in a robotic workcell.

Initially, this chapter will describe what an industrial robot is and how it is programmed. Next, a discussion of how to implement a robotic system will be presented, followed by a brief description of some robotic applications. Finally, future trends in robotics will be presented.

Dale E. Palmgren is an Assistant Professor, Manufacturing Engineering Technology, at Arizona State University in Tempe, Arizona.

WHAT IS AN INDUSTRIAL ROBOT?

An industrial robot can also be called a *reprogrammable multifunctional manipulator*. Put in other terms, this means that an industrial robot is a machine that can be given instructions that allow it to repeatedly perform a well-defined task. Such tasks might be loading a part on a machine, performing a cutting operation, or inspecting a completed part for adherence to specifications.

A robot that is **reprogrammable** is capable of acquiring more than one set of instructions, so that more than one task can be accomplished. The set of instructions (program) that the robotic controller interprets consists mainly of *locations* (points in space) and the specific directions (*paths*) the robot's toolholder or gripper should take to move between those points. These points and movements are usually programmed into, or *taught* to, the robot controller before the robot begins operation. In some applications, however, suitable sensors may be utilized to change the locations of the points in *real time* (while robot is in operation.)

The term *multifunctional* describes the robot as a machine that is capable of performing a number of *different* tasks. An example of a multifunctional robot would be one that is used primarily for gas metal arc welding (GMAW) on an assembly, but that also is used, at different times, to apply sealant around windows.

The word **manipulator** describes the robot's ability to move parts or tooling through space. To accomplish this task, the manipulator must be mechanically and structurally sound and able to handle static and dynamic loads while moving a part or tool along a prescribed path to the desired location.

ROBOT CLASSIFICATIONS

The industrial robot is made up of a variety of mechanical, electrical, and electro-mechanical components. The robot is configured in such a way to provide for a predetermined range of reliable and repeatable motions through space. The robot can be classified on the basis of four different considerations:
- Geometric configuration.
- Number of axes of motion.
- End effectors.
- Type of control.

Geometric configuration
There are four possible geometric configurations of a robot manipulator. They are:
- Cartesian.
- Cylindrical.
- Spherical.
- Articulated.

Selection of a particular manipulator geometry depends on the size of the work envelope and complexity of the task to be performed. The **work envelope** of a robot is a volume of space defined by the maximum reach of the robot arm in three dimensions.

The **Cartesian coordinate robot** manipulator has three mutually perpendicular axes of motion, as shown in Fig. 11-1. There are two advantages of having a Cartesian coordinate robot. The first is that the algorithms within the robot's computer controller to generate path motion are relatively simple. The relationship between the coordinates of a point in space and an axis of the robot is easier to describe in Cartesian coordinates than in other coordinate systems. The second advantage is that a large work envelope can be covered if a gantry robot, Fig. 11-2, is used. However, the ability of a Cartesian coordinate robot to reach around obstacles in its path is severely limited.

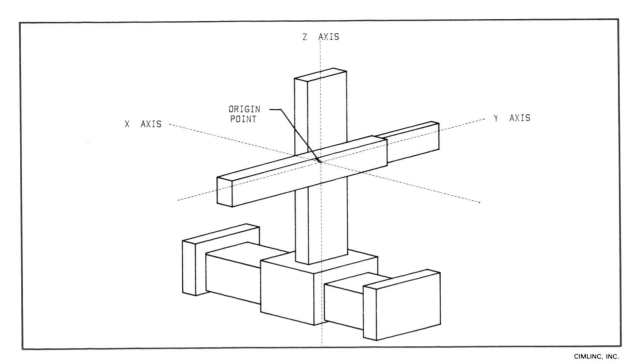

Fig. 11-1. An example of a Cartesian coordinate robot. The motions of the robot arms is parallel to the X, Y, or Z axes.

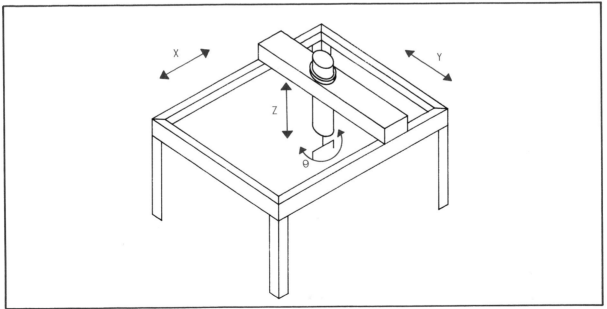

Fig. 11-2. An example of a gantry robot. Except for rotation about the Z axis, the motions of this robot are similar to those of the Cartesian coordinate robot.

The *cylindrical coordinate robot* manipulator, Fig. 11-3, has the advantage of being easily controlled. However, it is limited by the size of its work envelope and by its inability to reach around obstacles.

The *spherical coordinate robot* manipulator, Fig. 11-4, is also easy to control. For a given maximum reach of the robot, the work envelope is smaller than that of a cylindrical robot. The ability of a spherical coordinate robot manipulator to reach around obstacles is also limited.

An *articulated*, or *jointed-arm*, robot manipulator is a little more difficult to control. This is due to the fact that there is no simple relationship between the geometry of the robot and the coordinate or the point in space.

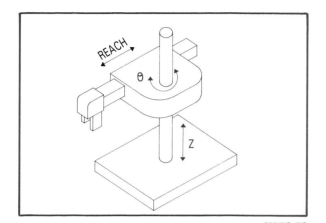

CIMLINC, INC.

Fig. 11-3. A cylindrical coordinate robot is easy to control, but applications are limited by the size of its work envelope.

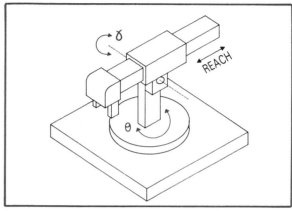

CIMLINC, INC.

Fig. 11-4. A spherical coordinate robot has a work envelope even smaller than that of the cylindrical coordinate robot.

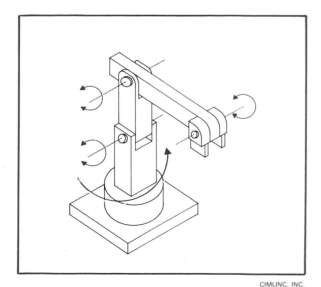

CIMLINC, INC.

Fig. 11-5. An articulated robot is able to reach around obstacles. It has a range of motions similar to those of a human being.

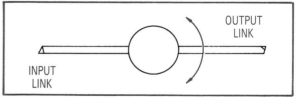

CIMLINC, INC.

Fig. 11-6. The motion of a rotational joint. This type of joint is most common on an articulated robot.

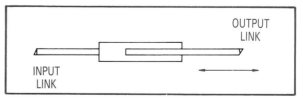

CIMLINC, INC.

Fig. 11-7. The motion of a translational joint. A Cartesian coordinate robot would use this type of joint for its motions in the X, Y, or Z directions.

The advantage of the articulated robot manipulator is its ability to reach around obstacles. It does this in much the same way that the human shoulder, elbow, and wrist work together to position the hand in space. Refer to Fig. 11-5.

Axes of motion

The number of axes of motion and the degrees of freedom of a robot are important because they dictate the *flexibility* of the robot. Axes of motion also affect the positions and orientations that a robot's *end effector* (tool or gripper) can have in space. The number of axes of motion and the number of degrees of freedom available in a robot do not have to be equal, although in general they are related.

An *axis of motion* can be defined as the rotation of one link about another link or about a stationary base. See Fig. 11-6. The translational motion of one link relative to another link or stationary base, Fig. 11-7, is also considered to be an axis of motion.

A **degree of freedom** is defined as an independent motion that can be either rotational or translational. To better understand this, consider a point in the X-Y plane. As shown in Fig. 11-8, a point in this plane can *translate* in a direction parallel to either the X or Y axis, or it can *rotate* about the origin (intersection of the two axes). Since the point can move in three independent directions, this plane is said to have three degrees of freedom.

Now, consider a point in three-dimensional space of the X-Y-Z coordinate axis. A point in three-dimensional space has a maximum of three translations and three rotations, as shown in Fig. 11-9. They are:

- Translation parallel to the X axis.
- Translation parallel to the Y axis.
- Translation parallel to the Z axis.
- Rotation about the X axis.
- Rotation about the Y axis.
- Rotation about the Z axis.

Thus, three-dimensional space has a maximum of *six* degrees of freedom.

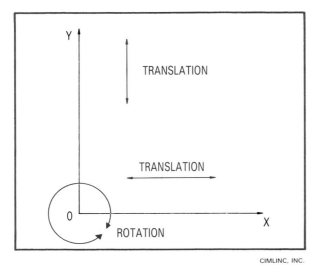

CIMLINC, INC.

Fig. 11-8. The degrees of freedom in a plane. There are two translations (linear movements) and one rotation about the origin, for a total of three degrees of freedom.

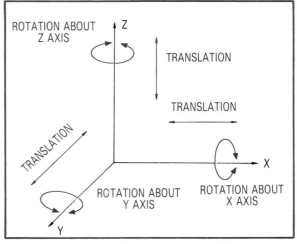

CIMLINC, INC.

Fig. 11-9. The degrees of freedom in three-dimensional space. There are six degrees of freedom in three-dimensional space: three translations and three rotations.

Now consider a Cartesian coordinate robot with three translational axes of motion. How many degrees of freedom does it have? The answer is three—the only motions possible are *translations* parallel to the X, Y, or Z axes, as shown in Fig. 11-10.

If a *rotational* motion of the end effector about the X axis of the wrist is added, a robot with four degrees of freedom will result. Clearly, it is possible to get a robot with *six* degrees of freedom by adding two more rotations (about the Y and Z axes).

Now consider adding another axis of rotation about the X axis of the end effector (which already rotates around the X axis of the *wrist*). Will this increase the number of degrees of freedom of the robot? *No*—the maximum number of degrees of freedom for a point in space is six, regardless of the number of axes of motion (in this case, seven). However, the extra axis of motion *does* give the robot greater mobility. A robot with greater mobility is more useful because it can reach around obstacles into areas that normally could not be reached. See Fig. 11-11.

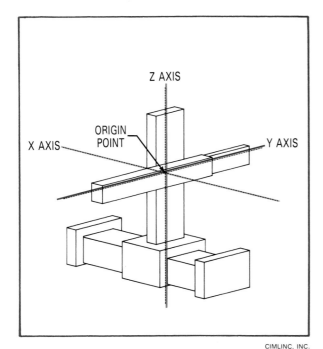

CIMLINC, INC.

Fig. 11-10. A Cartesian coordinate robot with only three degrees of freedom. Each degree of freedom is a translation parallel to the X, Y, or Z axis.

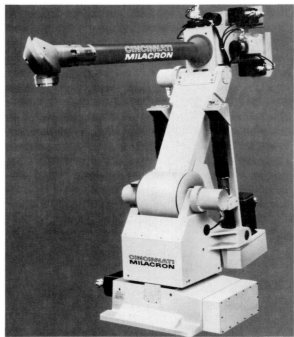

CINCINNATI MILACRON

Fig. 11-11. A robot with a high degree of mobility. The increased mobility of the articulated design allows the robot to reach into difficult locations.

End effectors

Also called *end-of-arm tooling*, end effectors are attached to the wrist of the robot and provide it the means to physically interact with its environment and perform designated tasks. The end effector may be any of a number of devices, from a spray painting gun or a welding electrode to an inspection probe to a gripper for picking up integrated circuits. Because of the variety of tasks performed by robots, it is not uncommon for this tooling and associated equipment to be designed and built for particular applications. A number of "off-the-shelf" end effectors are available.

Examples of some different types of end-of-arm tooling are shown in Fig. 11-12.

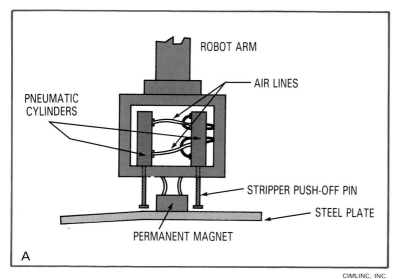

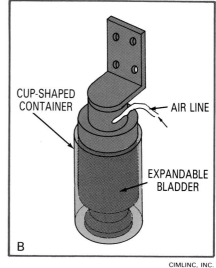

CIMLINC, INC.

CIMLINC, INC.

TYPICAL TWO FINGER MODIFICATION

TWO FINGER RIGID DISC GRIPPER

TYPICAL THREE FINGER MODIFICATION

THREE FINGER RIGID DISC GRIPPER

ROBOT ARM

AIR LINES

PNEUMATIC CYLINDERS

STRIPPER PUSH-OFF PIN

STEEL PLATE

PERMANENT MAGNET

A

CUP-SHAPED CONTAINER

AIR LINE

EXPANDABLE BLADDER

B

C

MACK CORPORATION

Fig. 11-12. Various types of end-of-arm tooling. A—Tooling that uses a permanent magnet to pick up a steel plate. B—Tooling that uses an expandable bladder to pick up a hollow cylindrical container. C— Several types of finger grippers.

Type of control

The control of a robot is accomplished by providing correct information from the controller to actuators. The information that is provided from the controller to the actuators is both position and velocity (speed) information. It allows the end effector to move to a particular point or along a pre-scribed path at a designated velocity.

Depending on the design of the robot, the controller may or may not know the absolute position and velocity of each actuator as the robot goes through the sequence of operations. This gives rise to the two basic types of control, open-loop control and closed-loop control.

Open-loop control. If the robot controller does *not* know the position and/or velocity of each actuator during the operation of the robot, the robot is said to be under *open-loop control*. When the robot is operating under open-loop control, the controller must *assume* that each link of the robot is at its correct position and has moved there at the correct velocity, so that the end-of-arm tooling has moved to the appropriate position at the correct velocity. However, the controller does not receive any feedback of actual position and velocity information, so it cannot check and correct itself. See Fig. 11-13. Thus, under open-loop control, it is possible to deviate from prescribed positions and velocities and not be able to correct a problem.

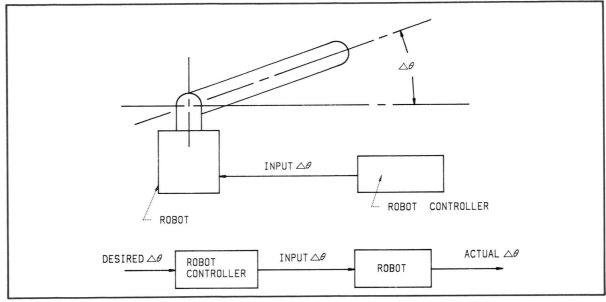

CIMLINC, INC.

Fig. 11-13. A block diagram of open-loop robot control. In this particular case, the controlled variable is position.

Closed-loop control. When the robot controller *does* know the position and/or velocity of each actuator during the operation of the robot, the robot is operating under *closed-loop control*. Under this type of control, information from position and velocity *sensors* (feedback) is provided to the robot controller. The sensor information is compared to the values that have been calculated by the controller for position and/or velocity. If there is a difference between the actual value and the calculated value, an error signal is generated. The robot controller responds to the error signal and corrects the position and/or velocity of the link or links. This keeps the end-of-arm tooling moving along the prescribed path and at the correct velocity. See Fig. 11-14.

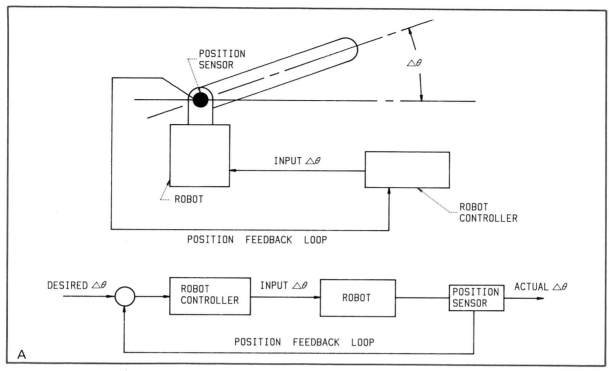

POSITION
SENSOR

△θ

INPUT △θ

ROBOT

ROBOT
CONTROLLER

POSITION FEEDBACK LOOP

DESIRED △θ   ROBOT CONTROLLER   INPUT △θ   ROBOT   POSITION SENSOR   ACTUAL △θ

POSITION FEEDBACK LOOP

A

CIMLINC, INC.

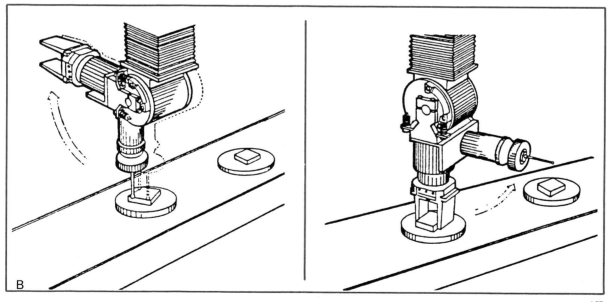

B

ACT

Fig. 11-14. *Closed-loop robot control uses a feedback loop to adjust positioning. A—Block diagram of closed-loop robot control. The position feedback loop ''closes'' the control system block diagram. B—At left, a position sensor establishes the position of a randomly placed part by sensing its X,Y, and Z coordinates. At right, gripper rotates into position to pick up the part, using the coordinate information fed back to the robot control by the sensor.*

ROBOT MOVEMENT

Actuators are pneumatic, hydraulic, or electrical devices that are used to move a robot's manipulator links. They provide the torque necessary to move rotational joints or the force necessary to move translational joints.

Pneumatic systems use a *compressible* fluid, usually air. Air under pressure is readily available throughout most manufacturing facilities, making this medium very attractive for robot actuators. See Fig. 11-15. Pressure variations can create erratic operation of robot actuators. Other problems, such as water in the system and noisy operation, can make pneumatically actuated robots undesirable.

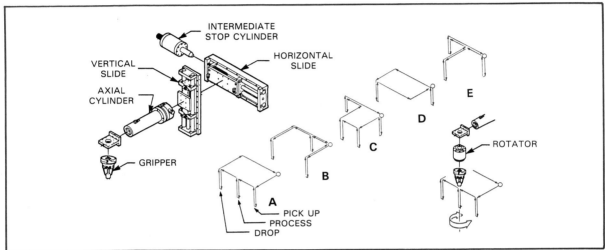

Fig. 11-15. Example of components that can be used with pneumatic actuators for movement in the X, Y, and Z axes. Diagrams A-E show typical patterns of movement. A rotator, lower right, permits rotational movement around the Z axis.

Hydraulic robots use an *incompressible* fluid, such as oil, for a working medium. System pressure is generated by a hydraulic pump, with oil supplied to the robot via high-pressure hydraulic hoses. Since high pressures can be generated with hydraulic actuators, it is possible to generate high torques and forces. Thus, hydraulic actuation is usually considered when heavy loads must be moved by the robot.

Pneumatic and hydraulic actuators can be considered together, since they operate on basically the same principles. It should be noted that actuators may also be used to operate robot end-of-arm tooling.

Pneumatic or hydraulic cylinders usually provide force in one direction (either forward or backward, depending on the configuration of the cylinder). In some applications, a double-acting cylinder with pushrods at both ends is used. The force that is developed by the cylinder depends on the pressure of the fluid and the cross-sectional area of the piston, Fig. 11-16. Note that the two areas (A1 and A2) are not equal; because of the attachment of the pushrod to the piston, A2 is less than A1.

Pneumatic or hydraulic *rotary* actuators provide torque for actuation of a rotating joint. Fig. 11-17 shows two of the available types of rotary actuators. The speed at which a pneumatic or hydraulic cylinder or rotary actuator operates is determined by the flow rate of the fluid into the actuator. Speed of operation of the actuator is obviously important, since a robot often has to track a particular path at a specified speed.

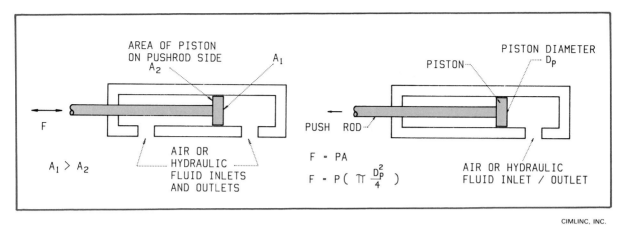

Fig. 11-16. Examples of pneumatic or hydraulic cylinders. These cylinders or actuators can be used to move robot arms or to operate grippers or other end-of-arm tooling.

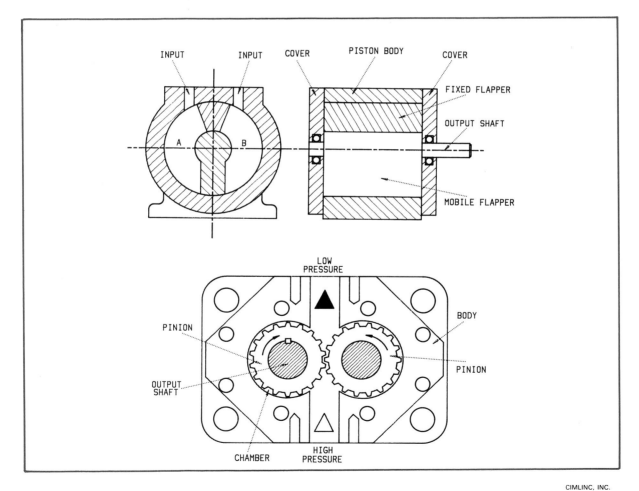

Fig. 11-17. Typical hydraulic rotary actuators include, at top, a flapper motor and, at bottom, a gear motor.

Electrical actuators are usually either DC electric motors, Fig. 11-18, or stepping motors.

The **permanent-magnet motor** is a common type of DC (direct current) motor. The motor speed can be varied by changing the applied voltage; direction of rotation can be changed by reversing the polarity of the motor input leads.

The **stepping motor** derives its name from the fact that the motor receives discrete electrical impulses that cause it to *step* (rotate) a specified number of degrees. To compute the angle of rotation, you can use the following formula:

Degrees of rotation = Number of steps x Degrees per step

The rotational speed of a stepping motor is varied by changing the frequency of the electrical impulses or (steps) flowing into the motor. The higher the step frequency the faster the rotation and vice versa.

JORDAN CONTROLS, INC.

Fig. 11-18. A rotary actuator consists of a DC servomotor and gear train. It provides precise closed-loop motion control.

Position and velocity feedback

If a robot is operated under closed-loop control, it is necessary to provide information (feedback) on position and velocity to the robot controller. This is accomplished by using sensors attached to the robot joints. Various types of sensors can be used for position or velocity feedback. The most common position sensor is an **optical encoder**; the most common velocity sensor is a **DC tachometer**.

The two types of optical encoders currently available are the absolute encoder and the incremental encoder, Fig. 11-19. The choice of which encoder to use for a robot depends on the design of the controller, the resolution desired, and the cost.

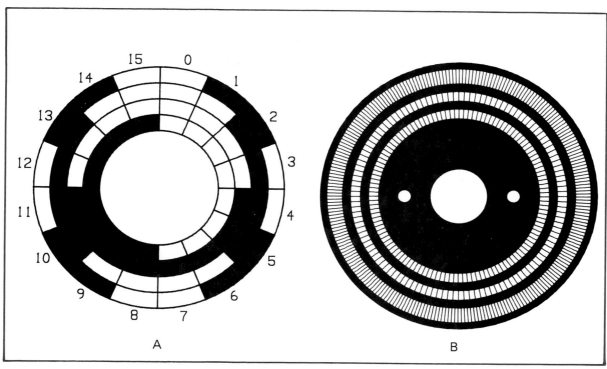

Fig. 11-19. Two types of optical encoders are used as position sensors. A—An absolute encoder. B— An incremental encoder.

An **absolute encoder** gives *exact* position of the shaft. This information is obtained from the binary-encoded information on the encoder plate. As the encoder plate rotates, light is allowed to pass through its clear areas. This generates an electrical impulse, so the output takes the form of digital data, or voltage states corresponding to *ones* (higher voltage) and *zeros* (lower voltage).

The resolution of an absolute encoder is given by:

> Resolution in degrees = $360/2^K$,
> where K = The number of channels or bits on the encoder plate.

Absolute encoders are available with 15 bits or a resolution of about 0.01°. Because of their high resolution, absolute encoders are quite expensive.

The **incremental encoder** gives positional information about the joint by counting the number of increments that the encoder plate has rotated, then multiplying this value times the angle between increments. Look at Fig. 11-19.

The formula for determining positional information is:

> Amount of rotation = Increments counted x Angle between increments

Incremental encoders may have as many as 1700 increments around the circumference of the plate, producing an angle of 0.21° between increments. Incremental encoders are less costly than absolute encoders, because they are easier to manufacture.

Velocity feedback is accomplished by using a DC tachometer. A *DC tachometer* is nothing more than a direct-current generator attached to the rotating shaft of the motor. The voltage output of the DC tachometer is proportional to the shaft's rotational speed. See Fig. 11-20.

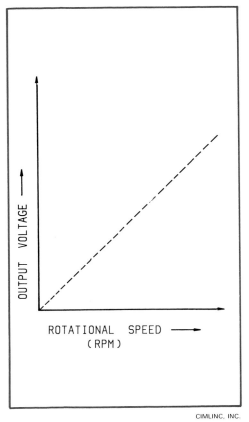

Fig. 11-20. The output voltage-rotational speed relationship for a DC tachometer.

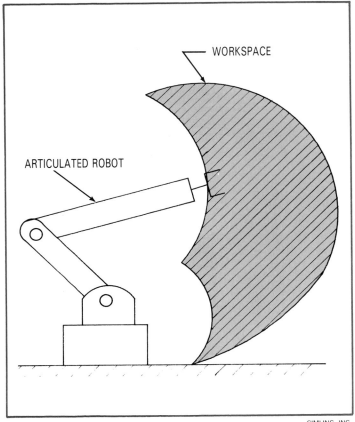

WORKSPACE

ARTICULATED ROBOT

Fig. 11-21. Planar view of a robot's workspace. The shape is typical of a articulated robot.

ROBOT PROGRAMMING

As noted earlier, the robot workspace or work *envelope* is the area surrounding the robot that contains all points that can be reached by the end-of-arm tooling. Consideration of the work envelope is very important during the selection of a robot, because the robot must be able to perform the specified task on a part without need to relocate either the robot or the part.

The workspace of a robot is usually presented in a planar view, as shown in Fig. 11-21. An important point to remember when looking at drawings of a robot's work envelope is that the workspace is a three-dimensional *volume*, not a two-dimensional *area*, as the planar view suggests.

Robot coordinate system

The *robot coordinate system* is defined as the motion geometry of the end-of-arm tooling, and may or may not be the same as the *robot geometry*. When programming the robot, the robot coordinate system to be used can often be selected by pressing a button that instructs the robot controller to move the end effector according to the specified coordinate system.

Fig. 11-22 shows an example of an articulated robot that can move in Cartesian, cylindrical, or polar coordinates.

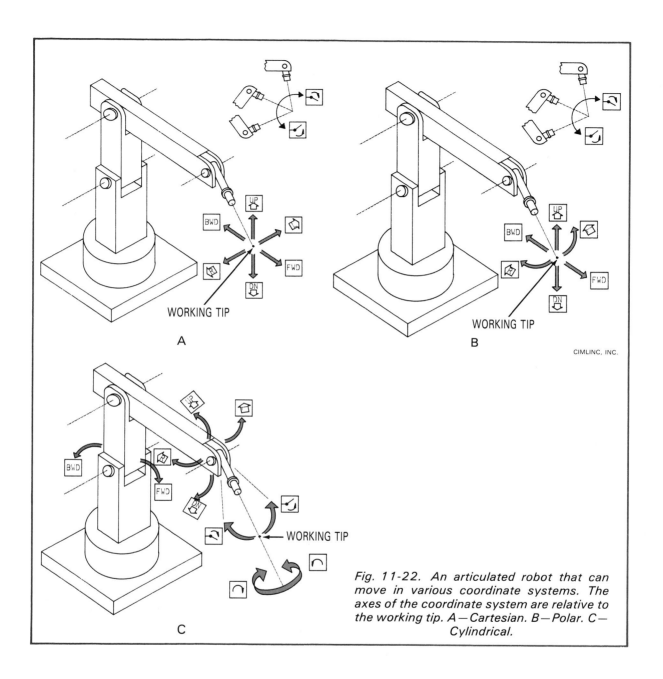

Fig. 11-22. An articulated robot that can move in various coordinate systems. The axes of the coordinate system are relative to the working tip. A—Cartesian. B—Polar. C—Cylindrical.

CIMLINC, INC.

Path programming

Applications determine whether control of the robot's path is important. For example, the path may not be important if the robot is spot welding, but is *critical* if the robot is spray painting. Therefore, a robot may be programmed in the point-to-point mode or for movement along a continuous path.

Point-to-point programming. When this programming mode is used, specific points that are input or taught by the operator are stored in the robot controller and played back when the robot is operating. Motion of the end-of-arm tooling, in the point-to-point programming mode, is said to be *path-*

independent, because only the taught points themselves are important locations. From a programming standpoint, it is very important to remember that the path of the robot between the taught points is not necessarily straight. Intermediate points may have to be taught so that the robot does not collide with the part, tooling, or another robot. See Fig. 11-23.

Continuous path programming. Robots may be programmed for continuous path operation through use of either linear interpolation or circular interpolation. Sometimes, a combination of the two is used.

When using *linear interpolation*, the beginning and end points of the robot path are taught by the operator. When the robot program is executed, the controller generates the intermediate points between the taught beginning and end points, so that the robot moves along a straight line.

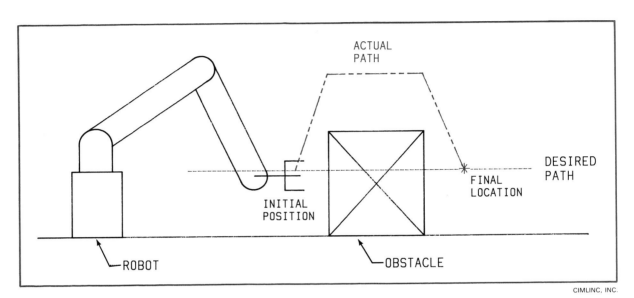

CIMLINC, INC.

Fig. 11-23. Robot programming for obstacle avoidance. Since an obstacle prevents the robot from moving in a straight line between the initial and final positions, two intermediate points must be programmed to avoid the obstacle.

Circular interpolation is done by teaching at least *three* points. The robot controller then defines a circle that will pass through those three points. On playback, the controller will generate a circular path through the three points. (In practice, a fourth point is taught very close to the first point, so that the circle will be closed.) Circular interpolation can also be used for defining segments of circles, as shown in Fig. 11-24.

It should be noted that the path the robot travels along during linear or circular interpolation is not a perfectly straight or smoothly curving line. Rather, the movement is within a tube-shaped boundary with a radius that is equal to the robot's dynamic repeatability (accuracy of positioning while in motion), Fig. 11-25. This may be an important consideration when selecting a robot, if the application requires the robot to track a path quite closely.

Programming methods

Programming of robots can be done using one of three methods. These three methods are:
- Active robot teaching (teach pendant).
- Passive robot teaching (lead-through).
- Off-line robot programming.

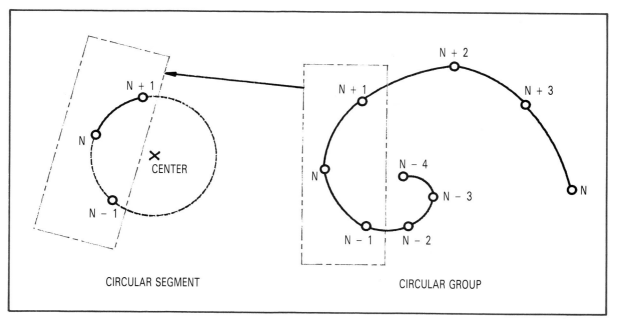

Fig. 11-24. *The use of circular interpolation in robot programming. Circular interpolation can be used for circles or arcs of circles.*

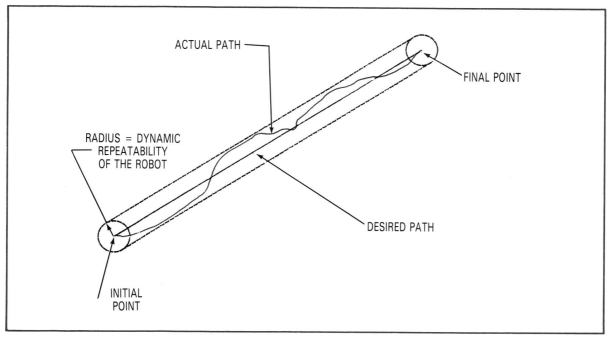

Fig. 11-25. *The actual robot path resulting from the use of linear interpolation. The actual path is not quite a straight line, but only deviates from the desired path by an amount equal to the robot's dynamic repeatability.*

Which programming method is used may depend on the manufacturing process and the work envelope. Unfortunately, most robot controllers only allow the robot to be programmed using one of these methods. In general, a robot that is programmed using a teach pendant (active robot teaching) cannot be taught by the lead-through method (passive robot teaching), and vice versa.

In the ***active robot teaching*** method of programming, the robot's actuators and feedback sensors are activated. Using a teach pendant, Fig. 11-26, the operator "drives" the robot's end effector (end-of-arm tooling) to the desired location and orientation. When the correct location/orientation is achieved, the operator "teaches" that point to the robot controller, which stores it in memory for playback. The pendant allows the operator to teach the robot controller how to perform other tasks, such as:

- Setting the speed for end-of-arm tooling movement.
- Establishing path of movement (using linear or circular interpolation).
- Turning outputs on or off.
- Receiving information from inputs.
- Setting welding parameters.

Approximately half of all robots in use today are programmed using the active method.

CINCINNATI MILACRON.

Fig. 11-26. Active robot teaching. The operator uses a teach pendant to move the end effector to the desired location. Movements, positions, and orientations are retained in memory and become the program for the robot's task.

Passive robot teaching is a robot programming method in which the feedback sensors are activated, but the robot actuators are deactivated. To program a robot using this method, the operator uses hands or manipulators to grasp the end-of-arm tooling. The operator then leads the robot along the desired path at the correct speed. The feedback sensors at the robot's joints send position and velocity information back to the controller where it is stored. When the robot is placed in operation, the controller plays back the stored information to generate the desired path and functions. Passive robot teaching is used for about one-third of all robots in present use.

Off-line robot programming is a programming method that does not involve use of either the actuators or feedback sensors in the teaching process. Instead, the robot program is developed by typing in the locations and orientations of the end effector at each stage of the process. The program is then stored in the robot controller for playback.

A variety of off-line programming languages are available which allow the operator to input locations/orientations. It also allows the operator to perform logical operations, control inputs and outputs, adjust operating parameters, and define coordinate axes around different parts or tooling. Off-line programming accounts for about 15 percent of all robot programming.

The use of **computer graphics** (both wire-frame and solids modeling), has made it possible to simulate robot motion. When the desired robot motion is complete, the computer software writes an off-line program of the simulation and downloads this information to the robot controller for playback.

DESIGN, JUSTIFICATION, AND IMPLEMENTATION OF ROBOTIC SYSTEMS

A robot is usually used with other machines and tooling in a robotic workcell. To successfully implement a robotic workcell, consideration must be given to:

- Layout of robotic and associated equipment.
- Design of the end-of-arm tooling (end effector).
- Robotic peripheral equipment.
- Communications between the robot and other machines within the workcell.
- Justification of the system.
- Bringing the system on-line in an already existing manufacturing environment.

DESIGNING THE ROBOTIC WORKCELL

To lay out a robotic workcell, you must have a thorough knowledge of the manufacturing operations necessary to perform the specified task on the part. With this knowledge, you can then make a preliminary assessment of the requirements for the robot, such as: number of axes of motion, work envelope size, and the method of programming. This information permits you to seek information from manufacturer on specific robots that might be used for this application.

A preliminary layout of the robotic workcell can now be made, using all equipment that is necessary to complete the desired manufacturing operations. It is important to realize that the workcell layout involves not only the physical location of the equipment, but the type and number of machines needed, and the flow path of the part through the workcell. Laying out of the workcell, like any design process, is interactive in nature, but a few guidelines can be useful to minimize the number of iterations. These guidelines include:

- *Identify the rate-limiting step in the workcell.* This is important, so that an attempt can be made to reduce the time of the longest operation to about the same time as any other operation in the workcell.
- *Maximize operational time of the robot and associated equipment.*
- *Be aware that collisions can occur between the robot and associated equipment.*

- *Make provision for the safe and efficient movement of parts to and from the workcell.*
- *Design the layout for ease of access to machines to simplify maintenance.*
- *Make all possible provisions for operator safety.* After making a first pass at the design of the robotic workcell, you can address such specifics as end-of-arm tooling, robotic peripheral equipment, communications within the workcell, cost justification, and workcell implementation. Remember, this is an interactive process; it may require rework of any specific area before the design of the robotic workcell is complete.

Design of end-of-arm tooling varies from one application to another. However, *grippers* are very common end effectors, and may be designed for specific applications. Grippers take on a variety of shapes and sizes, but are basically similar in design.

Consider a very simple single-action gripper like the one shown in Fig. 11-27. This gripper is considered to be a single-action type, because only one finger moves while the other remains stationary. Single-action grippers are simple and can be used for picking up rectangular objects.

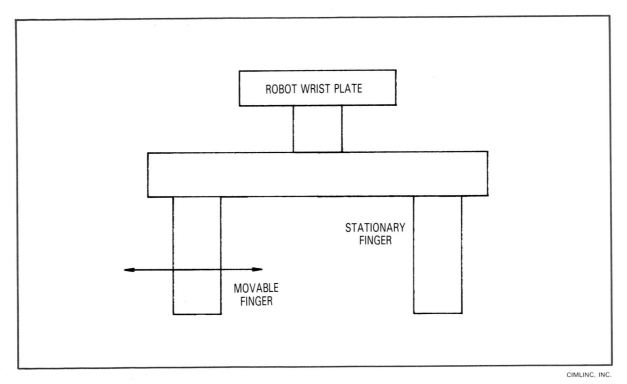

Fig. 11-27. An example of a single-action gripper. In this case, only one finger moves.

Another simple end effector is the double-action gripper, shown in Fig. 11-28. It is defined as a double-action gripper because the fingers move toward each other simultaneously. This particular gripper is good for picking up rectangular objects and will center the rectangular object in the middle of the gripper.

Other grippers have pads attached to the fingers to compensate for any misalignment during operation of the robot. These pads can also be shaped like v-blocks to pick up cylindrical objects, as shown in Fig. 11-29.

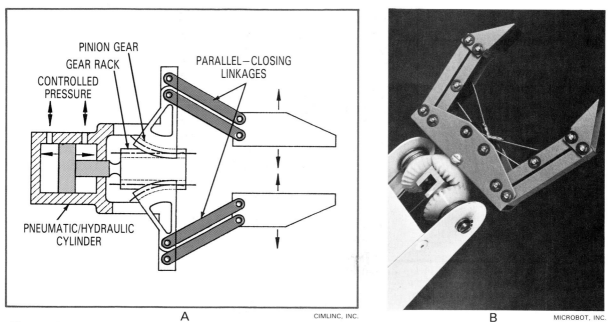

Fig. 11-28. Examples of double-action grippers. A—When the double-action gripper operates, both fingers move. B—A double-action, cable-actuated gripper used on a robot intended for classroom use.

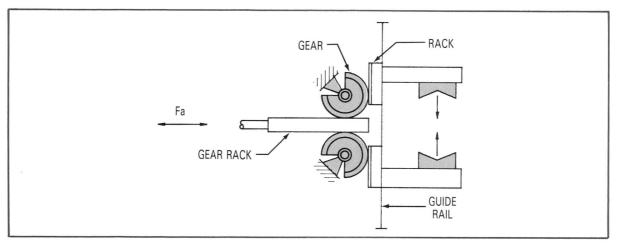

CIMLINC, INC.

Fig. 11-29. A double-action gripper with pads. The pads are attached to the fingers to accommodate parts of different shapes.

End-effector considerations

When purchasing or designing an end effector, it is very important to match total weight of the end-of-arm tooling and the part to be handled to the *maximum payload* that can be carried by the robot. It is also important that the torque on the wrist plate of the robot not exceed the manufacturer's specified limits.

Determination of the maximum weight of the end-of-arm tooling, the torque generated by the end-of-arm tooling, and the payload is a fairly simple problem. Consider the following example:

Robot Specifications

Maximum weight of the robot payload = 25 lbs.
Maximum torque on the robot wrist plate = 125 in./lbs.
Weight of part to be carried = 9 lbs.

What is the maximum weight of the end-of-arm tooling?

Maximum robot payload = total weight of the end-of-arm tooling + weight of part.
25 lbs. = total weight of end effector + 9 lbs.
Total weight of end effector = 25 lbs. − 9 lbs. = 16 lbs.

Therefore, the maximum weight allowable for end-of-arm tooling is 16 lbs.

The torque created by the end-of-arm tooling and part must also be checked, so that it does not exceed the maximum permitted by specifications.

The torque on a wrist joint, like the one shown in Fig. 11-30, can be computed as follows:

12 lbs. (3 in.) + 9 lbs. (6 in.)
= 36 in./lbs. + 54 in./lbs.
= 90 in./lbs.
90 in./lbs. ≤ 125 in./lbs., so torque is O.K.

Now check that the end-of-arm tooling (for example, a gripper) and the part weight and location are within the specifications of the robot.

12 lbs. + 9 lbs. = 21 lbs.
25 lbs. ≥ 21 lbs., so weight is O.K.

If either the weight or the torque had exceeded the robot's maximum limits, the tooling would have to be redesigned, the part location in the end-of-arm tooling moved, or a different robot selected for the workcell.

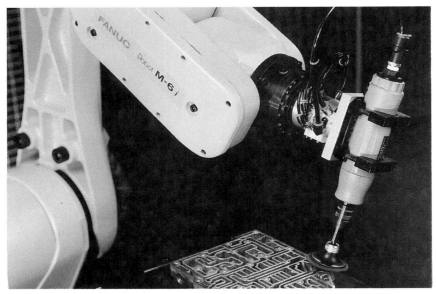

FANUC ROBOTICS NORTH AMERICA, INC.

Fig. 11-30. End-of-arm tooling attached to the robot wrist. The robot wrist allows for a variety of end effectors to be attached for different tasks such as the deburring illustrated above.

In addition to grippers, other types of end-of-arm tooling, such as gas metal arc welding torches, spray painting guns, or grinders, are available. Specialized tooling can be designed, as well, for use in various applications. However, the tool designer must be careful to not exceed the maximum limits of the robot.

Compliance. End-of-arm tooling may be designed so that it is *compliant*, to prevent damage from small, unwanted loads that are placed on the end-of-arm tooling. These small loads may come from misalignment of the parts, or other unplanned obstacles in the robot's path.

Compliance may be designed into the system in either passive or active form. ***Passive compliance*** is achieved by using an elastic medium at some point along the end-of-arm tooling, so that deflections can occur when loaded. This is shown in Fig. 11-31.

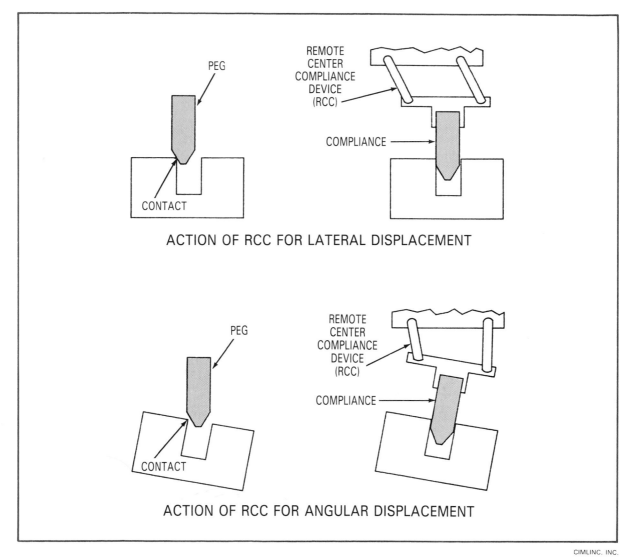

Fig. 11-31. *Passive compliance for end-of-arm tooling allows for small amounts of misalignment in the mating parts.*

Active compliance makes use of sensors in the end-of-arm tooling to detect any unwanted loads that are present. Information from the sensors is passed to the robot controller, which moves the robot in real time to remove the load.

Force sensing. *Force sensing* on the end-of-arm tooling can be accomplished by mounting strain gauges at desired locations. Strain gauges measure strain, which is related to stress. If the value of stress can be measured, and the geometry of the end-of-arm tooling is known, the *force developed* can be calculated, using concepts from strength of materials. Force sensing is necessary when using active compliance.

Fortunately, there are commercially available sensors that can be attached between the end-of-arm tooling and the wrist plate of the robot. These force sensors measure forces and torques in the X, Y, and Z axes.

Tactile sensing. Tactile sensing can also be incorporated into the end-of-arm tooling. *Tactile sensing* enables the robot to "feel" the object in much the same way that humans feel through their finger-tips.

The simplest tactile sensing device is a microswitch that provides the robot controller with binary information when contact has been made with an object. Another type of tactile sensor, pressure-sensitive conductive rubber, provides analog information that must be digitized before being sent to the robot controller. Other materials that are *piezoelectric* (capable of generating an electric current as a result of pressure) can be used for tactile sensing, as well.

Regardless of the type of tactile sensing device used, the reasons for sensing include:
• Determining local shape of the part.
• Finding orientation of the part.
• Detecting gripping pressure.

Tactile sensing allows parts that are soft, weak, or delicate to be handled by a robot.

Machine vision. The use of machine vision with robotic systems is becoming more widespread. Machine vision can be used to identify, locate, inspect, or measure parts, Fig. 11-32. For a successful machine vision installation, attention must be given to six areas:
• Part presentation.
• Part lighting.
• Camera selection.
• Optics.
• Interface to computer.
• Computer algorithm used.

Part presentation is important so that the camera views the part in the center of its field of view. The camera can look at the part from the top or look at it from the side. In some cases, the part may be viewed with two cameras at once for *stereoscopic* (3-D) viewing.

Lighting of the part is extremely important. By backlighting the part, a silhouette results. This gives a good outline of the shape, because of the high contrast. When *front* lighting is used, care must be taken to minimize shadows, since they could mask the part's shape and make the interpretation of the part by the computer more difficult.

Cameras use either a vidicon tube or a solid state element, such as a CCD (charge-coupled device), to register the image. A vidicon-tube camera uses a photoconductive vacuum tube for imaging and is relatively inexpensive.

Solid state cameras are arrays of photosensitive elements that generate an electrical charge in response to light falling on them. Solid state cameras are relatively expensive, but more rugged than vidicon cameras.

The camera *lens* is an important part of the machine vision optics. The focal length of a lens will change the field of view, while the aperture setting changes the amount of light that enters the camera, thus changing the depth of field (portion of the field of view that is in sharp focus).

Fig. 11-32. A machine vision system being used to aid placement of electronic components on circuit boards.

The *interface* between the camera and the computer must perform three functions:
- Timing.
- Digitization.
- Buffering.

The timing circuit coordinates the picture being processed with the digitizing hardware. This is important so that the object being viewed is in harmony with what the computer thinks it is looking at, Fig. 11-33.

The digitizing hardware takes analog signals and converts them to binary information. This is done using an ***analog-to-digital converter***.

When a ***buffer*** is used in a machine vision system, digitized information is stored in the buffer until the computer can process it. This is often necessary because of the length of time that is required by the software to process this information.

There are two computer ***algorithms*** (sets of rules used to solve problems step-by-step) that can be used to determine the shape of the part. The simplest algorithm is template matching.

With ***template matching***, the shape shown by the camera is compared with known shapes stored in the computer. This method is useful when you want to determine if the whole observed part corresponds to what it is expected to be.

Connectivity analysis is the second algorithm that is used in machine vision software. The information is stored as binary images, allowing rapid processing when attempting to determine the shape. The part shape can be segmented into connected components and analyzed one component at a time.

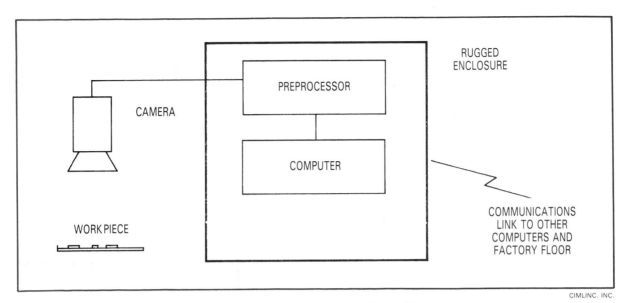

Fig. 11-33. The components of typical vision system.

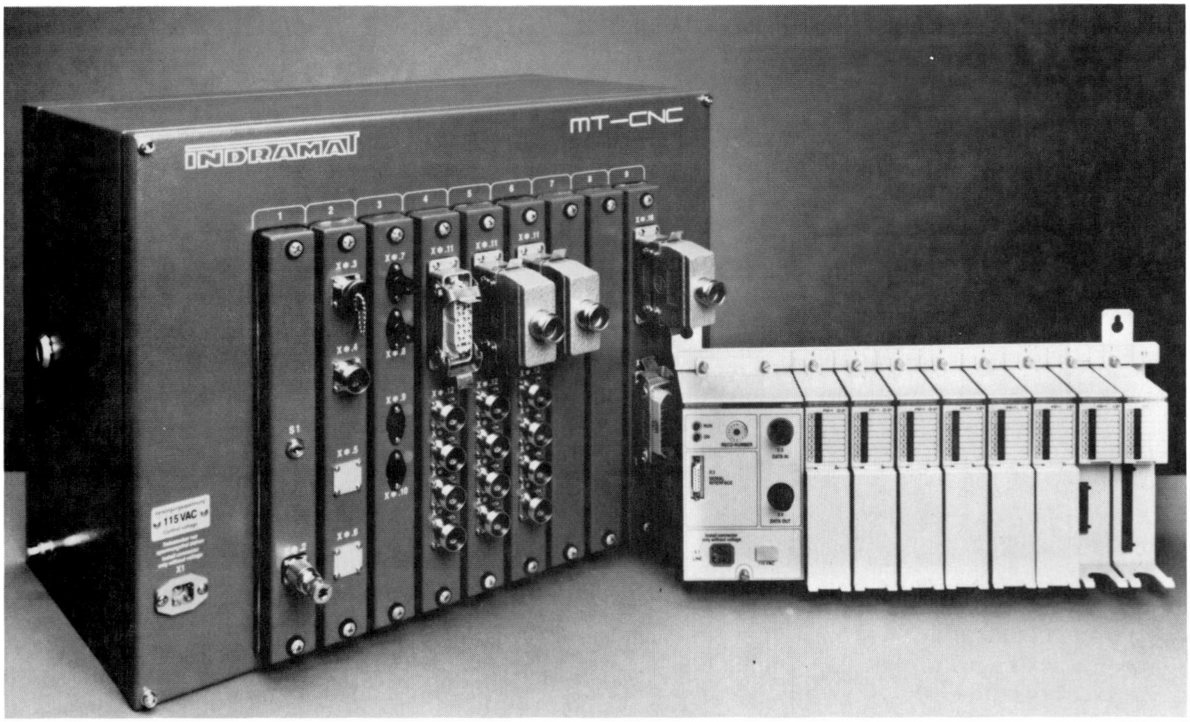

Fig. 11-34. A programmable logic controller and CNC functions are integrated in this multi-tasking, multi-axis controller. It can simultaneously control and coordinate several robots and machine tools (up to 20 axes of motion) through a fiber-optic interface.

294 *Computer-Integrated Manufacturing*

Workcell control

Coordinating communication of the robot with other robots, machine tools, machine vision systems, and material handling systems is a complex task. Although it is desirable to monitor the robotic workcell from a factory-wide host computer, a controller at the workcell level is necessary. This could be accomplished by using a microcomputer or even a *programmable logic controller* (PLC) to oversee the workcell. A sophisticated robot controller could also control the workcell, Fig. 11-34.

The advantage of using a dedicated microcomputer as the workcell controller is that it controls the operation of the workcell alone, and not a particular machine or process. A robot controller, however, would have to control the robot as well as the workcell.

JUSTIFYING A ROBOTIC SYSTEM

When justifying a robotic system, you must consider the economic implications. Each organization has its own method of determining the payback period or return on investment (ROI). It is used to determine whether or not a piece of equipment should be purchased.

The simplest formula used to determine the payback period, which is also known as the *capital recovery period*, is:

$$P = \frac{I}{S\text{-}E}$$

Where: P = number of years for payback.
  I = total investment in robot and tooling.
  S = total annual labor savings.
  E = total annual expenses of the robot.

If the number of years for payback is less than the number of years determined by the organization, the robot is economically justified.

IMPLEMENTING A ROBOTIC SYSTEM

Implementation of a robotic workcell involves the purchase of a robot and other necessary equipment and tooling, education of personnel, and scheduling of existing manufacturing operations to allow for installation and start-up of the new equipment.

The purchase of a robot and other equipment involves obtaining quotations from vendors. In dealing with robot vendors, a performance specification is often used.

The performance specification is an agreement between the buyer and vendor that states the robot or robotic system will not be purchased by the buyer until the vendor proves that the robot can perform the task designated by the buyer.

When implementing a robotic system, education is essential for two different groups of employees. The first group consists of those people who will deal directly with the operation and maintenance of the robot. The second group includes those who work in the factory, but whose jobs are not *directly* related to the operation of the robotic system.

The people whose work will be directly related to the operation of the robot can obtain specific information from the manufacturer. The robot manufacturer provides instructional manuals and training courses in maintenance or operation of that specific robot model.

People who are not directly involved with the operation of the robotic system must understand why it is being installed and how it will operate. This is very important to the development of favorable employee attitudes toward installation of the equipment.

Some downtime of the existing manufacturing facility may be experienced during installation of a robotic system. The effects of such downtime can be minimized with proper planning. Creating higher-than-normal inventories or making use of other production facilities are ways of minimizing the effect on the manufacturing facility during robotic system installation.

ROBOTIC APPLICATIONS

One of the earliest robot applications was in resistance spot welding. This particular early robotic application used point-to-point robots. Since resistance spot welds are made at specific points, a continuous path robot was not necessary.

The payload capacity of a spot welding robot has to be rather large. The large capacity is needed because, even though the welding transformer is remotely located, the weight of the spot welding tips, power cables, and air hoses is quite formidable.

Spray-painting robots are usually taught in the passive (lead-through) mode by an operator. Obviously, the best robot operator in this case would be a person who is quite proficient in spray painting. A spray-painting robot must be able to handle moderate payloads, and should have its exposed components (such as joints and motors) covered so that paint overspray will not contaminate them.

The use of robots for electronic assembly is growing rapidly. These robots must be fast and have a high repeatability, typically about ±0.004″ (0.010 mm). Fortunately, this application has low payload requirements, which enhances repeatability. Programming of the robot can be done in the active mode (using the teach pendant) or by off-line programming.

Machine loading can be done by robots that can handle large payloads at reasonably fast speeds. The repeatability of the robot can be relatively low. The repeatability may range from ±0.008″ to ±0.100″ (0.020 to 0.254 mm), depending on the specific machine and application.

Future trends

The use of robots in industry is expected to show continued growth. The *applications* for which the robots are used are shifting somewhat. Use of robots for assembly and inspection tasks is expected to increase, while welding and painting applications will decrease. Material transfer and machine tending applications will continue at approximately the same level.

SUMMARY

An industrial robot can be defined as a *reprogrammable multifunctional manipulator*. *Programmable* means that a device has the capability of acquiring more than one set of instructions. This means that the robot can perform more than one task. A *multifunctional* machine is capable of performing a variety of tasks. *Manipulator* implies the movement of parts or tooling through space. The industrial robot consists of various mechanical, electrical, and electromechanical devices that permit predetermined movement of a tool or workpiece through space.

The robot's four major areas of function are the *robot geometry*, the *end-of-arm tooling*, the *number of axes* or *motions*, and the *control method*.

The four basic *robot geometries* are Cartesian, cylindrical, spherical, and articulated.

The *axes of motion* are important because they dictate the flexibility of the robot, as well as the positions the end effector on the robot arm can occupy in space.

The *end-of-arm tooling* attaches to the wrist of the robot and provides the means for the robot to interact with the environment. The tooling can vary from a paint gun, to a weld head, to a parts gripper, to a heat-treat rack.

The *robot control* is the electronic means of providing information and instructions to the actuators that control physical movement of the robot. Robot controllers are built to be either *open-loop* or *closed-loop*. Usually closed-loop systems are preferred, because they give the robot controller continuous knowledge of position and velocity of movement during operation.

The movement of the robot joints, arm, and end effectors is accomplished by pneumatic, hydraulic, or electromechanical actuators. These actuators provide torque to move or rotate the robot.

The methods used to program robotic systems include *active teach pendant programming*, *passive lead-through programming*, and *off-line programming*.

Active robot teaching is defined as a robot programming method where actuators and feedback sensors are activated using a teach mode pendant. The operator, using the teach pendant, simply drives the robot around the desired path one step at a time. Then the location and orientation of each step is stored in the robot controller memory for use in production operations.

Passive robot teaching is a programming method in which the feedback sensors are activated but the robot actuators are deactivated. In this method, the operator holds the end of the robot tooling and leads it through its desired path, at the desired velocity. The coordinate locations and the travel speeds are fed back from the sensors to the robot controller. Under production conditions, the robot controller plays back the stored information to generate the robot path.

The *off-line robot programming* system is a programming method in which both the actuators and feedback sensors are deactivated and the robot program is typed into the robot controller or downloaded to the controller. Some of the latest systems use computer graphics to simulate the robot motion, and download this path into the controller automatically.

In the design of the robot, there must be thorough analysis of such items as type of gripper, type of end effector or end-of-arm tooling, wrist torque load, robot arm weight load, types of sensors, type of robot controller, and the use of vision. If these items are not properly calculated, the robot system will not be successful.

When justifying a robotic system, the economic implications must be considered. Each organization has its own method of determining the payback period or return on investment (ROI). It is used to determine whether or not a piece of equipment should be purchased.

Implementation of the robotic system must consider plant layout, other support equipment and tooling, the performance specification of the supplier, employee training, maintenance instructions and spare parts, and the employee time needed to develop the programs for the production runs. If this phase is not well-planned, serious problems may be experienced in achieving successful implementation.

IMPORTANT TERMS

absolute encoder
active compliance
active robot teaching
actuators
algorithms
analog-to-digital converter
articulated (jointed-arm) robot
axis of motion
buffer
capital recovery period
Cartesian coordinate robot
circular interpolation
closed-loop control
computer graphics
connectivity analysis
cylindrical coordinate robot
DC tachometer
end effector
force sensing
incremental encoder

linear interpolation
manipulator
off-line robot programming
open-loop control
optical encoder
passive compliance
passive robot teaching
permanent-magnet motor
programmable logic controller
real time
reprogrammable
robot coordinate system
sensors
spherical coordinate robot
stepping motor
stereoscopic
tactile sensing
template matching
work envelope

QUESTIONS FOR REVIEW AND DISCUSSION

1. Name the possible geometric configurations of a robot manipulator. Describe the advantages and disadvantages of each configuration.

2. Consider a Cartesian coordinate robot with three translational axes of motion and one rotational motion (about the X axis). How many degrees of freedom are possible with this robot? List them.

3. Contrast the *open-loop* and *closed-loop* control methods used with robots.

4. What are actuators? How are they used in a robotic manipulator? Which type would you choose to handle a heavy load? Why?

5. Describe the two types of interpolation used to program a robot for continuous path operation.

6. Name the most common method of robot programming, and describe how it is performed.

7. What guidelines should be followed in designing a robotic workcell?

HYPERTHERM

Advances in plasma arc cutting technology have resulted in oxygen arc systems that have a higher arc current density and a narrow kerf, permitting precision cutting of metal components.

Advanced Manufacturing Processes

by Russell Biekert

Key Concepts

☐ Most types of advanced (nontraditional) manufacturing processes could be integrated with some form of automation into a total CIM system.

☐ Electrical Discharge Machining uses a carefully controlled electric spark for precise material removal. It produces extremely smooth, burr-free finishes.

☐ Reversing the common process of electroplating, Electrochemical Machining removes a controlled amount of material from a workpiece by electrolysis. The tool (electrode) is shaped as a mirror image of the desired workpiece contour.

☐ Lasers are achieving wide use for such processes as drilling, cutting, welding, and marking. The ND:yag (neodymium/yttrium aluminum garnet) laser is the most powerful and widely used type.

☐ The kinetic energy of moving electrons is transformed to heat in the joining method known as Electron Beam Welding.

Overview

The term ***nontraditional manufacturing processes*** is not a new one. It has been defined many different ways in the last two decades. There has been much debate regarding whether some processes should be considered nontraditional or conventional.

This chapter will consider those advanced manufacturing processes that have computer-assisted nontraditional applications. Most of the processes discussed in this chapter could be integrated with some form of automation into a total CIM system.

Russell Biekert has held a number of key positions in the manufacturing and engineering areas for Garrett Engine Division, Allied Signal Aerospace Company in Phoenix, Arizona.

ELECTRICAL DISCHARGE MACHINING

Electrical discharge machining (EDM) is an advanced production process that uses a fine, accurately controlled electrical spark to erode metal. The frequency of sparks, or *discharges*, is usually many thousands of cycles per second. Each spark removes a tiny amount of material from the workpiece. A very small amount of material is also eroded from the electrode. The *spark gap* (space between the electrode and the workpiece) is flooded with dielectric oil.

The EDM process is similar to what happens if the clamps at one end of a set of jumper cables accidentally touch after the clamps at the other end have been connected to terminals of an automobile storage battery. A powerful spark is produced, and the two clamps show signs of cratering, pitting, and metal removal where the spark has jumped. It is this same cratering phenomenon that makes EDM possible. However, with EDM, both the spark and the arc gap are carefully regulated for precise material removal.

THE EDM ELECTRODE

Fig. 12-1 is a schematic representation of the EDM process. The tool, or electrode, is shaped in the mirror image of the desired configuration of the workpiece. It is mandatory that the cutting gap be maintained accurately at all times for proper operation. Tool feed is accomplished through the use of a special servomechanism that acts as a sensing device. Based on the condition of the spark and gap, the servomechanism plunges or retracts the tool.

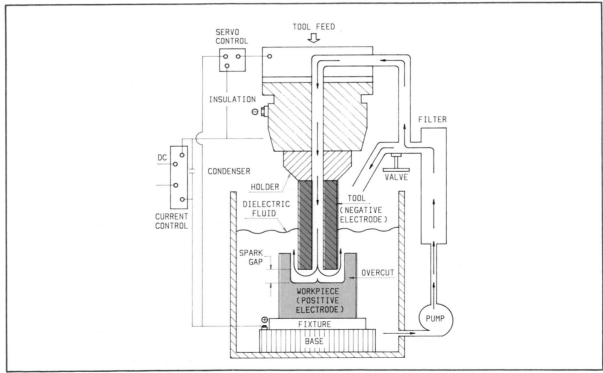

CIMLINC, INC.

Fig. 12-1. An EDM system is shown in cross section. The main components are the electrode, the workpiece, the dielectric servo control head control, spark gap, and filter system.

The tool usually, but not always, has a negative electrical charge. The tool pictured in Fig. 12-1 is typical of cylindrical electrodes. Wherever possible, the electrode is either hollow or contains passages to allow dielectric oil to be pumped through it. This is done to help flush the sediment of the cutting operation from the gap area. In some situations, dielectric oil is used to cover only the area being machined; in others, the whole workpiece may be submerged in dielectric oil, as shown in Fig. 12-1.

The dielectric oil is usually stored in holding tanks inside the frame of the machine itself. The oil is pumped through a filter, then introduced into the spark gap to flush away particles resulting from the EDM process.

The typical spark gap ranges from 0.0005" to 0.020" (0.0127 mm to 0.508 mm). When the electrical potential between the workpiece and the tool is sufficiently high, the dielectric is partially ionized in the gap area, allowing a spark to jump (arc across) the gap.

The power supply current output may range from 0.5 amps to 400 amps; the voltage output may vary from 40 volts to 400 volts. DC or direct current is used, typically at a frequency of around 200 Hz. This produces pulses of about 20,000 sparks per second.

The circuit capacitor is variable in ranges from less than 0.001 microfarad (mfd) to higher than 400 mfd, and serves to alter and intensify the discharge. Generally speaking, the higher the capacitance, the faster the cutting and the more stable the servo feed.

Tool feed designs

There are two basic designs for tool feed: ram and quill. Ram-feed machines are typically heavier duty than the slightly less expensive quill-feed machines.

Ram-feed machines use a hydraulic cylinder to control the movement of the head, *quill-feed machines* use a hydraulic motor to drive a *leadscrew*. Both systems allow for controlled advancing and retracting of the tool. The servo control compares a selected reference voltage set on the control panel with the actual gap voltage, and causes the tool to either advance or retract to adjust for precise machining. If a short-circuit occurs (caused by metal chips bridging across the gap between the electrode and workpiece), the servo will retract the tool until arc control is restored.

Most EDM machines allow for rotation of the electrode to achieve more even tool wear. This feature is useful, of course, only when the electrode is cylindrical in shape. When irregular shapes are being machined, the spindle is locked to prevent accidental rotation of the electrode.

Since EDM tooling is expendable, the electrode may have to be replaced several times to finish a given job. It is therefore necessary to provide a tool-holding method that reliably relocates the new electrode. Special chucks and fixtures may be required for this purpose.

Electrode materials

Typical electrode materials are copper, brass, graphite, and copper tungsten. Since the electrode is expendable, the rule of thumb is to try to find the cheapest electrode material that will do the job.

Graphite is, by far, the most widely used type of material for electrodes. It is easily machined, low in cost, light in weight, and has excellent wear characteristics. However, graphite requires better dielectric flushing than most other electrode materials. In addition, care must be taken not to exceed the current-carrying ability of a particular electrode because of its size.

Under certain conditions, with the proper machine settings, *no-wear machining* can be accomplished using graphite electrodes. The term "no-wear" means that workpiece erosion is accomplished with practically no loss of electrode. The frequency setting for no-wear conditions is very low. The no-wear technique produces a rough surface finish, so it is generally used only in roughing operations.

Copper tungsten is a very dense, heavy, brass-colored material. Electrodes of copper tungsten provide excellent wear rates and good detail, and are extremely useful in cutting tungsten carbides. Copper tungsten is easily machined and allows for more minute detailing than graphite (which is more brittle). Likewise, it is not as easily damaged by rough handling before and during use, as is graphite.

Brass is ideal for some electrode applications, since it is readily available and easily machined. Brass is relatively inexpensive and is produced in many standard shapes and sizes. For production hole sinking, brass rod or tube is often preferred because of its low cost and off-the-shelf availability in finished sizes. Brass is a poor choice for complex cuts, however, because of its relatively slow removal rates and low wear ratio.

Low-wear (and sometimes *no-wear*) conditions can be obtained when using *copper* electrodes. Electrodes made from copper produce extremely smooth finishes on the workpiece. Free-machining grades of the metal are easily cut into complex shapes. Free-machining grades are not readily available in all common sizes, however. Copper is slightly more expensive than brass, but has better wear characteristics.

In EDM, the electrode does not touch the workpiece. Therefore, *workpiece hardness* does not control the choice of electrode material. For example, brass may be used to machine anything from soft steel to tungsten carbide.

Regardless of the electrode material, it usually will have to be altered in shape to produce the desired hole or cavity in the workpiece. An exception to this statement would be the already-mentioned example of hole sinking with off-the-shelf brass tubing or bar stock. Thus, another factor governing the choice of electrode material is its machinability.

Electrode wear is a major problem with EDM. Minimizing electrode wear to increase tool life is a prime concern. Most newer power supplies allow for adjustment of the duration, intensity, and polarity of the discharge. The proper adjustment of these parameters for each job can help keep electrode wear at a minimum. Long discharges of medium intensity create less electrode wear than short discharges of high intensity. However, this also decreases machining speed. (The final selection of working parameters depends largely on the nature of work and the complexity of fabricating the electrodes.

Material removal rates

In EDM, material removal rates depend primarily on the amount of current supplied across the gap. Fig. 12-2 depicts the results of various current-level settings for single discharges.

Fig. 12-2A shows the relatively small cratering of a 1-amp discharge. Fig. 12-2B shows a 2-amp discharge and the resultant cratering. Notice that the crater produced by the 2-amp discharge is approximately twice the size of the 1-amp discharge. Likewise, the 4-amp discharge shown in Fig. 12-2C produces a crater twice as large as the 2-amp discharge. Essentially, the material removal rate in-

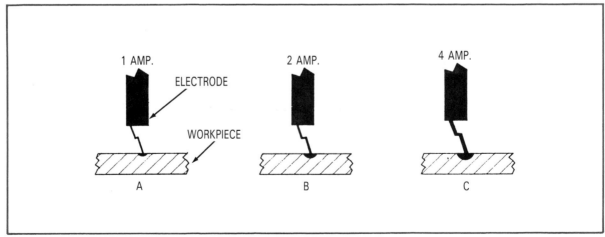

CIMLINC, INC.

Fig. 12-2. The current flow affects the material removal rate, as shown in these three views. A—With 1 Amp. current. B—With 2 Amp. current. C—With 4 Amp. current.

creases almost proportionately to the machining current rate increase for a given frequency.

An *average* rate for material removal when cutting steel with a graphite electrode is 0.05 cubic inch per hour per ampere. This figure is based on ideal cutting frequencies with ideal flushing abilities. Actual conditions will vary depending on electrode configuration, depth of cut, and dielectric flushing pressure.

Under ideal conditions, and with the proper control settings on the power supply, EDM can produce finishes in the neighborhood of 8-10 rms. It may be necessary to use several electrodes to obtain finer finishes in EDM. For instance, a roughing electrode may be used to remove large amounts of material in as short a time as possible. It is then the job of the finishing electrode, or electrodes, to remove the small amount of material remaining and impart the smoother finish.

The accuracy obtainable with modern EDM equipment, such as the machines shown in Figs. 12-3 and 12-4, is in the range of approximately 5% of the spark gap. This represents only the integral accuracy of the cavity being machined. The linear locations of the machined cavities on a workpiece are dependent upon the mechanical accuracy of the machine tool positioning system.

Fig. 12-3. A modern computer numerically controlled EDM system. The power supply is in the cabinet at left.

Fig. 12-4. A specially built two-spindle EDM system. CNC control is accomplished through the controller at right.

Since the spark tends to arc across the point that represents the shortest distance from the electrode to the workpiece, it is necessary to make the electrode smaller than the desired finished size of the hole. Standard tables are available, giving the amount to undercut electrodes to assure the proper finish size in the workpiece. Factors that must be considered in these recommended values include electrode material, workpiece material, electrical frequency, and applied gap current.

EDM advantages and disadvantages

EDM is a proven method of machining parts that cannot readily be produced by conventional means, either because of the part shape or because of the material from which it is made. A major benefit of this machining process is that electrically conductive materials of any hardness can be machined readily. This allows EDM to be performed as a final operation even after heat treating, thus eliminating distortion resulting from the hardening process.

Since the tool does not touch the workpiece, there are no cutting forces generated; therefore, very fragile parts can be machined. EDM can also produce thin slots, intricate shapes, and sharp internal corners without sectioning. Heating is localized and controlled by the dielectric fluid, eliminating stresses that would otherwise be induced into the part by the machining process. EDM produces smooth surface finishes with an absence of burrs, eliminating or minimizing the need for hand finishing.

Savings of as much as 60% have been realized in the sinking of dies and 40% in the production of new dies. Additional savings are gained through EDM by reducing the extent and number of secondary operations required on molded and cast parts.

The only limit on the complexity of cavity shapes is the ability to machine the electrode to the needed shape. Therefore, almost any shape can be formed in both the male and female members of die sets. Clearance for forming dies and drawing dies can be maintained evenly and as minutely as necessary. Complex fine blanking dies made of solid tungsten carbide have been constructed with as little as 0.0006" (0.152 mm) total clearance.

While EDM has many advantages, it is definitely not a mass production or replacement process. If the same work can be done economically in any other manner, that method is usually the most practical.

The average EDM material removal rate for most machines is in the range of 1.5 to 3 cubic inches per hour. Along with this relatively slow material removal rate, another possible limitation of EDM is that it will machine only electrically conductive materials.

The biggest disadvantage is that most EDM machining results in rapid tool wear. Since tools are often expensive, this might be the deciding factor in whether or not EDM is used. In many instances, a *combination* of machining processes can best accomplish the required work. For example, the biggest portion of material removal in a forming die may be performed by conventional methods before heat treating. After the part is hardened, the finishing touches can be applied by EDM.

Although EDM has its uses, it should not be thought of as a "cure-all." Because of its higher cost, EDM should not be substituted when conventional methods can be used.

EDM applications

As described earlier in this chapter, EDM is quite often used in punch and die fabrication. Brass, zinc, or other metal electrodes make possible the economical machining of heat-treated forging dies, die-casting dies, and plastic-molding dies. Since the finished die can only be as accurate as the electrode used in its manufacture, a series of electrodes can be used to finish the final part. The first few are used to rough out the metal, with the last one finishing the detail of the die. The same electrodes can be used in sinking second, and subsequent, dies by just adding one each time for final finishing. The process may be used until the roughing electrodes are no longer usable.

Electrical discharge machining is ideally suited to diemaking. The die set consists of two pieces: the female member (the *die*) and the male part (the *punch*). A unique operation is used to make the die set. First, a duplicate of the male electrode is placed into the machine and erodes the material to be used for the female electrode. The two electrodes are then used, in separate operations, to make the die and punch.

Thin, flexible workpieces, such as sheet-metal parts for jet engines, can be readily machined using EDM techniques. Since there is no tool pressure imparted to the workpiece during the process, there is no worry that the workpiece might buckle or collapse.

Complex extrusion dies and wire-drawing dies are easily machined by EDM techniques. Although wire drawing dies are basically very simple in design, the material from which they are made (usually tungsten carbide), makes it almost impossible to machine them using any other method.

NC wire EDM. The introduction of **numerically controlled wire feed EDM** revolutionized electrical discharge machining applications. The system uses a spool of wire for the electrode, as shown in Fig. 12-5. To best picture the process, consider a band saw blade. If the wire has an electrical charge opposite that of the workpiece, EDM material removal will occur when the moving wire is brought in close proximity to the workpiece. If the workpiece is clamped to a movable machine table operated by servomechanisms, the cutting rate and direction can be controlled. See Fig. 12-6.

In NC wire EDM, the coordinate moves of the table are controlled by a numerically controlled positioning device. Only straight cuts in the axis of the wire can be made, but since the table moves, any number of complex shapes can be generated. Some advantages of NC wire EDM are:

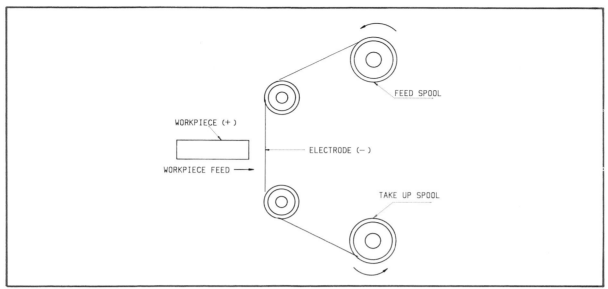

Fig. 12-5. This schematic shows the essential elements of wire-feed EDM: the negative-polarity wire electrode being fed from the feed spool to the takeup spool, while the positive-polarity workpiece is fed perpendicular to the wire.

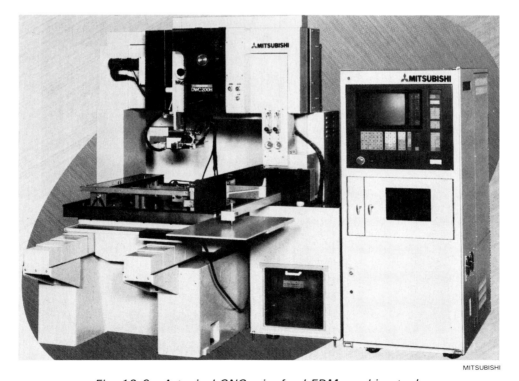

Fig. 12-6. A typical CNC wire-feed EDM machine tool.

- elimination of expensive EDM electrodes.
- more productive use of skilled help through use of unattended machining.
- reduction of lead time.
- generation of small radii and intricate shapes without special equipment.

The use of EDM is expanding rapidly, as a result of its ability to perform jobs that are difficult or impossible to do by conventional means. Research and development is expected to contribute greatly to the further growth of this process. Wider recognition and understanding of EDM and its applications will lead to increased importance and acceptance. EDM has truly removed many design barriers and has added new dimensions to the technology of manufacturing.

ELECTROCHEMICAL MACHINING

Electrochemical machining (ECM) is a specialized production process that relies on the principle of *electrolysis* (producing chemical change by passing electric current through a conductive liquid called an *electrolyte*). Electrolysis was first studied by Michael Faraday in the early 19th century. He showed that if two pieces of metal are placed in a solution of electrolyte and energized by a direct current, metal will be displaced from the *anode* (positive pole) and plated onto the *cathode* (negative pole). Initially the phenomenon of electrolysis was valued solely for the cathodic side of the reaction. Initially, it was used to electroplate parts for material buildup, wear coating, or decorative purposes. The anode was always an expendable, wasted component of the reaction.

Electrochemical machining simply *reverses the priorities* of the anodic and cathodic reactions; through a controlled erosion rate, it removes material from the workpiece that serves as the anode. See Fig. 12-7.

SPEEDFAM CORPORATION

Fig. 12-7. Electrochemical machining can be used to automatically deburr complex workpieces, especially those that have normally inaccessible areas. This ECM machine includes the deburring work unit, an electrolyte and tank, and an ultrasonic workpiece washing unit.

ECM has become one of the most rapidly expanding processes in modern industry. There are several reasons for its remarkable growth rate, the most important of which are:
- The need to machine harder and tougher materials.
- The need to machine complex configurations beyond the scope of traditional machining processes.
- The increasing cost of manual labor.

Obviously, to make effective use of ECM, more is required than to simply connect a conductor to the workpiece and submerge it in an electrolytic solution.

THEORY OF ECM

ECM is sometimes called *deplating*, a name which stems from the fact that as the cathode is being plated, the anode is being consumed or deplated. If both the cathode and anode are completely submerged in the electrolyte and a direct current potential is introduced between them, the anode would be evenly consumed and the cathode would be evenly plated with material displaced from the anode. In ECM, the object is to *control* the deplating in such a way that the anodic side, or the workpiece, is eroded in a *predictable* manner. To accomplish this, the cathode is shaped into an approximate *mirror image* of the configuration desired on the workpiece (anode). This is shown in Fig. 12-8A, which portrays the initial conditions in a typical ECM application. Fig. 12-8B illustrates the final conditions and tool-to-part relationship.

Typically, the tool-to-workpiece gap may be about 0.005" (0.127 mm), but can vary from 0.001" to 0.030" (0.025 mm to 0.767 mm).

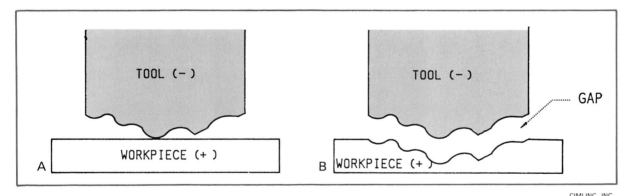

CIMLINC, INC.

Fig. 12-8. The process of electrochemical machining reverses the traditional plating process of metal deposition. Instead, it removes metal. A — The initial relationship of shaped electrode and workpiece. B — The final stage of the process, in which the workpiece cavity reflects the shape of the tool (electrode).

ECM operation

Fig. 12-9 is a schematic diagram of a typical ECM machine. The power is supplied as shown, with the tool being charged negatively. The workpiece is fastened to the holding fixture, which in turn is fastened to the table of the machine through an insulation barrier.

Electrolyte is pumped through the hollow electrode and flows out through the gap between the electrode and the workpiece. The electrode, or tool, is insulated along its upper section to eliminate unwanted ECM action that would tend to make the cut tapered or bell-mouthed.

The electrolyte discharges into a sedimentation tank, where sludge resulting from ECM action is removed through a combination of sedimentation and centrifugal separation. The electrolyte is next pumped through a filter to remove all remaining sludge and impurities. A pressure relief (bypass) valve located between the pump and filter allows control of the electrolyte pressure to the electrode.

The whole work area is enclosed by exhaust ducting. The ducting collects the hydrogen gas created by the electrolysis reaction, and a fan exhausts it to the atmosphere. Servomechanisms maintain a constant tool feed and keep the tool from coming in contact with the workpiece. If actual contact is made, the tool will usually be destroyed, because of the high current density required by ECM operations.

The basic electrochemical reaction occurring in Fig. 12-9 is:

$$2Fe + 4H_2O + O_2 ---\rangle 2Fe(OH)_3 + H_2$$

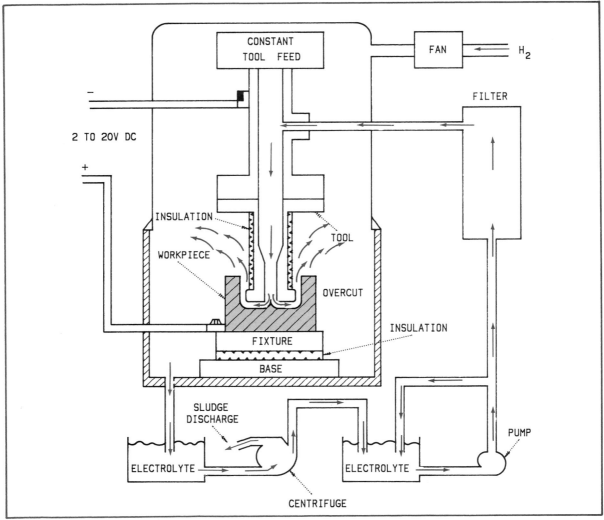

CIMLINC, INC.

Fig. 12-9. The main components shown in this cross-sectional view of an ECM system are the tool, the workpiece, electrolyte, insulation, flushing system, hydrogen fan system, and filtering system for sludge.

This equation is typical of applications where the workpiece is iron or steel. Note that the electrolyte (NaCl) is not shown, because it does not enter into the reaction. It is not consumed, and therefore does need not to be replaced. The $Fe(OH)_3$ is a red-brown sludge that must be continuously removed from the electrolyte solution.

The electrolyte is forced through the tool and into the gap between the tool and the workpiece by high-volume pumps. There are two reasons for this procedure. First, the electrochemical reaction generates a great deal of heat and the electrolyte in the gap must not be allowed to boil. If boiling occurs, the electrolysis rate will be greatly reduced. Second, the sludge produced by the reaction must be continuously removed from the gap to prevent electrical shorting between the tool and the workpiece.

Electrodes

The ECM electrode or tool is perhaps the most important single part of the ECM machine. A potential tool material must meet many requirements. First and foremost, it must be a good conductor of electricity. It must resist the chemical activity of the electrochemical reaction. It should be readily obtainable and easily machined. It should have good thermal conductive properties. Finally, it should be capable of accepting a highly polished finish. This finish will allow electrolyte to flow uniformly over it.

Materials most commonly used for ECM electrodes are copper, brass, bronze, and certain stainless steels. Conditions will often dictate the selection of the tool material. For example, if the electrolyte solution is a strong acid, stainless steel will probably be chosen. Although stainless steels have much lower thermal conduction properties and are typically much harder to machine and finish than copper, their resistance to acid corrosion may be the governing factor in the choice of tool material.

Since the tool is virtually nonconsumable in the ECM process, great attention to detail can be justified in its manufacture. Theoretically, only one tool is needed. However, since the possibility of tool damage or destruction is always present, it may be advisable (in some cases) to produce two or more identical tools for a job. An important aspect of tool design is the placement of the passages for the electrolyte flow. It is vital for proper operation that there be an even flow of the electrolyte over the whole area of the workpiece in the gap. Besides electrolyte flow, another reason for a smooth tool finish is the fact that imperfections on the surface of the tool will be transferred to the workpiece.

It is quite possible that the tool will come in close proximity to parts of the workpiece that do not require material removal. Wherever there is a possibility of unwanted electrochemical machining, it is necessary to insulate either the tool or the workpiece. The tool is usually coated with an insulating material, such as an epoxy-type resin, to eliminate any unwanted reaction. The insulating material must be bonded securely to the tool surface—if it cracks or works free from the tool surface, the ECM reaction will occur in areas where erosion is not wanted. The result may be the need to scrap the workpiece. An added hazard is the possibility of a piece of the insulating material lodging itself between the workpiece and the tool, causing mechanical damage to the tool.

Current and material removal rates

The power supply is usually the most expensive single component in an ECM installation. It must be capable of supplying low voltage (2-20v) direct current to the gap at current ratings of 5000 amps/in., or higher. Obviously, one factor affecting the material removal rate is availability of current from the power supply: the higher the current, the faster the removal rate. Actual current ratings vary from less than 100 amps to more than 10,000 amps.

Electrolyte solutions

The electrolytic solution's function is to provide ions to carry the current across the gap from the anode to the cathode, thereby causing the electrochemical reaction and erosion of the workpiece. The selection of electrolyte material is based on several factors. Electrolytes may be aqueous (water) solutions of salts, acids, or bases. They may also be nonaqueous liquids, molten salts, or gases. The most common electrolytes are water solutions of salts, predominantly chloride.

Because of the high current densities involved with ECM, the electrolyte must be *highly conductive* to minimize power losses. Even the most conductive electrolytes are much less conductive than metals. For this reason, the electrolyte is much more susceptible to resistance heating than the workpiece or the tool. Obviously, the electrolyte should have a low viscosity to allow it to flow as rapidly as possible from the working gap (thereby removing the heat as it leaves). An electrolyte should also have as high a specific heat as possible, high thermal conductivity, and a high boiling point.

The electrolyte must remain chemically stable during use even though it is in molecular contact with the ions being removed from the workpiece. If the electrolyte were to combine with the released metallic ions, it could form insoluble products and adversely affect the machining process.

The electrolyte temperature is precisely controlled in most ECM applications, since the chemical activity and the electrical conductivity of the electrolyte are both affected by temperature changes. Therefore, it is not uncommon to equip the electrolyte holding tanks with heat exchangers. These heat exchangers control the temperature of the electrolyte to an accuracy of $\pm 1°F$.

Initially, the electrolyte may need to be heated to the proper operating temperature. After the process begins, however, the heat must be continually removed from the electrolyte, since the reaction generates a great deal of heat. This is accomplished by refrigeration or the use of evaporative water towers.

Surface finish and tolerances

There are three main factors that govern surface finish in the ECM process. These are the *workpiece material*, the *type of electrolyte*, and the *machining conditions* (density of current and flow of electrolyte).

Workpiece material. If the base material of the workpiece is not completely homogeneous, some components may be eroded more readily than others, especially at low current densities. Even if a material is completely homogeneous, structural variations at grain boundaries may lead to surface finish problems and cause varying removal rates.

In general, if surface finish is poor because of material structural variations, it can be improved by *increasing* the current density. Another method of improving surface finish is lowering the conductivity of the electrolyte and increasing the voltage potential between the anode and cathode. However, these adjustments to the process parameters are useful only when small improvements are needed.

Type of electrolyte. The type of electrolyte used is also a very important factor governing quality of the surface finish. There are guidelines to the selection of the proper types of electrolytes for a given type of base metal and tool material. However, the actual composition of the electrolyte is usually selected by trial and error and is custom-fitted to the job. The electrolyte should have a stable electrical conductivity. If conductivity varies, the removal rate will vary, causing lack of conformity to the tool configuration. Such problems usually cause shape variations, rather than poor surface finish, but should be considered nevertheless.

Machining conditions. In general, the higher the current density, the better the surface finish. Of course, the limiting factor to this statement is the actual heat-dissipating abilities of the tool, workpiece, and electrolyte. Not only must the current density be as high as possible, it must be distributed as evenly as is practical over the work area. If the current is more dense in some areas and less dense in others, the removal rates in these areas will differ, causing surface finish problems.

The flow of the electrolyte in the work area can also affect the surface finish. If turbulence or eddies are created, small amounts of electrolyte will stay in the work area longer than others. Because they are in the work area longer, these small amounts of electrolyte tend to become more ionized, increasing their electrical conductivity. This locally increased conductivity will create a higher removal rate, causing the eddy pattern to be transferred to the workpiece.

About the best surface finish available with electrochemical machining is in the neighborhood of 0.000005" per inch. The surface finish is dependent on the above-mentioned parameters, of course, and can be much worse than 0.000005" per inch.

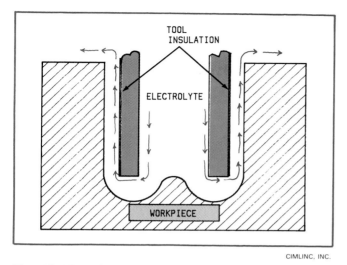

Fig. 12-10. ECM drilling is often done with a tubular cathode that is insulated on its outer surface to prevent overcutting the hole diameter. When completed, this will be a through hole.

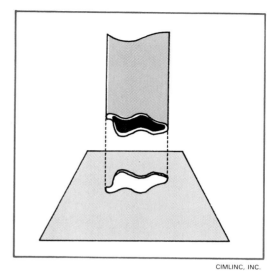

Fig. 12-11. ECM is ideal for drilling irregularly shaped holes, since the finished hole will be shaped the same as the tool used.

Accuracy and working tolerance with ECM are generally limited to within ±0.0005" (0.013 mm). If the machined part is not within the specified tolerances, the problem may be due to inaccurate tooling.

Two common causes of inaccurate tooling are poor manufacturing procedures and improper stress relief of tooling material. If the tooling is not produced with the required design configuration and tolerances, it will not produce accurate work. This is generally true of all types of material-removal processes. Also, if the tooling material has been stressed during its manufacture, it may stress-relieve itself during the process of ECM because of the heat generated. It may assume a new (but incorrect) shape.

ECM advantages

The advantages of ECM over traditional machining methods are varied. Although ECM is a very powerful tool in today's manufacturing industry, it must not be considered a remedy for *all* complex machining problems.

A specific advantage over standard machining methods is the longevity of tool life. If the tool is treated properly, and the electrolyte is chosen to minimize attack on the tool, it will last virtually forever. Another favorable characteristic of ECM is the fact that a completely machined surface is produced in one pass. Conventional machining methods require that a small cutter make many passes over the workpiece. This, coupled with the inevitable burrs resulting from conventional machining that must be removed, makes ECM a profitable approach to the solution of many material removal problems. In certain cases, ECM may be much faster than conventional machining.

Another favorable characteristic of ECM is that it does not require a great deal of operator skill. Once the tooling and process parameters have been established, the actual operation of the machine itself can be handled by a relatively unskilled operator who does nothing more than push buttons and change parts. Also, once a procedure has been established, and assuming that it is followed, there is only a slight possibility of creating defective parts.

Since workpiece material generally has no effect on ECM, the hardness or softness of the part material does not matter. Skyrocketing labor prices and exotic alloy parts with complex shapes, a combination that all but eliminates other removal processes, seem to insure a strong future for ECM.

ECM applications

Present applications of ECM encompass such areas as drilling, shaping, turning, and milling.

The process of ECM drilling is illustrated in Fig. 12-10. The cathode is usually tubular in shape, with insulation on its sides to eliminate ECM reactions between the side of the tool and the workpiece. The electrolyte is supplied to the work area through the center of the tube and is expelled along the outside between the tool and the workpiece.

ECM drilling is not restricted to round holes. The process is limited only by the complexity of the tooling required to do the job. If the tooling can be made, no matter how long it takes, ECM processing will be as rapid for the most complex shapes as it is for round holes. Fig. 12-11 shows a typical application of ECM producing an irregularly shaped hole.

One problem encountered with ECM drilling is a hole that must be drilled completely through the workpiece. As the tool starts to penetrate the back side, the electrolyte flows out the hole. As a result, pressure is lost, causing a variation in removal rate. To overcome this problem, all that is required is to back up the workpiece with another piece of metal, or with a nonconductor such as plastic. Nonconductive backups must be flexible, so the tool can pass completely through the workpiece without mechanical binding.

One of the great advantages of ECM drilling is its ability to drill very deep small-diameter holes. It is not uncommon to drill holes as small as 0.050" (1.27 mm) in diameter to depths of two feet (0.61 m) or more. Straightness of the hole depends on the straightness of the tool and the relative positioning of the tool to the workpiece. ECM drilling is used extensively in the gas-turbine industry for drilling cooling holes in turbine blades, and producing virtually burr-free shaped holes in thin sheet metal parts. A related use is in the electronics industry, which uses ECM to produce burr-free holes in thin sheets or wafers of semiconductor materials such as silicon.

Shaping is another use of ECM that is becoming more and more acceptable as a manufacturing process. Again, the gas-turbine industry has capitalized on the burr-free qualities of the ECM process to produce turbine blades faster and more economically than any other process available. Fig. 12-12 shows a typical application of ECM to produce complex aerodynamic contours on parts such as a turbine blade.

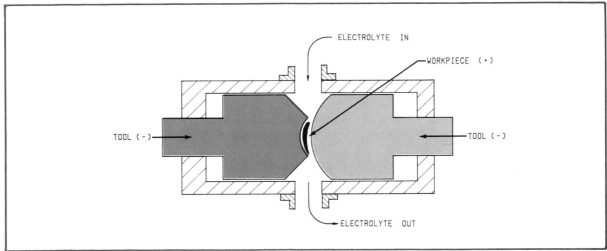

CIMLINC, INC.

Fig. 12-12. A two-axis ECM machine tool can be used to shape a contoured part, such as a turbine blade for use in an aircraft engine. The tools move toward each other until the workpiece achieves its final shape.

LASER PROCESSES

The term *laser* is an acronym formed by the initial letters of the words making up its description: **L**ight **A**mplification by **S**timulated **E**mission of **R**adiation. A more practical definition is: a device that utilizes the natural oscillation of atoms for simplifying or generating extramagnetic waves in the visible region of the light spectrum. This key principle is not new. It was first published by Albert Einstein in 1917.

Laser devices have revolutionized the manufacturing industry in recent years. Applications are limited only by our ingenuity. Lasers are used in drilling and cutting rubber, fabrics, paper, plastics, metals, and ceramics, Fig. 12-13. Lasers are also used for marking, welding, cladding, and heat-treating. Laser machines have also combined applications with robots for use in the automotive industry where parts are too large to be placed within the limits of a machine tool. The laser beam is transferred from the laser source to the workpiece at the extremes of the robot work envelope through a series of mirrors. This is pictured in Fig. 12-14.

NTC LASER MACHINE GROUP, MARUBENI AMERICA CORPORATION

Fig. 12-13. This cantilever laser cutting system does not use a conventional traveling cutting table. It allows the user more flexibility with the option of using a stationary material pallet in its place. Its open, simple design means sheet and plate material can be easily loaded/offloaded from the front or side, and larger stock can be positioned for processing.

Fig. 12-14. This laser machining center uses a series of mirrors to transfer the laser beam from its source to the robot's end effector, where it can act on the workpiece.

LASER THEORY

The *stimulated emission* concept of the laser process is the most complex to understand. First, an atom must be stimulated by an outside source of energy such as heat, light, or some chemical reaction. This is done to increase the *energy level* of that atom.

As the energy level of the atom is permitted to decay back to its original level, it releases a ***photon***, or particle of light energy. When the photon hits another atom, that atom is stimulated and reaches a higher energy level. Then, as the atom's energy level decays, another photon is emitted. This process starts a chain reaction that forms a concentrated light beam.

A second important concept is that each photon or light unit released is identical in terms of wavelength, phase, direction, and energy. The light beam is therefore known as a ***coherent light*** beam. The coherent light beam is very ***collimated*** (consisting of parallel rays) and can be focused at an object with great intensity or energy for work.

Properties of laser light

The properties of a laser light beam that make it very effective for processing are coherence, parallelism, monochromaticity, and brightness.

- *Coherence* means having one wavelength. Therefore, the light travels in phase.
- *Parallelism* can be described as traveling in a column or straight-line path versus converging (coming to a point) or diverging (spreading out in a cone shape). Laser light can therefore travel a greater distance without losing its intensity.
- *Monochromaticity* means having a single color frequency. Therefore, a simple lens can be used to focus the light beam to a pinpoint spot size.
- *Brightness* really results from the other stated properties. For comparison, laser light can produce 100,000 times more power density than a regular light bulb at the same distance.

Types of lasers

Lasers can be gaseous or solid state, depending upon the medium used to produce the laser beam. The most common gas lasers are xenon, krypton, and CO_2 (carbon dioxide). The CO_2 laser, Figure 12-16, is best applied to processing applications. Solid state lasers are: ruby, Nd:glass (neodymium/glass), Nd:yag (neodymium/yttrium aluminum garnet), and alexandrite. The Nd:yag laser is the most powerful of the solid state lasers and therefore has the greatest applications for processing.

Mode of operation

The laser *mode of operation* can be defined as the way that the light beam is applied to the work source. The laser beam can be either continuous or pulsed. All gas lasers can be operated as continuous process lasers. Nd:yag and alexandrite lasers, among the solid-state types, can also be operated in a continuous mode. Ruby and Nd:glass lasers can only be operated in pulse mode, since they exhibit low thermal shock resistance. This could cause the laser rod to crack. In summary, the CO_2 and Nd:yag lasers are the most powerful and the best suited to industrial applications. Nd:yag laser systems are available in power ranges up to 400 watts. CO_2 lasers can generate up to 10-12 kilowatts of power.

Laser applications

Drilling. Laser drilling can be defined as the process of piercing or penetrating through a material with multiple pulses of a stationary laser beam. This process is usually referred to as **percussion drilling**. Actually, the process partly melts and partly vaporizes the material. The hole is not perfectly round and has slightly tapered sides. Typically, percussion drilling is limited to holes with diameters of less than 0.030" (0.762 mm). Larger holes use a cutting technique described under cutting applications.

Marking. Laser marking uses a rapidly pulsing computer-controlled scanning system. As the laser beam travels along its path, the beam scans over the workpiece and vaporizes the material with a series of overlapping blind holes that form letters and numbers. The laser groove is typically 0.010" (0.25 mm) wide. This marking method is usually fifty times faster than conventional means and is much more readable.

Laser markings can also be placed on parts at locations almost impossible to mark by conventional means, due to contour or limited space on the workpiece.

Cutting. Laser cutting combines the high energy level of the laser beam with a high velocity jet of gas to vaporize the material being cut.

Travel (cutting) speeds of 100" per minute (IPM) or greater are common on thin sheet metal sections. The cutting process creates no cutting forces and leaves a smooth surface texture within close tolerances.

Laser cutting can eliminate expensive dies or special EDM electrodes and fixtures. The holding fixture for the laser can be very inexpensive and made mostly of aluminum, since the fixture does not have to resist tool forces. The cost savings from laser cutting over conventional cutting processes can be as great as 10:1 reduction. Gases used with laser cutting are compressed air, or oxygen.

Cutting can be accomplished in either pulsed or continuous operating modes. A continuous mode results in the highest cutting speeds. Pulsed modes result in the least heat and the least workpiece distortion.

Laser cutting of *holes* can be generated two ways: by trepanning and by CNC circular interpolation.

Laser *trepanning* can be defined as a means of generating a hole by an optical-mechanical device that rotates the laser beam around a center point. This is accomplished by rotating the focus lens around the centerline of the laser beam with a small servo-controlled drive. Typical hole sizes for laser trepanning range from approximately 0.020" - 0.100" (0.51 mm - 2.54 mm).

Laser-cut holes using ***CNC circular interpolation*** make use of the same technique as milling a circular pocket. Drives move the laser beam in two axes simultaneously to create a circular cut around the perimeter of a hole. This technique is usually done with holes over 0.100" (2.54 mm) in diameter, and can be carried out at cutting rates as high as 100 IPM.

Naturally, cuts are not limited to round holes; any shape that is programmable can be laser-cut. Most laser machining centers are available with three, four, and five axes of control. Figs. 12-15 through 12-18 show examples of current multi-axis laser machining centers.

Welding. Lasers are starting to show cost savings over conventional techniques for spot, seam, and full-penetration welding.

Spot and seam welding can be accomplished with either pulse or continuous laser modes. Spot welding is simply heating a small spot with the laser beam until the two members melt and fuse together. Seam welding is similar, except the weld bead is continuous. Usually, spot and seam welding depths are limited to 0.100" (2.54 mm) or less.

Full-penetration laser welding is achieved through a technique known as the *keyhole* method. The laser has to first vaporize the metal, producing a hole through the workpiece. Then, as the laser beam travels, the hole continues along the joint. The molten metal solidifies on the rear edge of the hole, forming the weld nugget.

With high-powered lasers, weld depths of 1.00" (25.4 mm) can be achieved. Experimental CO_2 lasers have demonstrated penetration depth in excess of 2.0" (51.8 mm). Therefore, laser welding can sometimes be cost-effective when compared to electron beam welding and to conventional welding processes.

EXPLOSIVE FORMING

Explosive forming is classified as a high-energy-rate forming process (HERF). ***High-energy-rate forming*** uses a relatively large amount of energy, applied for a very short period of time, to cause rapid deformation of the workpiece in a controlled and predictable manner.

The term "high-energy-rate forming" is somewhat ambiguous, because the *rate* of energy applied to a system is defined as **power**. Therefore, the name seems to imply HERF is a high-power forming process. Although HERF is, indeed, a high-power forming process, more is required to effect the HERF process than just the input of great quantities of power. Since HERF relies on the second-degree velocity term in the classical kinetic energy equation, ($E = 1/2mv^2$, which means energy increases by the square of the velocity), it is more correctly called ***high-velocity forming*** (HVF).

For instance, there are in use today many drop-hammer forging presses that develop more power than some HERF systems. However, the deformation velocity of these presses is in the order of 30 feet per second (fps), while HERF is capable of deformation velocities of 500 fps or higher. The *power input* to the system may be the same, but the *rate of deformation* must be in this higher range for a process to qualify as a HERF process.

GARRETT ENGINE DIVISION/ALLIED SIGNAL AEROSPACE

Fig. 12-15. The large cylindrical chamber is the generating source for this powerful CO_2 laser. The system has a five-axis control unit.

GARRETT ENGINE DIVISION/ALLIED SIGNAL AEROSPACE

Fig. 12-17. A five-axis YAG laser system smaller than the one shown in Fig. 12-16. This one has an 8'' rotary tilt table.

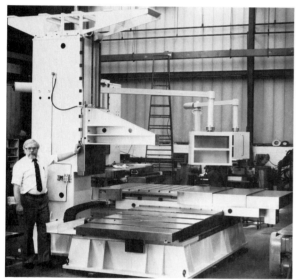

GARRETT ENGINE DIVISION/ALLIED SIGNAL AEROSPACE

Fig. 12-16. A five-axis YAG laser machining center with a rotary tilt table capable of handling a workpiece 36'' in diameter.

GARRETT ENGINE DIVISION/ALLIED SIGNAL AEROSPACE

Fig. 12-18. This three-axis laser system is used for precision drilling of small holes and cutting of small parts.

Explosive forming is a method of high-energy-rate forming that uses high-explosive charges as the energy source. The workpiece to be formed is generally clamped over an evacuated die cavity, as is illustrated in Fig. 12-19. The *energy transfer medium* is usually water or air. When the explosive charge is detonated, a severe shock wave is induced into the energy transfer medium. The shock wave forces the workpiece into the evacuated cavity, causing it to conform to the shape of the die. Explosive forming systems are capable of producing forming pressures well in excess of one million pounds per square inch (psi).

TYPES OF SYSTEMS

The system shown in Fig. 12-19 is a standoff system, in which the explosive charge is detonated somewhere off the surface of the workpiece. The forming energy is propagated through the energy

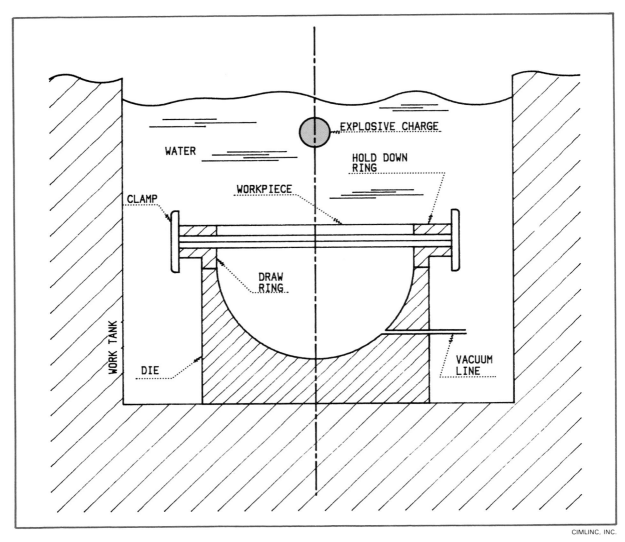

Fig. 12-19. Explosive forming uses high explosives as an energy source. This cross-sectional view shows a stand-off system, in which the explosion takes place above the surface to be formed.

transfer medium to the surface of the workpiece, resulting in a very evenly distributed force application. Such a system is capable of deformation rates of several hundred feet per second.

A variation of this method is the contact system, in which the explosive charge is placed in actual contact with the workpiece. When detonated, the charge violently deforms the workpiece away from the explosion. This method is much faster-reacting than the standoff system, and is more likely to fracture the workpiece. Both systems use solid explosive charges as energy sources.

Fig. 12-20 is a schematic representation of a combustible-gas-mixture system. It is similiar to the explosive forming systems just described, but uses an explosive gas for the energy input. Although the results obtained from gas detonation systems are more predictable, the pressures obtained are considerably lower than those obtained with solid explosive charges. Forming pressures in excess of 100,000 psi are obtainable, with deformation velocities of less than 200 feet per second.

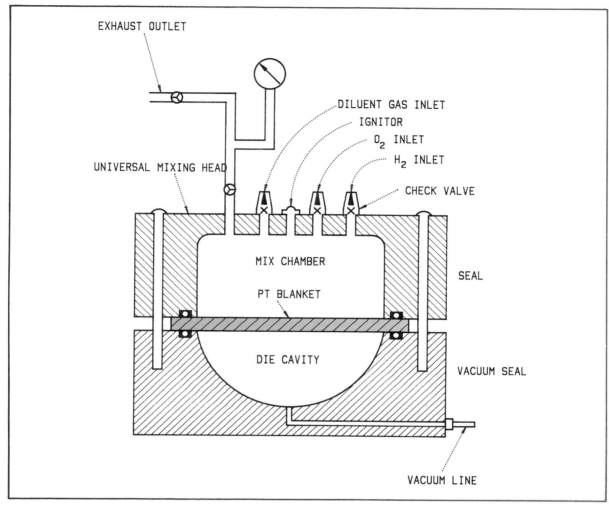

Fig. 12-20. A combustible gas mixture can be used to provide the force needed for explosive forming, as shown in this view.

ELECTRON BEAM WELDING

Electron beam welding (EBW) is a material-joining process in which a fusion weld is produced when a narrow beam of accelerated electrons strikes the work. The electrons' *kinetic energy* (energy of motion) is transformed into heat. This process is unlike other welding processes, in that it usually takes place inside a vacuum chamber. Within the vacuum chamber, electrons are accelerated to a very high voltage and are forced to travel in the direction of the material to be joined. Each speeding electron possesses a large amount of kinetic energy. The amount of energy is determined by the equation $E = 1/2mv^2$, where $\mathbf{m}$ = electron mass and $\mathbf{v}$ = velocity. The mass of an electron is 9×10^{-31} kg. While the *mass* of the electron is small, its *velocity* approaches the speed of light. As the electrons strike the work surface, the material's molecules are greatly excited. This causes the molecular bonds to break, and the material to melt.

Most materials, when heated, give off free electrons. In the electron gun, a filament (usually of tungsten) is resistance heated to approximately 3600°F. (2000°C) to provide the free electrons. Once the free electrons are available, a high potential (voltage difference) between the cathode and the anode accelerates the electrons to approximately 142,000 miles per second (228,500 kilometers per second). This acceleration is caused by the charge on the electrons.

Since electrons carry a negative charge, they are repelled by the cathode, which also has a negative charge. The electrons are, at the same time, attracted to the anode and workpiece, which are both at ground potential.

Fast-moving electrons traveling in random directions are of little use. They must be concentrated to increase the number of electron/workpiece collisions in a small area. This focusing of the electrons takes place in two areas of the electron gun. As the electrons first escape the filament, they are traveling in all directions. The cathode and anode are shaped to repel the electrons in a specific direction (toward the anode opening). This shapes the electrons into a cone, shown in Fig. 12-21.

The second stage of focusing takes place at the focus coil. The focus coil is usually an electromagnet or set of charged plates that further channels the electron beam into a narrow stream. See Fig. 12-21.

In order to control the beam current (number of electrons in the beam), a bias grid is often used. This portion of the gun also carries a negative charge. By varying the applied voltage, it will repel or pass electrons moving toward the anode.

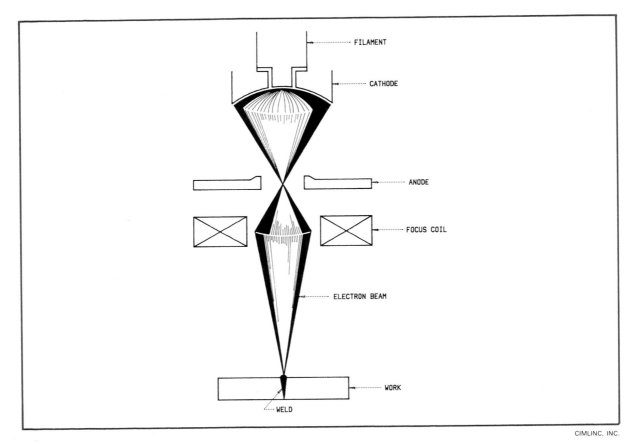

CIMLINC, INC.

Fig. 12-21. Electrons moving at high speed are shaped into a beam and focused to provide the heating necessary for the Electron Beam Welding process.

EBM APPLICATIONS

Electron beam welding has found many uses in modern manufacturing including precision welding of electronic parts and delicate measuring equipment; welding of exotic alloys; drilling of extremely small holes; heat treating; and vacuum melting of metals for diffusion plating.

EBW has many distinct advantages over conventional welding processes. These include high weld purity, a small heat-affected zone (only 1/25 width of conventional welding processes), the capability of making welds as narrow as 0.005″ (0.13 mm), very deep penetration (up to 20 times deeper than the weld width), and low workpiece distortion. EBW can be used to weld dissimilar metals and exotic high temperature alloys.

Like any other manufacturing process, EBW does have drawbacks. It must be done in at least a partial vacuum, so pump down time is long and production must be done in batches. The equipment is expensive and usually bulky, and fixturing often must be very elaborate. In addition, the process poses health hazards, since X rays are produced, it is often difficult to view the process as it takes place, and high operator skill is required.

SUMMARY

All of the advanced specialized high-technology processes and machine tools available today can be incorporated into the CIM philosophy of today's manufacturing plants. Although some of the advanced processes can function manually, most require computer control due to their complexity and the need to have fixed process controls to produce high-quality products.

Electrical discharge machining (EDM) is an advanced production process that uses a controlled electrical spark to erode or remove metal. The spark or frequency of discharges is usually in thousands of cycles per second. Each spark removes a tiny particle of metal. The electrode to produce the spark is usually copper, brass, graphite, or copper tungsten alloy. There are also numerically controlled wire-feed EDM systems in which the electrode is a spool of very fine copper wire. This application can be used for cutting contour shapes.

Electrochemical machining (ECM) is a specialized process that relies on the principle of electrolysis to remove metal. The process uses direct current between the *cathode* and the *anode*. A fine gap is maintained between the workpiece and the tool, with the entire process done in an electrolyte solution that carries away the eroded particles from the workpiece.

Laser devices have revolutionized the manufacturing industry. The most common applications are drilling holes, cutting contour shapes, marking parts, and fusion welding.

Laser is an acronym for Light Amplification by means of Stimulated Emission of Radiation. A laser can be defined as a device that uses the natural oscillation of atoms to generate electromagnetic waves in the visible region of the light spectrum. This light beam can be focused and used to cut, machine, or weld materials. There are two types of lasers, *gaseous* and *solid state*. Common gas lasers are xenon, krypton, and carbon dioxide. Solid-state lasers are ruby, neodymium glass, neodymium/yttrium aluminum garnet, and alexandrite.

Another specialized process in industry is *explosive forming*, a *high-energy-rate forming process* (HERF). HERF processes use relatively large amounts of energy applied for very short periods of time to cause rapid deformation of the workpiece. Explosive forming is a HERF method that uses high explosive charges as the energy source in a very controlled process to form parts from sheet metal or plates.

Electron beam welding (EBW) is a fusion welding process in which a narrow beam of highly accelerated electrons strikes the work, with the kinetic energy transformed into intense heat to fuse the material together. EBW has found many applications in modern manufacturing facilities. Most systems are directed by some level of computer control.

IMPORTANT TERMS

anode
brightness
cathode
CNC circular interpolation
coherence
coherent light
collimated
die
electrical discharge machining
electrochemical machining
electrolysis
electrolyte
electron beam welding
energy transfer medium
explosive forming
high-energy-rate forming
high-velocity forming

kinetic energy
laser
leadscrew
mirror image
mode of operation
monochromaticity
nontraditional manufacturing processes
no-wear machining
numerically controlled wire feed EDM
parallelism
photon
punch
quill-feed machines
ram-feed machines
spark gap
trepanning

QUESTIONS FOR REVIEW AND DISCUSSION

1. What kinds of material are used to make EDM electrodes? Why is electrode material cost often a consideration?

2. How does the process of NC wire EDM differ from conventional EDM?

3. Describe the basic operation of the Electrochemical Machining process.

4. Describe some of the advantages of ECM over traditional machining methods.

5. How has the laser affected manufacturing?

6. Name and describe the four properties of a laser light beam that make it effective for processing materials.

7. How does explosive forming use an energy transfer medium to shape the workpiece?

Factory of the Future

by Donald G. Kelley

Key Concepts

☐ Islands of automation can be cost-effective, but full-scale integration is preferable for the long term.

☐ Establishing a standard to permit sharing of computer data is vital to achieving integration of manufacturing with other systems required to produce a product.

☐ Planning for the *factory of the future* involves developing *as is* and *to be* factory scenarios.

☐ In Just-In-Time manufacturing, product is *pulled* through the factory by demand, eliminating most inventory.

☐ To survive rapid change, manufacturers must form partnerships with employees, suppliers, and customers.

Overview

Manufacturing is an extremely complicated affair, requiring a tremendous amount of knowledge to produce even simple component parts. It is obvious that no one person has the ability to keep in mind the details of producing a complex product, such as the modern aircraft engine. Historically, the solution has been to divide the manufacturing process into smaller, more manageable pieces, allowing a person to understand the small piece for which he or she is responsible.

For the most part, manufacturing industries today follow the practices of the past. Engineering designs a product, often with little regard for how products are actually manufactured or assembled. The shop floor personnel, industrial and manufacturing engineers make the product—frequently ignoring how it was actually designed. Fig. 13-1 shows typical lines of communication and responsibility in a manufacturing enterprise. Output from one department is funneled to the next department, with little or no feedback.

The *factory of the future* will have a different communication model. All departments will have direct lines of communication and responsibility for the finished product, as shown in Fig. 13-2. Sophisticated computers and modern manufacturing techniques are forcing the designer and the manufacturer to open new lines of communications.

Existing technology will be combined with new technology to produce products faster with shorter lead times and greater quality. For a company to evolve into the factory of the future, changes must be made in company philosophy, the quantity and type of personnel and of equipment.

The factory of the future, as presented in this chapter, may be somewhat oversimplified, but contrasting it with today's typical factory will serve to clarify and illustrate the key differences.

Donald G. Kelley is an Associate Professor, Department of Manufacturing Engineering Technology at Arizona State University in Tempe, Arizona.

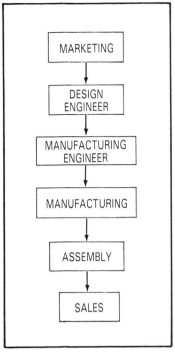

Fig. 13-1. The typical communication pattern in most manufacturing enterprises today is linear. Output from one department is sent to the next, with little or no feedback.

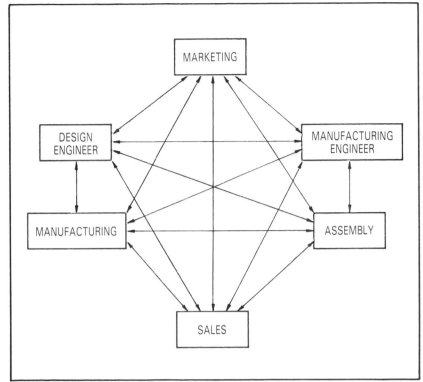

Fig. 13-3. Manufacturers continually seek new technologies to improve productivity and product quality. This machine tool can quickly and easily be converted from a horizontal milling machine to a vertical milling machine. It also provides infinitely variable spindle speeds.

MANUFACTURING TODAY

Greater demands for quality and pressure from the competition in the marketplace are forcing American manufacturers to continually select new technologies that offer more efficient approaches to product design and manufacturing. See Fig. 13-3. At present, most manufacturing companies view automation, implemented to a greater or less degree, as a means of improving productivity and remaining competitive in the marketplace.

In the context of manufacturing, *automation* is the automatically controlled operation of a machine or group of machines to produce parts or assemblies. The machines can include such devices as automated guided vehicles, cranes, robots, inspection and deburr equipment, and machine tools, all of which must be used with electronic or mechanical devices that simulate human intelligence and skill. Human intelligence is simulated by electrical or mechanical sensors on the manufacturing equipment with feedback to the computer.

ISLANDS OF AUTOMATION

It is common for manufacturing managers to implement emerging technologies as they become available. This approach has resulted in *islands of automation*, as shown in Fig. 13-4.

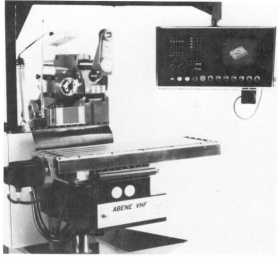

Fig. 13-3. Manufacturers continually seek new technologies to improve productivity and product quality. This machine tool can quickly and easily be converted from a horizontal milling machine to a vertical milling machine. It also provides infinitely variable spindle speeds.

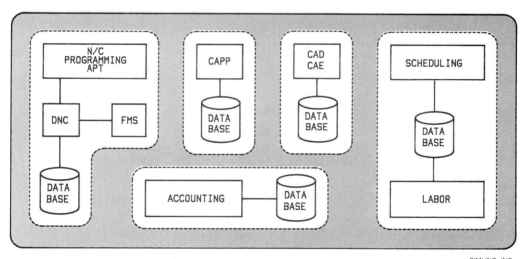

Fig. 13-4. Implementing new technologies as they become available has resulted in "islands of automation" in many manufacturing enterprises. Each island maintains its own database and functions with little reference to the others.

For example, computer-aided design and computer-aided manufacturing (CAD/CAM) was proclaimed to be the solution to integrating the design and manufacturing functions. Instead, CAD/CAM has become yet another island of manufacturing automation, and has in some cases, actually reduced productivity. This is due to the limited ability of CAD/CAM software to communicate with the other computer systems.

Another example of an island of automation approach has been the Flexible Manufacturing System (FMS), Fig. 13-5. In many manufacturing plants, the typical FMS operates independently of other FMS systems, without the benefit of hierarchical computer control.

One final example: at present, process planners will often use an isolated computer to create manufacturing instructions, then make drawings and sketches by hand to supplement the instructions. The reason for the handmade drawings and sketches is that these planners are producing work on computer systems that do not have graphic capabilities. To make matters worse, the process planning computer has not been interfaced to any other computer system. In this case, the ideal computer interface would be a CAD/CAM system capable to merging text and graphics.

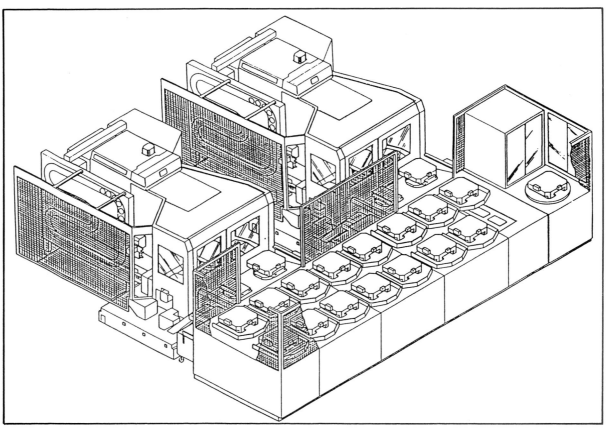

HITACHI SEIKI

Fig. 13-5. A flexible machining cell that consists of two horizontal machining centers with 60-tool changers, a 16-position pallet system to serve both machining centers, and a pallet washing system. Such cells often operate independently, without interconnection to the factory's mainframe computer.

Implementing automation

Islands of automation can be cost-effective if they are well planned and implemented. Without a plan, the islands can be linked together, but will not be truly integrated—some of the old problems will still exist. To solve the problems that result from linking islands of automation, it will be necessary to develop specifications for the company and suppliers of equipment and software.

There is no *one* product on the market that will satisfy all requirements of Management, Engineering, Planning, and Manufacturing. To satisfy these requirements, several products will need to be purchased or developed independently. However, the products *must* have common communication standards and goals. CIM is an integration of each separate system to a common database so that the data can be shared, transferred, and modified with ease. See Fig. 13-6.

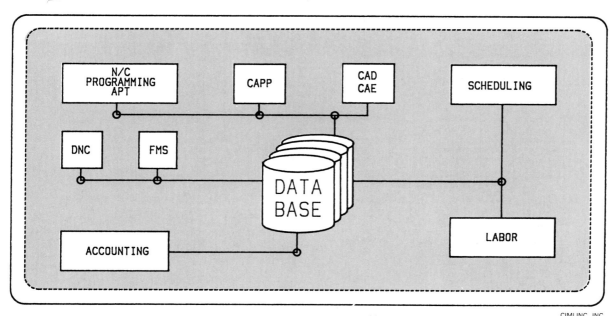

Fig. 13-6. In a Computer-Integrated Manufacturing situation, the individual databases built by departments or ''islands of automation'' are integrated into a common pool of information accessible to every department and function.

Integration projects must be broken down to components of manageable size, then prioritized. It is necessary to attack each area of integration separately, but with a common goal using common standards. Plans must include surveys of existing systems and equipment, training, costs, schedules, standards, and most important, methods to measure productivity gains.

The need for standards

To integrate the different products and systems it is very important to set company standards. These standards should not conflict with *national* or *international* standards. For instance, the huge number of different CAD/CAM systems used by industry forced the development of the *Initial Graphics Exchange Specification* (IGES).

IGES is used to transfer data between computer graphics systems. The graphics data is translated into a standard format, allowing the transfer of drawings between CAD systems from different manufacturers. IGES can also be used to transfer data between different CAD and CAM systems.

Some manufacturing experts believe that as much as half the total cost of shop floor equipment can be attributed to additional charges for making possible communications between the equipment and a computer. The reason for this is the large number of equipment manufacturers who have proprietary protocols and interfaces. General Motors Corporation developed a uniform set of communication standards, called *Manufacturing Automation Protocol* (MAP), that has been widely accepted.

The purpose of MAP is to define a multi-vendor communications network. See Fig. 13-7. The MAP users group is responsible for documenting and promoting an internationally accepted network environment for factory floor equipment. This equipment includes NC machine tool controllers, robots, computers, computer terminals, and programmable controllers.

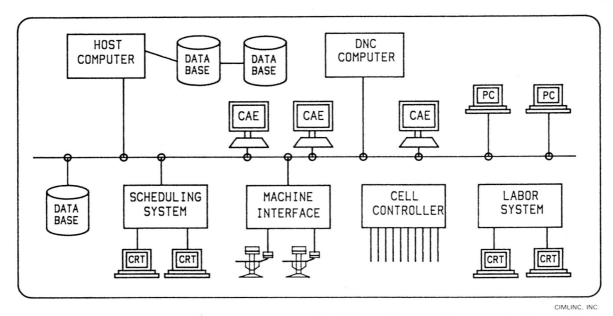

Fig. 13-7. The Manufacturing Automation Protocol (MAP) is a uniform set of communication standards. It is intended to make possible communication between systems originating with many different manufacturers.

ROLE OF THE COMPUTER

The computer will play an increasingly important role in the factory of the future and will continue to impact sales, marketing, factory planning, purchasing, process planning, scheduling, and assembly. As companies implement new computer technology, all aspects of the manufacturing process from product design to the assembly line will be affected.

The goal of many companies has been to produce "paperless manufacturing" through the use of computers. In the past, the cost of hardware and software has made this impractical, but as costs decline, this "paperless" environment is becoming a reality. In the factory of the future, it should not be surprising to find at every workstation a graphics terminal, Fig. 13-8, with the ability to access graphics and text database.

Computer-integrated manufacturing

Computer-integrated manufacturing (CIM) is more than just integrating computers with manufacturing. CIM is actually the sharing of common computer data between manufacturing disciplines and other systems required to produce a product. See Fig. 13-6.

Each company will need to develop its own CIM philosophy, strategy, and plan for implementation. CIM needs to be "top-down planning and bottom-up implementation" and must be a company-wide management philosophy for planning, integration, and implementation for automation. To suc-

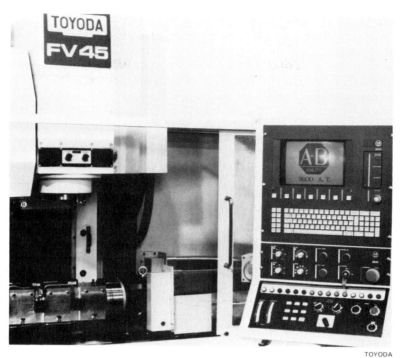

Fig. 13-8. A graphics terminal like the one on this vertical machining center's CNC controller will be a standard feature of workstations in the factory of the future. Such stations allow the operator to interact with the system and (when appropriate) modify a program to meet the requirements of a job.

cessfully integrate computers with manufacturing, management must commit to allocate the resources and develop an overall plan for automation.

Among the benefits of CIM are improved productivity, improved quality (reduced scrap and rework), reduced new part lead time from design to process planning, and increased factory utilization. To realize these benefits, each area of the manufacturing process from design to assembly needs to be reviewed. Plans for integration need to be developed and implemented.

MANAGING CHANGE

The effective manager or supervisor, as a function of his or her job, tries to create within the enterprise an environment that makes it possible to achieve the company's objectives. In doing this, the effective manager will select and train subordinates, organize their role relationships, direct their work, and evaluate the results. In other words, the duties of a manager include planning, staffing, organizing, directing, and controlling. Probably the most basic function a manager can perform is planning, a prerequisite to *managing change* in an organization.

Planning is a decision-making process involving selection among alternatives for future courses of action. It involves strategies, programs, policies, objectives, and procedures for achievement in the enterprise.

The ideal factory manager should exercise leadership. *Leadership* is the exercise of influence over others in situations where the influenced person believes it appropriate to follow. Leadership differs from authority. *Authority* is the formally delegated right to set goals and direct employee efforts to achieve them.

Rather than exercising leadership, however, factory managers and supervisors often adopt a reactive approach. They see themselves as constantly being required to react to emergency situations, and hindered in performing the job by unnecessary paperwork. They feel that they are involved in a decision-making process to overcome upper-level management failures, and maintain that they can accomplish required tasks if provided with adequate tools, material, and people.

To change from a reactionary approach to one that emphasizes planning, the manager must learn to make effective use of the computer to extract data that could be helpful in making management decisions. These important decisions can be based upon current events and processes, providing the manager is capable of retrieving needed information from the computer.

PLANNING FOR THE FACTORY OF THE FUTURE

Manufacturing in the future will see increasing competition from the Pacific rim countries like Japan, Korea, and China that have the manpower and raw materials, and Europe with its economic open border policy. With the increased competition there will be a "shakeout" of manufacturing companies. American companies have to begin to plan for the factory of the future *now*.

As noted earlier, to successfully integrate computers with manufacturing, management must commit to allocate the resources and commit to an overall plan. The plan should include the use of current technology and future technology that is being developed. To develop the plan, several major items need to be considered. These include such important steps as developing the "as is" and "to be" factory scenarios, selection of the projects, and justification. To generate a plan, there are several requirements:
- Develop the plan and allocate resources.
- Develop the "as is" factory scenario.
- Develop the "to be" factory scenario.

For any plan to succeed, the planner must have a clear understanding of the company strategy and goals. The plan needs to include a schedule for the execution of the defined tasks, reachable milestones, and resource requirements. There are a number of questions that have to be answered, such as:
- What is the overall company strategy?
- What are the goals of the company?
- What is the philosophy for human resources?
- Is the right department in the company doing the plan?
- Are there enough resources to complete the plan?

Once the planner has obtained answers to the relevant questions, it should be possible to make and implement the plan with only minor problems.

THE "AS IS" FACTORY SCENARIO

The *"as is" factory scenario* is a visualization (pictorial view) of the current conditions of the factory. It is required to help set the base from which to work. The following paragraphs describe some of the systems and conditions that need to be evaluated for the "as is" factory scenario.
- NC programming, CAPP, DNC, CAD/CAE, and scheduling are probably in development or are implemented, but the systems are not totally integrated. An example would be a company with DNC and NC programming systems that are integrated, but those systems are not integrated with CAD/CAE.
- Factory automation has been started, but only for higher volume simple parts; there is not an overall integration plan. Automation of the manufacturing support areas, such as scheduling, has not been truly addressed. Some form of scheduling software is probably installed on a com-

pany computer, but it is not intelligent, and cannot perform "what if" analyses based on specific conditions (*simulation*). Even if the scheduling system *can* perform "what if" simulations, that capability is certainly not available to the user on the factory floor.

- NC machine tools are being used effectively, but the "one person/one machine" syndrome still prevails, Fig. 13-9. A problem with NC equipment is that the machine controls are becoming easier to program manually, so the machine operator wants to make NC programs. This practice can be counterproductive, unless provision is made to assure that any changes made to the NC program at the machine level are reflected in the source program in the central database.

The "as is" factory scenario is valuable because it allows you to perform a diagnostic analysis of current problems. A direct benefit of the activity is that some of the problems will be easily corrected with little use of resources, thus producing a greater payback.

PEDDINGHAUS

Fig. 13-9. Many manufacturing operations continue to use their NC equipment in the traditional "one person/one machine" fashion.

THE "TO BE" FACTORY SCENARIO

The factory of the future will be totally integrated through use of the computer. To progress from the "as is" factory scenario to an actual computer-integrated manufacturing operation, it is necessary to generate a "to be" factory scenario. This scenario is a pictorial view of the factory 5, 10, or 20 years in the future. The following is a brief description of some of the systems and manufacturing strategies that need to be evaluated for the factory of the future.

Two prominent manufacturing strategies in the United States are *Group Technology* (GT) and *Just-In-Time* (JIT). These strategies are not limited to the manufacture of parts; they are also applied to assembly of the product. An example of an optimum application would be an automobile assembly line, where trunk lids would be manufactured complete and supplied one-at-a-time for assembly as needed (zero inventory).

JIT manufacturing

Basically, JIT is a philosophy that has two major points:
- Manufacture parts as they are needed for assembly, without building an inventory.
- Manufacture the parts in the general vicinity of the assembly facility.

Since there is not an inventory, the *customer* dictates the manufacturing schedule through *demand*. When a product is completed and ready to be delivered to a customer, another product must be manufactured. Therefore, Just-In-Time is a manufacturing technique that approaches a *continuous flow* process in which the piece parts are "pulled" through the manufacturing operation by demand.

GT manufacturing

Group Technology, on the other hand, is a different concept and is a variation of *batch* processing. With Group Technology and batch processing, piece parts are "pushed" through the manufacturing process. Identical piece parts are grouped in batches or *lots*. An operation must be performed on all parts in a lot before that lot can be moved to the next step, therefore, the parts are said to be *pushed* to completion. An example of the optimum GT application is a family of parts manufactured in a cell of machines.

In Group Technology, an important goal is to identify and group together parts similar in shape, size, composition, and required processing steps, then produce them using the same machines, routings, tooling, and personnel. This production approach makes use of the *manufacturing cell*, a group of machines that can perform most of the machining operations needed to manufacture a particular part. Usually, several manufacturing cells are required to make a factory.

Types of manufacturing systems

The type of manufacturing system used is usually determined by the *quantity* of parts or complete assemblies to be processed. See Fig. 13-10. With very low part quantities, it is sufficient to manufacture parts "one at a time." As the quantity increases, a *batch* process is generally used. For high volume output, *transfer lines* (such as the typical automobile plant assembly line) are usually employed. For very high volume production of an essentially uniform product, *continuous flow processing* is necessary. An example would be an oil refinery.

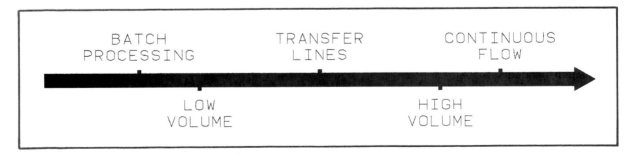

Fig. 13-10. The type of manufacturing system that is used by a company is normally determined by the volume of work to be done. As production quantities increase, the manufacturing operation moves from batch production through transfer lines to continuous flow production.

The JIT concept actually is applicable at all volume levels, from the very low to the very high. It is most effective when dealing with small quantities (the JIT ideal "a lot size of one"). Thus, as the time required to change the setup from one part to another approaches zero, "batch processing" becomes JIT. The final philosophy of the factory of the future will include elements of both JIT and GT manufacturing techniques, Fig. 13-11.

As has been stated before, the Just-In-Time and Group Technology concepts are not limited to the manufacture of parts, but also can be applied to assembly of the product. The key to efficiency of any assembly operation is making sure that parts and sub-assemblies are delivered to the workstation when needed.

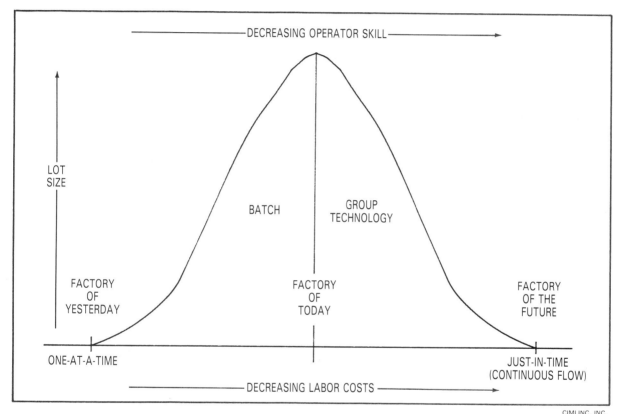

CIMLINC, INC.

Fig. 13-11. As the factory of yesterday evolved into today's manufacturing operation, lot sizes increased (from one-at-a-time to batch production), while labor costs and the level of operator skill needed decreased. As evolution continues, the labor cost and skill levels will continue to decrease, as will lot sizes. Group technology and JIT will result in a continuous flow of products in small lots.

Automated material handling

For efficient parts manufacturing or assembly, the factory of the future will have to make effective use of an ***automated material handling system***. The automated material handling system goes far beyond the obvious stage of using automated guided vehicles (AGVs) or an Automated Storage and Retrieval System (ASRS) to move parts to machines, as shown in Fig. 13-12. Instead, it must be a fully computer-integrated system involving both hardware and software. The ideal automated system would include a computer scheduling and control system, computerized warehouse (using a stacker crane or automated storage and retrieval system), AGVs, and a tooling and parts loading station. See Fig. 13-13.

The loading station should be the hub from which operations send parts and tooling to every workstation in the factory. The operators will inform the computerized scheduling system that they need a job or tool, and the computer system will then notify the computerized warehouse to deliver a new job (parts and tooling) to the loading station, using an AGV. The loading station operators will set the tooling and parts on a pallet for the specific operation. The AGV will then deliver the pallet, tooling, and parts to the proper workstation for machining. Tooling and parts may be delivered separately, or even by separate material handling systems.

TOYODA

Fig. 13-12. An automated storage and retrieval system is used to store and retrieve pallets and fixtures from shelving immediately adjacent to the horizontal spindle machining center at left.

When the machining operations are complete at one workstation, the AGV will either deliver the pallet to the next workstation or return it to the loading station. Loading station operators will then unload the parts from the fixture. They will load new blank parts, if necessary, or return the finished parts and tooling to the warehouse.

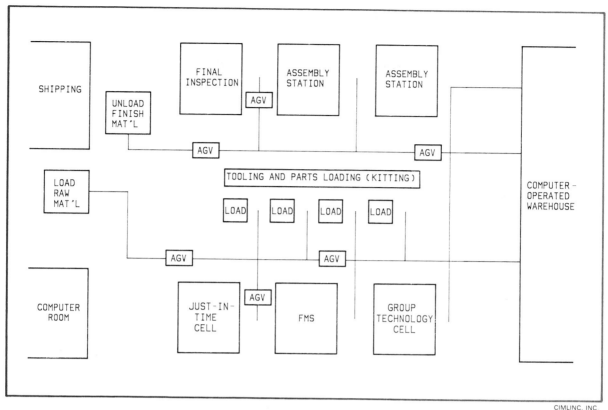

Fig. 13-13. A typical shop floor layout for the factory of the future. A vital link among the elements of the manufacturing operation is computer-controlled and scheduled material handling system.

Scheduling system

The most important link in the automated material handling operation is the computerized schedule and control system, which permits the many material and tool movement functions to be performed simultaneously in all areas of the factory.

The scheduling system must be able to adjust for personnel and equipment availability, machine downtime, changing customer requirements, and almost any other situation that might arise. To accomplish this task, an Artificial Intelligence (AI) computer scheduling system is usually needed. The scheduling system will be able to develop "what if" scenarios for differing situations, suggest to the user the optimum schedule, and even provide a graphic display. The AI scheduling system will be able to "learn" from various situations and successful solutions to problems. This will allow it to narrow the possible "what if" outcomes, even to the point of making decisions without user input.

Control functions

The control system must be integrated not only with the scheduling system, Fig. 13-14, but with each workstation or device (such as a robot, AGV, warehouse stacker crane). The control will need to send signals to a machine tool or robot to start a cycle (for example, to shuttle a pallet off an AGV to a machine tool) and also receive a signal from a machine or robot that it has completed a cycle. To do this, the control computer and all the machines and devices must be able to communicate, using MAP, Ethernet, or some other form of network.

Factory of the Future **335**

Fig. 13-14. Closed-loop control of this process monitoring and control system makes possible continuous monitoring and adjustment of processes and machines.

Machine tool requirements

The requirement that machine tools be capable of integration is apparent, but there are other less apparent characteristics that need to be considered. Critical requirements for machine tools are greater **uptime** (time available for actual use) and durability (long productive life). As manufacturing went from manual operation to NC operation, both uptime and durability had to increase. With the change from stand-alone NC to an integrated cell or FMS, these requirements are even more stringent. Machine tools that in the past were utilized only 30% of the time will be expected to be productive 60% to 90% of the time in an FMS. To accomplish this, machine tool builders are investigating new technology—such as the use of composite materials—to increase machine accuracy.

More and more, manufacturers are designing machine tools with parts made from composite materials instead of the traditional cast iron. The composites offer greater wear resistance and much better vibration-dampening characteristics. These enhanced machine tool characteristics are a requirement for unattended operation.

Unattended machine operation

The requirement to perform manufacturing operations unattended will adversely change personnel requirements. See Fig. 13-15. The level of skill needed for an operator to run a machine is decreasing, but the manufacturing engineering planning function has increased. There is a gap in industry between the need for skilled machine operators (machinists) and the number available from educational programs. To fill the gap, industry will need to employ more manufacturing engineers and to purchase machine tools that are able to operate unattended.

Fig. 13-15. This horizontal machining center, configured with a 13-position random-axis vertical pallet pool, is designed for unattended operation.

There are several means that can be used to simulate the input of a machine tool operator, thus permitting unattended operation. Vision and infrared systems, as well as vibration and tactile sensors are used to monitor the cutting time, tool forces, and power usage of the machine tool. Once the initial tool condition is established, one of two basic methods for predicting the need for tool replacement will be used. These are called the ***threshold method*** and the ***history method***. In the threshold method, a measured value is compared to a known machine limit (such as 50% horsepower usage), without regard to the actual condition of the tool. The history method compares the measured value to the value generated when the tool was new or sharp. Various automatic tool sensing systems and their functions are described below:

- A vision system consists of a television camera integrated with a computer (generally the machine tool controller) to visually monitor tool wear or breakage.
- A tactile (touch) monitor consists of a machine probe that is in physical contact with the part or the tool. It is used to measure dimensions of the part (to gauge material removal) or the tool (to monitor wear).
- A vibration monitor compares machine vibration with a set standard. When the standard is exceeded, a signal is sent to the controller causing it to initiate corrective action.
- An infrared monitor measures the heat generated by a tool, and compares that value with a set standard. When the standard is exceeded, a signal is sent to the controller.
- Tool force monitoring can be done by comparing the horsepower usage with a set standard. When the standard is exceeded, a signal is sent to the controller.

The objective for performing the "as is" and "to be" factory scenarios is to develop a detailed list of tasks that must be accomplished to move the company from the present to future manufacturing methods. The future factory must increase productivity by implementing systems and procedures identified by comparing the differences between the two scenarios. Once this important step is completed, the plan for implementation of the changes must be carried out.

EVOLUTION OF MANUFACTURING

Fig. 13-16 is a graphic representation of manufacturing evolution. During World War Two, quantity was the greatest concern to the user; manufacturing had to respond to the demand. From the 1970s through the 1980s, manufacturing had to respond to demands for improved quality. The 1990s bring a push for flexibility; by the year 2000, partnerships will be the emphasis.

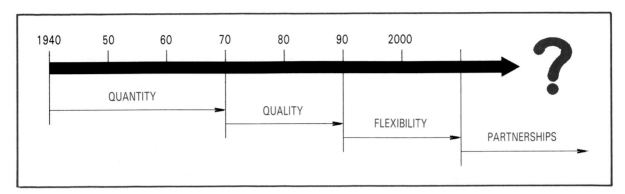

Fig. 13-16. The evolution of manufacturing from quantity emphasis through quality and flexibility to the need for developing partnerships at all levels.

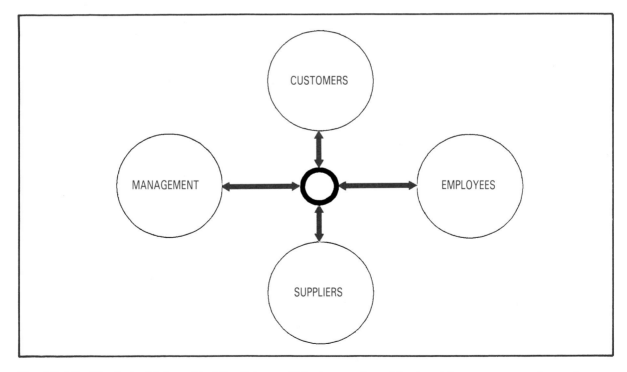

Fig. 13-17. The industrial world of the future will feature partnerships involving management, employees, customers, and suppliers. Such partnerships will be vital to survival of a business.

MANUFACTURING PARTNERSHIPS

For survival in a rapidly changing industrial environment, manufacturing company leaders must form partnerships with employees, suppliers, and customers. See Fig. 13-17. To achieve such partnerships, manufacturing companies must make organizational changes that will support the new company culture.

The availability of worldwide global communications also requires changes in basic business operations. Through the use of computers, modems, and phone lines, everyone in business enterprises—from top executives to customers—are linked together.

Role of the employee

In such a situation, the employee on the factory floor must take a leading role in automation from concept through implementation. Employee involvement in planning and implementing automation will require that companies carefully evaluate their organization structure and make changes to enhance the chances of company success. Future manufacturing organizations will permit effective communication and cooperation across all levels of management and manufacturing disciplines.

To achieve this type of organization, employees must perform multiple functions. Manufacturing employees will need to be trained to perform other manufacturing disciplines (to become flexible). Instead of just translating the product design into manufacturing plans and tooling, the manufacturing engineer will become more of an integrator. Managers will be leaders, not bosses. Essentially, they will manage the resources, not the employees.

Customer and supplier roles

Partnerships do not stop with the company's own management and employees, but must be extended to the company's customers and suppliers. The supplier will play an important role in the factory of the future. As manufacturing companies become more specialized, they will increasingly depend on suppliers for component parts, subassemblies, etc. Conversely, customers will have a comprehensive voice in product design and manufacturing. This new relationship will grow of necessity to control costs, delivery, and quality.

The degree to which companies form partnerships with their employees, suppliers, and customers will determine the future of those companies. Those who do it properly are more likely to survive.

SUMMARY

Automation has become a major part of manufacturing. It can be defined as the automatically controlled operation of a machine or group of machines to produce parts or assemblies.

As new technologies are implemented, *islands of automation* result. The creation of these *islands* is due to the limited ability of new software to communicate with other computer systems. These limitations brought about the creation of a uniform set of communications standards, called MAP (Manufacturing Automation Protocol). MAP is intended to define a multi-vendor communications network, making possible the interconnection of such equipment as NC machine tool controllers, robots, computers, computer terminals, and programmable controllers.

The introduction of the computer into manufacturing has allowed entire plants to be *integrated* with common computer systems from design to manufacturing. Process planning can be automated using a full (CAPP) system and an entire factory of machines can be linked to a Distributive Numerical Control (DNC) system network. Factory networks can also be integrated to a full MRP II system. Factory networks can also contain a full shop floor information data collection system for scheduling, machine monitoring, cost tracking, and labor reporting.

To be successful, management for the *Factory of the Future* must commit resources to the overall long-range plan. The plan should include the use of both current and future technology. The plan usually consists of three parts: *planning for resources, developing an "as is" scenario,* and *developing a "to be" scenario* for 5, 10, or 20 years into the future.

The *"to be"* scenario will probably contain phases for simultaneous engineering, group technology, manufacturing cells, automated guided vehicles, flexible manufacturing systems, Just-In-Time manufacturing, automated assembly, full MRP II system, and eventually, a complete automated and paperless factory.

The factory of the future must consider flexible manufacturing, in which highly versatile machines will eliminate much of the labor of producing goods. The new manager must be an excellent planning person. The management of the factory must also be trained in *total quality management* (TQM) principles, and will have to learn to exercise leadership instead of authority. Both direct and indirect employees must be empowered to make decisions. For example, operators will check their own work through operator certification and also will use statistical process control (SPC) at all levels within the factory. A company's success in forming partnerships may be directly related to its ability to survive and to prosper.

IMPORTANT TERMS

"as is" factory scenario
authority
automated material handling system
automation
demand
history method
Initial Graphics Exchange Specification

islands of automation
leadership
Manufacturing Automation Protocol
planning
simulation
threshold method
uptime

QUESTIONS FOR REVIEW AND DISCUSSION

1. What is meant by *islands of automation*? Give one or two examples.

2. Some people consider CIM to be nothing more than *wiring production machines to the computer*. Is this an accurate description? Comment.

3. What are the general categories of duties performed by a manager? Which of these is most important, in terms of managing change (such as CIM implementation). Explain.

4. Contrast the JIT and GT manufacturing philosophies, in terms of how they generate movement of product through the manufacturing process.

5. Describe the sequence of activities involved in moving a part from the loading station through an automated workstation and back to the unloading station.

6. List the various tool sensing systems typically used with unattended machine operations. Briefly describe their functions.

7. What changes in the role of the worker on the factory floor are likely to result as the factory of the future becomes a reality?

GLOSSARY

A

ABC ANALYSIS: A method of inventory control based on the general conclusion that a small percentage of a population accounts for the largest fraction of an effort or value.

ABSOLUTE ENCODER: An optical encoder that gives the exact position of the shaft.

ABSOLUTE POSITIONING SYSTEM: All dimensioned values programmed in the absolute system are referenced to the fixed point or origin of the Cartesian coordinate system.

ACTIVE COMPLIANCE: Compliance that uses sensors in the end-of-arm tooling to detect any unwanted loads that are present. This information is passed to the robot controller, which moves the robot in real time to remove the load.

ACTIVE ROBOT TEACHING: A method of programming in which the robot's actuators and feedback sensors are activated while a teach pendant is used to move end-of-arm tooling to the desired location and orientation.

ACTUATORS: Pneumatic, hydraulic, or electrical devices that are used to move a robot's manipulator links.

ADAPTIVE CONTROL: A CNC feature that optimizes cutting operations by changing feeds and speeds in response to feedback.

ADAPTIVE CONTROL: A system for altering (adapting) a process, as a result of changing conditions, to maintain optimum results.

ADAPTIVE CONTROL: Real-time response to changing process conditions, resulting in adaptation of the process (such as a change in cutting speed).

AD-HOC QUERIES: A method of database inquiry that lets users combine data to create reports and search for answers beyond what available applications provide.

AGV: Automated Guided Vehicle. Material handling equipment that functions without an operator. May be guided by a buried wire or other guidance system, and can be programmed to perform a variety of load/unload functions.

AI: Artificial Intelligence. A computer system (hardware and software) that can make decisions based upon given data.

ALGORITHM: A set of rules used to solve problems step-by-step.

ANALOG-TO-DIGITAL CONVERTER: Hardware that takes analog signals and converts them to binary information.

ANALYSIS TOOLS: Computer programs that can work with CAD models to evaluate stress, aerodynamic, thermal, and kinematic properties of a design before prototypes are built.

ANNUAL OPERATING PLAN: See AOP.

ANODE: The positive pole in an ECM system.

AOP: Annual Operating Plan. The near-term application of the strategic business plan.

APT: Automatically Programmed Tools. The most common language used for NC programming.

ARCHIVING: The term used to describe off-line storage for less-frequently accessed data files.

ARTICULATED: Term used to describe a jointed-arm robot manipulator.

ARTIFICIAL INTELLIGENCE: See AI.

"AS IS" SCENARIO: A visualization (pictorial view) of the current conditions of the factory.

ASRS: Automated Storage and Retrieval System. Computerized equipment used to store and access raw materials, components, finished goods, or critical tools and fixtures.

ATTRIBUTES: Significant characteristics of a part for GT coding purposes. Attributes are assigned alphanumeric symbols.

AUTHORITY: The formally delegated right to set goals and direct employee efforts to achieve them.

AUTOMATED GUIDED VEHICLE: See AGV.

AUTOMATED STORAGE AND RETRIEVAL SYSTEM: See ASRS.

AUTOMATIC SCHEDULE REGENERATION: A request from the FMS computer control system that will cause the FMS scheduling system to generate a new schedule to be placed in the Current Schedule File.

AUTOMATICALLY PROGRAMMED TOOLS: See APT.

AUTOMATION: The use of an automatically controlled machine or group of machines to produce parts or assemblies.

AXIS OF MOTION: The rotation of one link about another link or about a stationary base, or the translational motion of one link relative to another link or stationary base.

B

BACKSCHEDULING: A process used by MRP to determine a start date for each operation described in the routing, based on the due date of the work order for the parent being built.

BATCH MANUFACTURING: A system in which a discrete number of piece parts is stored in a lot awaiting processing; manufacturing done in small lots.

BILL OF MATERIAL: A list that defines which parts (and which versions of those parts) to use in assembling a product.

BILL OF RESOURCES: A listing of the production equipment, personnel, materials, and other resources needed to manufacture an individual unit of the specified product. This information can then be projected into a rough-cut capacity check.

BLOCK SCHEDULING RULES: A method used to help an industrial engineer develop quick "rule of thumb" standards and capacity estimates.

BUCKETS: Increments of weeks or months used to calculate order quantities.

BUFFER: In a machine vision system, a memory area where digitized information is stored in the buffer until the computer can process it.

BUFFERING: A procedure used with off-line data collection. The data is stored (buffered) until it is uploaded to the network.

BUSINESS PLAN: A document that describes the sales, profit, and capital required to achieve the objectives set by a company for growth, profitability, and return on investment.

BYTE: The unit of measure for computer memory size, roughly equivalent to one letter, number, or symbol.

C

CALS: Computer-Aided Acquisition and Logistics Support. A United States Department of Defense initiative aimed at improving access to CIM data between defense contractors and the Department.

CAM: Computer-Aided Manufacturing. In general, CAM refers to the use of computers to assist in any or all phases of manufacturing.

CAPACITY PLANNING: The function which has responsibility for establishing, measuring, monitoring, and sufficiently adjusting levels of capacity to properly execute all manufacturing schedules.

CAPACITY REQUIREMENTS PLANNING: Short-range capacity planning that compares actual to planned output, determines reasons for deviations, and initiates corrective action.

CAPACITY: The highest sustainable output rate that can be achieved with the current work schedule, product specifications, product mix, equipment, and worker productivity.

CAPITAL RECOVERY PERIOD: The length of time required for an investment in machinery or other capital goods to repay its cost through operating savings. Also known as the "payback period."

CAPP: Computer-Aided Process Planning. A computer system (hardware and software) used to produce a list of manufacturing and assembly steps (routing).

CARTESIAN COORDINATE ROBOT: A manipulator that has three mutually perpendicular axes of motion.

CARTESIAN COORDINATES:. A system in which each point in space is represented by a pair of numbers that correspond to its distance from the intersection of two axes at right angles to each other.

CASE: Computer-Aided Software Engineering. A system that uses computer programs (CASE tools) to help design components of an application and the information flows between them.

CATHODE: The negative pole in an ECM system.

CE: Concurrent Engineering. Also called Simultaneous Engineering, CE is a methodology for getting more departments in a company, and even close suppliers, to work together on a design. Before it is released for production, the group will assess the design for its ability to be manufactured, assembled, repaired, replaced, and inspected.

CELL CONTROLLER: The final distribution point for instructions to a group of 4 to 6 machine tools; in effect, the controller functions as a low-level DNC computer.

CENTERLINE PROGRAMMING: A feature of many current generation controls, in which centerline outputs in programming language, such as APT, are translated directly into machine moves.

CHECK SURFACES: In milling operations, pieces of geometry used as turning points.

CIM DATA ARCHITECTURE: The system design that defines database storage structures, security structures, network structures, and access structures, as well as establishing physical and descriptive data definitions.

CIM: Computer-Integrated Manufacturing. A company-wide management philosophy for planning, integration, and implementation of automation. CIM is an integration of each separate computer system (such as DNC, CAE, and CAPP), so that the data can be shared, transferred, and modified with ease.

CIRCULAR INTERPOLATION: A mode of contouring a two-dimensional curved shape by moving the tool through arcs of a circle.

CLASSIFICATION: A method of grouping similar things according to general likeness, then discriminating within those groups according to specific differences.

CLOSED-LOOP CONTROL: A type of control in which the robot controller does know the position and/or velocity of each actuator during the operation of the robot.

CLOSED-LOOP MRP: A Material Requirements Planning system that includes the function of feeding back performance measurement data.

CMM: Coordinate Measuring Machine. These computer-controlled machines permit the fast and accurate measurement of machined parts, with data recorded automatically for analysis and reporting.

CNC CIRCULAR INTERPOLATION: A hole-making technique in which drives move the laser beam in two axes simultaneously to create a circular cut around the perimeter of a hole.

CNC: Computer Numerical Control. A numerical control method in which one computer is linked with one machine tool to perform NC functions.

COHERENT LIGHT BEAM: One that has one wavelength and is very *collimated* (consisting of parallel rays).

COMMUNICATION PROTOCOL: A set of agreed-upon rules for data transfer and standards for physical linkages of computers and associated devices.

COMPARATOR: A device used to optically compare the profile of outline of a part against a template for inspection or process control.

COMPUTER NUMERICAL CONTROL: See CNC.

COMPUTER-AIDED ACQUISITION AND LOGISTICS SUPPORT: See CALS.

COMPUTER-AIDED MANUFACTURING: See CAM.

COMPUTER-AIDED PROCESS PLANNING: See CAPP.

COMPUTER-AIDED PROGRAMMING:. NC programming with the aid of software that makes it possible to easily develop programs for complex shapes and parts.

COMPUTER-AIDED SOFTWARE ENGINEERING: See CASE..

COMPUTER-INTEGRATED MANUFACTURING: See CIM.

CONCURRENT ENGINEERING: See CE.

CONFIGURATION CHANGE BOARD: A group of managers that meets to determine effectivity for engineering change.

CONFIGURATION MANAGEMENT: A means of managing engineering change that entails documenting of the configurations used throughout an entire product life cycle.

CONFIGURE-TO-ORDER MANUFACTURING: In this system, the product consists of a basic item with various options selected by the buyer to meet specific needs.

CONNECTIVITY ANALYSIS: A feature of machine vision software that allows the part shape to be segmented into connected components and analyzed one component at a time.

CONSTANT SURFACE FEET PER MINUTE PROGRAMMING: See CSFM.

CONTINUOUS-FLOW PRODUCTION: High-volume manufacturing of standard products.

CONTROLLER: See PLC.

CONVEYOR SYSTEMS: Fixed material handling devices that can be used to transfer pallets from load/unload stations to the machine tools of an FMS.

COORDINATE MEASURING MACHINE: See CMM.

COST ACCOUNTING: A cost control activity that tracks execution of the business plan, and is historical in nature.

COST MANAGEMENT: A cost control activity that is related to managerial performance and projects the future.

COST ROLL-UP: A costing method in which total cost is computed by adding together material, labor, and burden cost components.

COST/SAVINGS MEASUREMENT PLAN: A plan that shows what to measure, how to measure it, and how long to measure it. It permits management to see a project's success or understand its failures.

COUPLER CONTAMINATION: Adhering of chips or other foreign matter to the coupler; such contamination could cause imprecise indexing on the machine.

CRITICAL PATH: Another name for product cycle time.

CSFM: Constant Surface Feet Per Minute Programming. A control feature used to achieve constant cutting speed by adjusting the spindle motor speed (increasing or decreasing rpm).

CUSTOMER SERVICE PERFORMANCE: A measurement, expressed as a percentage, of efficiency in filling customer orders.

CYCLE COUNTING: An inventory method that identifies the causes of an inaccurate balance and allows appropriate action to be taken.

CYCLE TIME: The period of time from recognition of a customer need to shipment of actual product to meet that need.

D

DATA ARCHITECTURE: A system of data definitions and structures.

DATA DICTIONARY: A program element that defines how and where information is stored.

DATA ELEMENTS: The distinct types of information maintained in a database.

DATA INTEGRITY: Term describing the correctness of stored information. Data integrity is violated, for example, if duplicate files contain different values for a key field.

DATA SECURITY: A system used to prevent the making of unauthorized changes in information through limiting access to computer files and other strategies.

DATA SHARING: An essential feature of CIM. Computer systems must be capable of accessing data from various company databases.

DATA VALIDATION: The process of checking data for correctness. Data validation routines can be written into a computer program.

DATABASE SYSTEM: A structured file of information (data) that can be accessed for reference or for use in one or more applications.

DC TACHOMETER: The most common form of velocity sensor.

DEDICATED COMPUTER: A computer assigned strictly to a specific task.

DEEXPEDITE: The action of rescheduling an order for later production than planned.

DEMAND: Anything that reduces inventory.

DEMAND MANAGEMENT: Demand management is a process that identifies, for the factory, the items to be built.

DEPENDENT DEMAND: Demand that derives from a schedule of end products or service parts (the Master Production Schedule).

DESIGN FOR ASSEMBLY: See DFA.

DESIGN FOR MANUFACTURE: See DFM.

DESIGN LIBRARIES: Stored computer files that provide quick access to catalog parts (such as bolts and springs) that can be included in a model with a few keystrokes.

DESIGN REVIEW: A meeting in which representatives from the manufacturing areas can provide input to the product design.

DFA: Design for Assembly.

DFM: Design for Manufacture.

DIGITAL DATA: Information that has been converted to a form that can be stored and manipulated by the computer for manufacturing purposes.

DIGITIZING PROGRAMMING: A capability of today's control systems. The systems can convert part measurements to digital form, then use that stored digitized information to reproduce the part. This minimizes programming time.

DIGITIZING TABLET: A device that uses a

mouse-like or pencil-like device (stylus) that the CAD operator moves around on a digitizing surface to convert analog data to digital form.

DIRECT INPUT DEVICES: Devices such as sensors or code readers, that provide real-time instantaneous status data.

DISCREPANT LOTS: Parts lots that do not meet standards.

DISPATCH LIST: A document, produced on a daily basis from the planning system, identifies the priority of each job at each workcenter on the shop floor.

DISTRIBUTED PROCESSING: A system in which computing and directions come from the *lowest* level, closest to the action being performed. This frees upper-level computers to do overall planning, controlling, and reporting, and doesn't burden them with controlling and directing individual factory machines or processes.

DISTRIBUTIVE NUMERICAL CONTROL: See DNC.

DNC: Distributive Numerical Control. A control method in which one or more computers linked to a *number* of NC machine tools and capable of operating them simultaneously or separately.

DOCK-TO-STOCK PROCESSING TIME:. The amount of time it takes to remove a purchased product from the delivery vehicle and process it to the storeroom or point of use.

DUE DATE: Date of completion for a parent item.

E

EBM: Electron Beam Welding. A material-joining process in which a fusion weld is produced when a narrow beam of accelerated electrons strikes and heats the work.

ECM: Electrochemical Machining. A specialized production process that relies on the principle of electrolysis for material removal.

ECONOMIES OF SCALE: Cost savings provided by continuous-flow high-volume production.

EDM: Electrical Discharge Machining. An advanced production process that uses a fine, accurately controlled electrical spark to erode metal.

EFFECTIVITY: The date upon which an engineering change becomes effective.

EFFICIENCY: A performance measure equal to standard hours earned divided by actual hours worked on product.

ELECTRICAL DISCHARGE MACHINING: See EDM.

ELECTROCHEMICAL MACHINING: See ECM.

ELECTROLYSIS: A means of producing chemical change by passing electric current through a conductive liquid.

ELECTROLYTE: A conductive liquid used in electrochemical machining.

ELECTRON BEAM WELDING: See EBM.

EMBEDDED COMPUTERS: Units built into NC machine tools and other equipment to provide local processing power or memory, freeing departmental or mainframe computers for other tasks.

ENABLING TOOL: Term used to describe how a computer provides the potential for making information available, for protecting it, and for helping people create, summarize, and analyze it.

END EFFECTOR: A tool or gripper attached to the end of a robot arm.

ENERGY TRANSFER MEDIUM: Water, air, or another fluid used to transmit the shock wave from an explosive charge to the workpiece. The shock wave forces the workpiece into the evacuated cavity, causing it to conform to the shape of the die.

ENGINEERING DATA CONTROL: The process of creation, organization, and maintenance of records used by all segments of a business to identify, manufacture, and market its products.

EVEN PARITY: A tape coding system in which the rows of perforations contain only even numbers of punched holes.

EXPEDITE: The action of rescheduling an order for earlier production than planned.

EXPERT SYSTEMS: Computer programs that represent a particular domain of human knowledge. They permit a computer to gather information from a database and use it to draw a conclusion or select a course of action.

EXPIRATION TRACKING: A stock rotation principle that identifies a specific date after which an item can no longer be used.

EXPLODING A BILL OF MATERIAL: Breaking down the bill of material into quantities of needed components.

EXPLOSIVE FORMING: A high-energy-rate

forming process in which explosives are used to shape metal.

F

FACTORY NETWORK: A system of wires used to transmit program instructions and other data between computers and production machines.

FAMILY OF PARTS: As identified though the use of Group Technology, parts that have features in common or follow similar manufacturing processes.

FAS: Final Assembly Schedule. The process of bridging the MPS completion to the customer orders that consume it.

FEATURES-BASED DESIGN SYSTEMS: CAD programs that store geometric information as lists of features with parameters to define details of each feature type and how it connects to the overall part.

FEEDBACK: A situation in which sensors constantly monitor the machining process and send signals (feedback) to the computer about favorable or unfavorable conditions. The computer then changes feeds or speeds as needed.

FEEDRATE (F) COMMAND: A program command used to establish the proper feedrate (in inches per minute) for both machining and non-machining moves.

FIELD: A defined portion of a database record. A field holds one piece of data.

FILE MANAGEMENT SYSTEMS: Software that provides tools for making file backups, and for archiving, retrieving, and revising files.

FILTERING TOOLS: In MRP, filtering tools allow the use of ranges (percentages, quantities, and dates), rather than absolute values in determining performance.

FILTERS: In MRP II systems, tolerances or restrictions on action.

FINAL ASSEMBLY SCHEDULE: See FAS.

FINANCIAL PLAN: A company's sales plan translated into dollar amounts.

FINITE CAPACITY LOADING:. A work placement approach in which the planned orders developed by MRP are loaded into a workcenter up to the level of that workcenter's demonstrated capacity. Planned orders in excess of that capacity are rescheduled into future time periods.

FIRM PLANNED ORDER:. A type of order that allows the planner to control a specific order quantity or due date.

FIRST ARTICLE INSPECTION:. Thorough checking of the initial shipment of purchased parts to make sure that the vendor can actually build the part to specifications.

FIXED AUTOMATION: A collection of machines and tooling designed to achieve high-volume production of a single part at low unit cost.

FIXTURE: In a manufacturing system, the workpiece holder.

FLEXIBLE MANUFACTURING SYSTEM: See FMS.

FLOAT: The number of workpieces in the FMS that are not being directly worked on.

FMS: Flexible Manufacturing System. A group of processing or workstations connected by an automated material handling system and operated as an integrated system under computer control.

FORCE SENSING: A measurement, used in active compliance, of stress placed on the end-of-arm tooling can be accomplished by mounting strain gauges at desired locations.

FORECASTING: In market research, the task of predicting (often with the aid of a computer program) the required production rate and volume to meet anticipated demand.

FORMAL SYSTEM: The "official" system of communication in a company. It implies accountability and measurement.

FORWARD SCHEDULING: A means of determining when a part will be completed, developed by working forward from a known start date.

FULL REDUNDANCY: The practices of using identical machines to prevent unexpected downtime.

G

GENERATIVE PROCESS PLANNING:. A method in which a detailed description of the workpiece and its features is used by the CAPP system to automatically generate a process plan.

GRAPHICS: Images displayed on a computer screen.

GROUP TECHNOLOGY: See GT.

GT: Group Technology. A manufacturing technique in which "families" of similar parts are identified and grouped into lots that are then processed on dedicated machines. Piece parts are said to be "pushed" through the manufacturing process.

H

HARD COPY: A print on paper or Mylar® of a CAD drawing, usually produced by a plotter or electrostatic printer.

HIERARCHICAL STRUCTURE: Term used to describe CIM computer networks that generally form a pyramid-like structure. There are different levels (hierarchies) of control in the network.

HIGH-ENERGY-RATE FORMING: A process that uses a relatively large amount of energy, applied for a very short period of time, to cause rapid deformation of the workpiece in a controlled and predictable manner.

HISTORY METHOD: A method for predicting the need for tool replacement that compares the measured value to the value generated when the tool was new or sharp.

HORIZONTAL DEPENDENCE: The relationship of availability of components that are common to given parent.

I

IDLE TIME: Time when a machine tool is not actually removing material from the workpiece.

IGES: Initial Graphics Exchange Specification. A graphics standard established to make possible transfer of data between different computer graphics systems.

IMPLOSION: The *where-used* or "bottom-up" form of the bill of material.

INCREMENTAL ENCODER: An optical encoder that gives positional information about a joint by counting the number of increments that the encoder plate has rotated, then multiplying this value times the angle between increments.

INCREMENTAL POSITIONING SYSTEM: A system that uses a floating coordinate method instead of a fixed reference point.

INDEPENDENT DEMAND: Demand derived from an external source.

INDUSTRIAL ROBOT: A reprogrammable multifunctional manipulator.

INFINITE CAPACITY LOADING: A work placement approach in which MRP assumes floor capacity to be infinite, and loads work accordingly.

INFORMAL SYSTEM: The methods employees follow in the absence of formal information procedures.

INITIAL GRAPHICS EXCHANGE SPECIFICATION: See IGES.

IN-PROCESS INSPECTION: The process of verifying that parts are made to specifications at each operation.

INPUT-OUTPUT CONTROL: A method of balancing the shop floor load so that both the input and output cumulative deviations are zero.

INTEGRATION: Sharing of the same information from the same source by two or more software applications.

INTELLIGENT SUBSYSTEMS: Machining cells, or workcenters, that will process work as directed by the central factory computer.

INTERACTIVE: Term used to describe a computer system that provides the operator with an instant visual display for manipulating images quickly and effectively.

INTERCOMPANY ORDERS: Orders issued to a related business (another division of a company, for example).

INTERFACING: A process in which data needed by two systems is duplicated, but can be sent between them through some transfer or translation strategy.

INTERPOLATION: The process of calculating of the relative movement between the tool and the workpiece on an NC machine.

INVENTORY STRATIFICATION: See ABC ANALYSIS.

ISLANDS OF AUTOMATION: Isolated examples of automation as a result of manufacturing managers implementing emerging technologies as they become available.

ITEM MASTER DATA: Records that describe the characteristics of each unique manufactured and purchased part.

J

JIT: Just-In-Time. A system that eliminates work-in-process (WIP) inventory by scheduling arrival of parts and assemblies for an operation at the time they are needed and not before. Piece parts and assemblies are said to be "pulled" through the manufacturing process.

JUST-IN-TIME: See JIT.

K

KEYBOARD: A data entry device with letter, number, and special-purpose keys.

KINETIC ENERGY: The energy of motion.

L

LAN: Local Area Network. The system of cables connecting computers and controllers in a DNC situation. The LAN considerably simplifies the cabling job and minimizes the footage of cable that must be strung through the factory.

LASER: An acronym formed by the initial letters of the words making up its description: **L**ight **A**mplification by **S**timulated **E**mission of **R**adiation.

LEADERSHIP: The exercise of influence over others in situations where the influenced persons believe it is appropriate to follow.

LEAD TIME ELEMENTS: All of the increments that make up the total cycle time of a particular factory.

LIGHT PEN: An object, shaped like a pen wired to the computer, that can be used to locate points on the screen.

LINEAR INTERPOLATION: A mode used in contouring a two- or three-dimensional curved shape using a tool moved in straight-line segments. It involves simply designating the end coordinates of the shape. Also, a robot programming method in which the beginning and end points of the robot path are taught by the operator.

LOAD FACTOR: A performance measure calculated by dividing standard hours earned by the total available hours.

LOCAL AREA NETWORK: See LAN.

LOT TRACEABILITY: A system used to maintain the relationship between specific lots of parts and the products they are used on.

M

MACHINE COMMANDS: Program codes that tell the machine what to do.

MACHINE CONTROLLERS: Electronic devices that convert binary instructions into timed electronic pulses to move stepper motors on a machine the desired direction and amount.

MACHINE VISION: A quality control system that records the shape, size, and position of some object, compares it to a stored image, and provides an accept/reject decision based on established tolerances.

MACHINE-READABLE CODING: Symbols (such as letters, numbers, bars, etc.) that can be scanned and interpreted by a computer system.

MACHINING CELL: Two or more machines or workcenters that are served by a local material handling system and controlled by a dedicated computer. Also called a manufacturing cell.

MACHINING CENTERS: Multiple-purpose machine tools that can be programmed to perform a number of operations (drilling, milling, boring, reaming, tapping, or counterboring) on a single part.

MACHINING STATEMENTS: Programmed instructions used to direct a cutter around previously defined geometry, following a predetermined sequence.

MAD: Mean Absolute Deviation. A calculation method for measuring forecast error of a specific forecast trend line.

MAINFRAME: The largest type of computer. Mainframe computers are generally quite expensive, have vast amounts of memory, and have extremely fast processing speeds.

MAKE-TO-ORDER MANUFACTURING: A system in which the customer specifies the product before any resources are committed. It meets a need for specific, not normally available, product.

MAKE-TO-STOCK MANUFACTURING: A system used for products that are essentially identical and that are produced in large quantities.

MANAGEMENT INFORMATION SYSTEM: See MIS.

MANIPULATOR: A mechanical device that has the ability to move parts or tooling through space along a prescribed path to the desired location.

MANUFACTURING: A series of interrelated activities and operations that involve product design, and the planning, producing, materials control, quality assurance, management, and marketing of that product.

MANUFACTURING AUTOMATION PROTOCOL: See MAP.

MANUFACTURING CELL: A group of machines organized to produce one family, or several similar families, of parts. Also called a machining cell.

MANUFACTURING ORDERS: Documents issued for work that is completed inside the company.

MANUFACTURING REQUIREMENTS PLANNING: See MRP.

MANUFACTURING RESOURCE PLANNING: See MRP II.

MANUFACTURING RESOURCE RECORDS: Documents that describe machines, work centers, and other physical resources.

MANUFACTURING ROUTINGS: Process sheets that describe the sequence of operations to be performed on a specific part.

MAP: Manufacturing Automation Protocol. A standard that defines a multi-vendor communications network for all types of factory floor equipment.

MASTER PRODUCTION SCHEDULE: See MPS.

MATCHED SETS OF PARTS: Complete sets of components and parent items necessary to build a product efficiently.

MATERIAL HANDLING SYSTEM: A system used to move workpieces from machine to machine, or to move raw materials and components to and from storage areas.

MATERIAL REVIEW BOARD: A management group within a company that deals with material and supplier problems.

MATHEMATICAL CAPABILITY: The ability of a CNC machine tool to store mathematical functions within a machining statement.

MEAN ABSOLUTE DEVIATION: See MAD.

MEASUREMENT BASELINE: An assessment of the current status, established as a point for measuring performance improvement.

MEMORY: The amount of storage a computer has available for use with programs and data.

MIN/MAX SYSTEM: A method of inventory control in which the estimated demand during replenishment lead time, plus a reasonable buffer (the trigger point for reorder), forms the *minimum* level. The *maximum* level is the on-hand quantity that management does not want to exceed.

MIPS: Millions of Instructions Per Second. The unit of measure for the number of instructions per second a computer can execute.

MISCELLANEOUS (M) COMMANDS: Those functions that prepare an NC system for operation.

MIS: Management Information System. A database application designed to provide management with information needed for decisionmaking.

MODAL: Term used to describe action of program "M" commands: once in effect, they remain in effect until changed by another "M" command.

MODE OF OPERATION: The way that the laser light beam is applied to the work source, either continuous or pulsed.

MODEM: Acronym for "modulator/demodulator", a device for converting computer signals for transmission over telephone lines.

MONOCHROMATICITY: A property of laser light that means having a single color frequency.

MOUSE: A small hand-held device that provides precise and rapid cursor control.

MPS: Master Production Schedule. This schedule is a refining of the production plan into item-built priorities.

MRP II: Manufacturing Resource Planning. A system that back-schedules activities from the final ship date of a product, based on lead time of assemblies, components, and raw materials.

MRP: Manufacturing Requirements Planning. A scheduling system that is similar to MRP II, but considers a smaller organizational scope and does not consider production capacity when scheduling.

MULTIPLE-LEVEL BILL OF MATERIAL: A list of the parent and its components through all intermediate levels until the purchased items are reached.

MULTI-USE TOOLING: Tools that, with only slight modification, can be used for machining or fabrication of a number of similar parts.

N

NC PART PROGRAM: The complete series of codes necessary to produce a part on a numerically controlled machine.

NC: Numerical Control. A system of machine control by the output of a computer. Coordinates, expressed in numerical form, are used to establish tool locations and cutting paths.

NET REQUIREMENTS: Quantities determined by deducting available inventories and supply orders from the gross requirements.

NETTING: In MRP, the process of determining quantity.

NETWORK: An electronic interconnection of various devices, including computers, terminals, data collection equipment, and automated machine tools to permit communication among them.

NON-RECURRING LEAD TIMES: The delays needed to design, analyze, prototype, and test a design, or any other delays from the conception of a product up to the point it is ready for production.

NON-RECURRING PROCESSES: In manufacturing, events that take place a limited number of times for a product or component, no matter how many times that part is produced.

NONTRADITIONAL MANUFACTURING PROCESSES: Those advanced manufacturing processes that have computer-assisted nontraditional applications.

NO-WEAR MACHINING: In electrical discharge machining, graphite electrodes can be used so that workpiece erosion is accomplished with practically no loss of electrode.

NUMERICAL CONTROL: See NC.

NUMERICALLY CONTROLLED WIRE FEED EDM: An electrical discharge machining application that uses a spool of wire for the electrode.

O

OCR: Optical Character Recognition. A system that permits a computer to recognize letters or symbols read by a scanning device. OCR can be used to read labels for parts identification or finished goods inventory.

ODD PARITY: A tape coding system in which the rows of perforations contain only odd numbers of punched holes.

OFF-LINE: A term that describes an action, such as writing a program, that is done away from the machine, rather than during operation.

OFF-LINE ROBOT PROGRAMMING: A programming method in which the robot program is developed by typing in the locations and orientations of the end effector at each stage of the process. The program is then stored in the robot controller for playback.

OFFSET: In international business, the percentage of a product's components or parts that must be manufactured in a given location as a condition of a contract.

OPEN-LOOP CONTROL: A type of control in which the robot controller does not know the position and/or velocity of each actuator during the operation of the robot.

OPTICAL CHARACTER RECOGNITION: See OCR.

OPTICAL ENCODER: The most common type of position sensor.

OVERTRAVEL MONITORING: A control feature that can prevent damage to the machine tool and the part being processed. If the control senses that a programmed move will make the tool move beyond the limit set in a given axis, it will not permit the move to be made.

P

PACING RESOURCE CODE: In manufacturing control, a code that identifies whether a particular workcenter is machine-paced or labor-paced.

PALLET: A carrier or platform used with automated material handling systems.

PAPERLESS FACTORY: A "factory of the future" concept in which production information is available from computer-generated video display and hardcopy terminals, rather than on reams of historical computer printout.

PARABOLIC INTERPOLATION: A mode of contouring a curved shape by moving the tool though a parabolic (bowl-shaped) arc. Also called quadratic interpolation.

PARALLELISM: A property of laser light that can be described as traveling in a column or straight-line path versus converging (coming to a point) or diverging (spreading out in a cone shape).

PARAMETRIC DESIGN TOOLS: Computer programs allow creation of a generic model that can be automatically redrawn to scale after dimensions are entered.

PARAMETRIC PROGRAMS: Control programs used to produce parts in which only certain elements (parameters) change from one production run to the next.

PARENT ITEM: Product or components assembled from other components.

PARTS CLASSIFICATION AND CODING: A method of grouping parts in which various design and/or manufacturing characteristics of the part are identified, listed, and assigned a code number.

PASSIVE COMPLIANCE: Compliance achieved by using an elastic medium at some point along the end-of-arm tooling, so that deflections can occur when an unwanted load is applied.

PASSIVE ROBOT TEACHING: A robot programming method in which the feedback sensors are activated, but the robot actuators are deactivated. The robot's end-of-arm tooling is then moved along the desired path at the correct speed.

PEGGING: An MRP capability that displays, for a given item, the detailed sources that generate its gross requirements.

PERIODIC COUNT: The annual physical inventory.

PERIODIC REVIEW: An inventory control method in which records are reviewed at fixed intervals, and sufficient material is ordered to restore the on-hand plus on-order quantities to specified levels.

PERMANENT-MAGNET MOTOR: A common type of DC (direct current) motor in which the speed can be varied by changing the applied voltage, and the direction of rotation changed by reversing the polarity of the motor input leads.

PHOTON: A particle of light energy.

PICK DATE: The date when components are gathered for assembly of the parent item.

PICK LIST: A term often used for bill of material.

PLANNING: The process of deciding in advance what to do, how to do it, when to do it, and who is to do it. Product Planning and Process Planning are distinct functions in a manufacturing enterprise.

PLANNING HORIZON: In MRP, the time span for which planning has been done.

PLC: Programmable Logic Controller. A solid-state control system that has a user-programmable memory to store instructions. The basic executive software used with such controllers is designed so that new features can be easily added.

POINT-TO-POINT SYSTEM: System that allows tool movement only in straight line segments

POSITIONING SYSTEMS: Systems that identify locations numerically, in terms of a specific *reference point*, to ensure precision in tool and workpiece movement.

POST-PROCESSORS: Software programs used to interface NC programs with the controllers used in specific machine tools.

PREPARATORY (G) COMMANDS: Functions that change the mode of operation of an NC system.

PRISMATIC: Term used to describe box-like parts processed by an FMS.

PROCESS DESIGN: A description of the sequence of operations to make each component and/or assemble components into a final product.

PROCESS PLAN: A method for making component parts and assemblies that meet defined design criteria.

PROCURABILITY: An assessment of whether parts or materials needed to manufacture a product can be obtained.

PRODUCIBILITY: An assessment of how easy or difficult a product or component will be to manufacture.

PRODUCT DESIGN: The process of defining what a product should look like, what it should do, how long it should last, how it should work, and how well it should meet a defined need.

PRODUCT PLANNING: A systematic program intended to identify products that will generate maximum profits for a manufacturing enterprise.

PRODUCT SPECIFICATION: A clear, complete statement of the technical requirements of a product, and of the procedure for determining if those requirements are met.

PRODUCTION PLAN: A conversion of the Annual Operating plan in which rates of production and resource requirements are defined.

PRODUCTIVITY: A performance measure equal to standard hours earned divided by the total available hours.

PROGRAMMABLE LOGIC CONTROLLER: See PLC.

PURCHASE ORDERS: Documents issued to obtain goods or services outside the company.

PURCHASING CONTROL: Function that is responsible for activities involving contact with a company's external suppliers.

Q

QUEUING: Temporary storage of workpieces.

QUILL-FEED: A type of EDM machine feed that uses a hydraulic motor to drive a leadscrew.

R

RAM: Random Access Memory. The data in RAM is volatile, which means that it is lost when the computer is shut off.

RAM-FEED: A type of EDM machine feed that uses a hydraulic cylinder to control the movement of the head.

RANDOM ACCESS MEMORY: See RAM.

RANDOM LAUNCHING: A situation in which any workpiece among the parts family (or families) handled by the FMS can be introduced to the system without requiring downtime for machine set-up.

READ-ONLY MEMORY: See ROM.

REAL TIME: Term that refers to events happening now, at this instant, in contrast to *historical time* (events that happened hours or days ago).

RECORD ACCURACY: The value placed on any given data element.

RECORDS: Groups of related fields within a database.

RECURRING PROCESSES: In manufacturing, the series of events that will take place for a given item each time the item is produced.

REFERENTIAL INTEGRITY: A quality of a database design that will not permit data references and relationships to be damaged unintentionally by changes in data.

REPLICATION: A CAD feature that allows a part of a drawing to be copied as needed for use in the same or another drawing.

REPROGRAMMABLE ROBOT: A robot that is capable of acquiring more than one set of instructions, so that more than one task can be accomplished.

REQUEST FOR QUOTATION: A formal invitation to vendors to bid on supplying specific goods or services.

RESCHEDULING: The action of changing production priorities.

RESOURCE REQUIREMENTS PLANNING: Planning that considers the load on long-range resources, such as land, facilities, cash flow, capital equipment, and labor skills and availability.

RESPONSE TIME: The time needed for a computer to respond once a user enters some information.

RETURN ON INVESTMENT: See ROI.

REVISION CONTROL: A system of marking drawings or other files to indicate which information is most current and which revision should be used by each application.

ROBOT COORDINATE SYSTEM: The motion geometry of the end-of-arm tooling, and may or may not be the same as the *robot geometry*.

ROI: Return On Investment. An accounting measurement of the percentage of increased profit that can be attributed to a specific capital investment.

ROM: Read-Only Memory. ROM contains programming instructions and constants that the computer user cannot change.

ROTATION: A feature that lets the CAD operator move the design around to see if from different angles or perspectives.

ROTATIONAL: Term used to describe turned parts processed by an FMS.

ROUGH-CUT CAPACITY PLANNING: Intermediate-range capacity planning that considers adjustment of resources required for such activities as make/buy, subcontract, hire/lay off, or additional tooling.

ROUTING: The sequence of operations that a part will follow as it is being processed on the floor.

ROUTING SHEET INSPECTION: A GT grouping method that involves reviewing the process plans or routings to establish similar manufacturing processes.

S

SAFETY STOCK: Raw material and components protected from allocation by both formal and informal systems. Safety stock can be expressed as either a specific quantity or as a specified length of time.

SCALING: The ability of a CAD program to change the proportions or size of one part of the image in relation to the others.

SCRAP: Parts that are not manufactured to engineering specifications. Also, those that are produced to engineering specifications, but do not perform the intended function.

SENSORS: Electronic or electromechanical

devices designed to detect changes in certain variables, such as spindle deflection, torque, temperature, or vibration.

SERIAL NUMBER TRACKING: A variant of lot traceability most often applied to an end item, eliminating the need for maintaining multiple-level relationships.

SERVICE INDUSTRIES: Businesses such as banking, insurance, or merchandising, that provide a service rather than produce a physical product.

SERVICE LEVEL: A measure of customer service performance against requirements.

SERVOMECHANISMS: Devices that use pneumatic, hydraulic, or electromechanical means to achieve controlled motion.

SHOP FLOOR CONTROL: A function that executes the priority plan calculated by MRP while using the capacity determined by Capacity Requirements Planning (CRP).

SHOP FLOOR SCHEDULING SYSTEMS: Software that captures actual performance information and uses it to dynamically update the standard times. The same systems can monitor the move, wait, and queue times and adjust their values as improvements are made.

SHRINKAGE: The amount of an item that is not acceptable for use.

SHUTTLE: The transfer mechanism used to move a pallet from the transporter to the machine.

SIC: Standard Industrial Classification. A system used by the U.S. Bureau of Census to classify and segment industrial and business activities.

SIGHT INSPECTION: The simplest method of grouping parts into families for GT manufacturing. Sight inspection involves looking at actual parts, photos of parts, or blueprints of parts.

SIMPLE SCHEDULING RULES: A method used to help an industrial engineer develop quick "rule of thumb" standards and capacity estimates.

SIMULATION: The application of computer techniques to create models of either objects or processes to explore alternative solutions.

SINGLE-LEVEL BILL OF MATERIAL: A display listing the parent item and its immediate components only.

SMART TOOLING: Cutting and part-holding tools that readily reconfigure themselves to accommodate a variety of shapes and sizes within a given part family.

SOFT MOCKUP: A software simulation (3-D on-screen image) of a part.

SPARK GAP: In electrical discharge machining, the space between the electrode and the workpiece.

SPC: Statistical Process Control: The science of measuring a process to predict, as soon as possible, trends away from the expected process. Usually referred to as SPC.

SPECIAL SUPPLY ORDERS: Rework and subcontract orders.

STAND-ALONE NC MACHINE TOOLS: Those that are used individually, rather than being part of a machining cell or other configuration. They are appropriate for "job-shop" and small batch manufacturing.

STANDARD CODES: Agreed-upon coding systems used on perforated tape for NC programming.

STANDARD COST: A performance measurement baseline established (usually at annual intervals) as the standard.

STANDARD INDUSTRIAL CLASSIFICATION: See SIC.

STANDARD ROUTING: A system in which all parts in the family follow the same standardized sequence of operations.

STATISTICAL PROCESS CONTROL: See SPC.

STEPPING ACTION: The precise angular rotation of the motor shaft in response to application of an electronic signal.

STEPPING MOTOR: A motor that derives its name from the fact that discrete electrical impulses cause it to *step* (rotate) a specified number of degrees.

STEREOLITHOGRAPHY: A process that uses a CAD model to guide a laser through a polymer solution. The laser hardens the solution within the boundaries of the part, one thin layer at a time, to produce a real, three-dimensional mockup.

STRATEGIC INFORMATION MANAGEMENT: A process that includes developing a data architecture and providing for data integrity.

SUPPLY: Anything that replenishes inventory.

SYSTEMS INTEGRATION: The process of interfacing different types of computer and network hardware or in getting software applications to talk to each other.

T

TACTILE SENSING: A sensor application that enables a robot to "feel" the object in much the same way that humans feel through their fingertips.

TEMPLATE MATCHING: In a machine vision system, the process of comparing the shape shown by the camera with known shapes stored in the computer.

THRESHOLD METHOD: A method for predicting the need for tool replacement, in which a measured value is compared to a known machine limit, without regard to the actual condition of the tool.

THROUGHPUT: A term for the productivity of an FMS.

TIME FENCES: The critical decision points in supply management.

TIME-PHASED BACK-SCHEDULING: The process of using lead time elements for parents and components to determine the release date of an order.

"TO BE" SCENARIO: A pictorial view of the factory 5, 10, or 20 years in the future.

TOLERANCE STACK SHEET: A chart, developed through mathematical methods, that is used to determine how the engineering drawing dimensions and tolerances can be distributed throughout the machining operations to obtain the desired final results.

TOOL: In simplest terms, anything that allows something else to be done.

TOOL CHAIN: In an FMS workstation, the holder for tools to be used. Each tool chain may be capable of holding 60 or more cutting tools.

TOOL (T) COMMANDS: Program commands used to specify a desired tool from the tool magazine of a random-access automatic tool-changer.

TOOLING: Such items as perishable tools, fixtures, gages, and production supplies.

TOOLING SET-UP: The process of collecting the required tools from the tool crib and setting them in the machine.

TOPOLOGY: Term used to describe the configuration of a computer network.

TOTAL QUALITY CONTROL: See TQC.

TOUCH LABOR: Worker time and effort required to transport, position, adjust, reorient, and relocate material and parts as they pass through operations in the factory.

TQC: Total Quality Control. An approach based on the concept that it is possible to produce a perfect part every time. TQC places the responsibility for quality on the individual operator.

TRACEABILITY: A database characteristic that permits older data to be retrieved and tied to products by serial number, lot number, date, or other reference. Traceability is important to companies making products that may impact public safety.

TRANSFER LINES: The highest form of fixed, or hard, automation.

TRANSLATION: A feature that permits a CAD operator to move portions of the drawing around from one location to another.

TRAVELER: Paperwork accompanying an order through production.

TRAVELING COLUMN: A machine design in which tool positioning is done by moving the column over a stationary workpiece.

TREPANNING: A means of generating a hole by an optical-mechanical device that rotates the laser beam around a center point.

TWO-BIN SYSTEM: An inventory control system with one supply (bin one) containing inventory issued for use and another (bin two) reserved for future use.

U

UPTIME: Time that equipment is available for actual use.

UTILIZATION: A performance measure equal to actual hours worked divided by total available hours.

V

VALUE ENGINEERING: An activity that involves investigating reasons for supply and demand fluctuations, identifying waste, and recommending alternative solutions to the problem being addressed.

VARIANT-TYPE CAPP SYSTEM: The key to a variant-type system lies in being able to identify whether a part is similar to some previous part or family of parts. Also referred to as a *retrieval-type* CAPP system.

VENDOR: An external supplier of goods or services.

VERTICAL DEPENDENCE: The relationship between the parent item and its direct components.

VISUAL REVIEW: This inventory control method involves checking the actual physical inventory, not just inventory records.

VOICE INPUT: Technology that allows a worker to communicate with the computer, using a limited and specific vocabulary, to report status of observations of a production process.

W

WIP: Work-In-Process. Materials or parts that are actually being processed into final products.

WIREFRAME VIEW: In a CAD program, an on-screen representation in which all lines in a model are visible, as if it were made from glass.

WORK ENVELOPE: A volume of space defined by the maximum reach of a robot arm in three dimensions.

WORKAROUND: A means of overcoming a limitation or achieving a result that a system was not intended to provide.

WORKCENTER: A group of different machines that produce an item.

WORK-IN-PROCESS: See WIP.

WORKPIECE SET-UP: The process of adjusting the workholding fixture and getting raw materials ready.

WORKSTATION: An individual computer (usually a microcomputer) that provides the user with access to a set of programs designed to improve productivity.

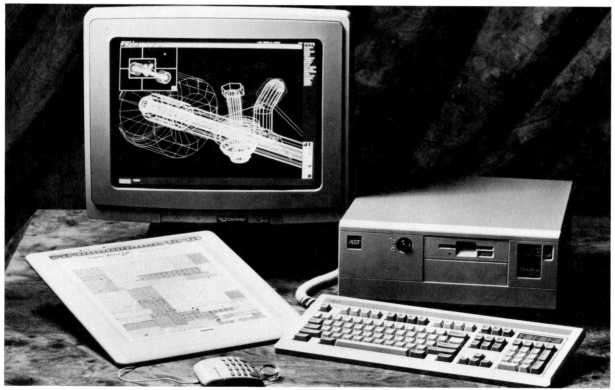

A typical CAD workstation today consists of a powerful desktop computer and high-resolution monitor, and such input devices as a keyboard and a graphics tablet.

INDEX

Coolant and chip handling systems, 257
Coordinate measuring machines,
  27, 195, 216
Coordinate system, robot, 282, 283
Cost accounting, 169
Cost control, 169-171
Cost management, 169-171
  costing methods, 169-171
  value analysis engineering, 169
Cost roll-up, 171
Costing methods, 169-171
Cost/savings measurement plan, 43
Coupler contamination, 253
Critical path, 140
Criticality, 145
CRT, 203
CSFM programming, 227
Current and material removal rates, 309
Current capabilities and applications,
  39-42
  embedded computers, 41
  hardware communications, 40, 41
  networks, 40
  systems integration, 42
Current factory technology, 190-197
Customer and supplier roles, 339
Customer order management, 104-106
Customer service performance, 104
Cutting tool delivery, 260
Cycle counting, 147
Cycle time, 134-136
Cylindrical coordinate robot, 272

D

Data architecture, 13
Data collection and reporting, 191, 192
Data control, engineering, 137-140
Data dictionary, 16
Data integrity, 37-39
Data integrity control, 168
Data security, 38
Data sharing, need for, 36, 37
Data validation, 38
Database architecture, 263, 264
  inspection history file, 263
  inventory file, 263
  machine queue file, 264
  machine tool status file, 263
  MHS status file, 263
  pallet/fixture status file, 263
  part master file, 263
  part programs, 263
  production history file, 264
  production sequence of current schedule
    file, 263
  standards file, 263
  system maintenance file, 263

  tool inventory file, 263
  tool requirements file, 264
Database system, 36
Databases, 36, 37
Databases, need for data sharing, 36, 37
DC and hydraulic variable speed
  drives, 207
DC tachometer, 280
Dedicated computer, 193
De-expedite, 113
Demand, 114, 332
Demand, characteristics of, 111, 112
  horizontal and vertical dependence, 112
  independent and dependent demand,
    111, 112
Demand management, 100-106
  customer order entry, 104-106
  forecasting, 101-104
Dependence, horizontal and vertical, 112
Dependent demand, 111, 112
Design,
  justification, and implementation of
    robotic systems, 287-296
  process, 20, 21
  product, 20, 22, 57, 59
  scheduling and control, CIM, 6
Design considerations, 62
Design libraries, 35
Design review, 59
Design systems, features-based, 50
Devices, output, 84, 85
DFA (Design for Assembly), 64
DFM (Design for Manufacture), 64
Digitizing, 228
Digitizing programming, 228
Digitizing tablet, 82
Direct input, 191
Discrepant lots, 146
Dispatch list, 158, 159
Distributed processing, 80, 81
Distributive Numerical Control, 203,
  229-232
  CNC "behind the reader system", 231
  conventional system, 229-231
  minicomputer system, 231, 232
DNC, 24, 94, 95
Dock-to-stock processing time, 163
Documentation, production, 68, 69
Drafting machines, 216, 217
Drilling machines, 211, 212
Drives, hydraulic and DC variable
  speed, 207
Drives, stepping motor, 207

E

EBM applications, 321

ECM,
  advantages, 311
  applications, 311, 312
  operation, 307-309
  theory of, 307-312
Economies of scale, 181
EDM advantages and disadvantages,
  303, 304
EDM applications, 304-306
EDM electrode, 300-306
  electrode materials, 301, 302
  material removal rates, 302, 303
  tool feed designs, 301
Education and training, 136, 137
Effectivity, 140
Efficiency, 153
EIA code, 209, 210
Electrical discharge machining, 216,
  300-306
Electrochemical machining, 306, 307
Electrolysis, 306
Electrolyte, 306, 309, 310
Electron beam welding, 319-321
Embedded computers, 41
Enabling tool, 10, 11
End effector, 273, 274
End-effector considerations, 289-294
  compliance, 291, 292
  force sensing, 292
  machine vision, 292-294
  tactile sensing, 292
Energy transfer medium, 317
Engineering,
  concurrent, 50, 51
  industrial, 77, 78
  manufacturing, 74-77
  manufacturing production, 73-95
  NC programming, 77
  process planning, 75-77
  production, 74-78
  tool design, 77
  value analysis, 169
Engineering data control, 137-140
Engineering/manufacturing communica-
  tion, 27, 28
English-like words, 220
Equipment, inspection and test, 216
Evaluation/producibility/performance,
  64, 65
Even parity, 210
Evolution of manufacturing, 338, 339
Evolving standards, 50, 51
Expandability, 244
Expedite, 113
Expert systems, 59-61, 91, 197
Expiration tracking, 168
Exploding, 111
Explosion, 138